ENVIRONMENTAL SCIENCE, ENGINEERING AND TECHNOLOGY

ENVIRONMENTAL MANAGEMENT

SYSTEMS, SUSTAINABILITY AND CURRENT ISSUES

ENVIRONMENTAL SCIENCE, ENGINEERING AND TECHNOLOGY

Additional books in this series can be found on Nova's website under the Series tab.

Additional E-books in this series can be found on Nova's website under the E-books tab.

ENVIRONMENTAL SCIENCE, ENGINEERING AND TECHNOLOGY

ENVIRONMENTAL MANAGEMENT

SYSTEMS, SUSTAINABILITY AND CURRENT ISSUES

HENRY C. DUPONT
EDITOR

Nova Science Publishers, Inc.
New York

Copyright © 2012 by Nova Science Publishers, Inc.

All rights reserved. No part of this book may be reproduced, stored in a retrieval system or transmitted in any form or by any means: electronic, electrostatic, magnetic, tape, mechanical photocopying, recording or otherwise without the written permission of the Publisher.

For permission to use material from this book please contact us:
Telephone 631-231-7269; Fax 631-231-8175
Web Site: http://www.novapublishers.com

NOTICE TO THE READER

The Publisher has taken reasonable care in the preparation of this book, but makes no expressed or implied warranty of any kind and assumes no responsibility for any errors or omissions. No liability is assumed for incidental or consequential damages in connection with or arising out of information contained in this book. The Publisher shall not be liable for any special, consequential, or exemplary damages resulting, in whole or in part, from the readers' use of, or reliance upon, this material. Any parts of this book based on government reports are so indicated and copyright is claimed for those parts to the extent applicable to compilations of such works.

Independent verification should be sought for any data, advice or recommendations contained in this book. In addition, no responsibility is assumed by the publisher for any injury and/or damage to persons or property arising from any methods, products, instructions, ideas or otherwise contained in this publication.

This publication is designed to provide accurate and authoritative information with regard to the subject matter covered herein. It is sold with the clear understanding that the Publisher is not engaged in rendering legal or any other professional services. If legal or any other expert assistance is required, the services of a competent person should be sought. FROM A DECLARATION OF PARTICIPANTS JOINTLY ADOPTED BY A COMMITTEE OF THE AMERICAN BAR ASSOCIATION AND A COMMITTEE OF PUBLISHERS.

Additional color graphics may be available in the e-book version of this book.

LIBRARY OF CONGRESS CATALOGING-IN-PUBLICATION DATA

Environmental management : systems, sustainability, and current issues / editor, Henry C. Dupont.
p. cm.
Includes bibliographical references and index.
ISBN 978-1-61324-733-4 (hardcover)
1. Environmental management. I. Dupont, Henry C.
GE300.E566 2011
363.7--dc23
2011017627

Published by Nova Science Publishers, Inc. † *New York*

CONTENTS

Preface vii

Chapter 1 Integrating Dynamic Social Systems into Assessments of Future Wildfire Losses: An Experiential Agent-Based Modeling Approach 1
Travis B. Paveglio and Tony Prato

Chapter 2 The European Experience on Environmental Management Systems and the Third Revision of the "Eco Management and Audit Scheme" (EMAS) 43
Fabio Iraldo, Francesco Testa, Tiberio Daddi and Marco Frey

Chapter 3 Minimizing Environmental Impact from Applying Selected Inputs in Plant Production 83
Claus G. Sørensen, Dionysis D. Bochtis, Thomas Bartzanas, Nikolaos Katsoulas and Constantinos Kittas

Chapter 4 Natural Adsorbents from Tannin Extracts: Novel and Sustainable Resources for Water Treatment 103
J. Sánchez-Martín and J. Beltrán-Heredia

Chapter 5 ISO 14001 Research: An Academic Approach 133
G. Lannelongue, J. Gonzalez-Benito and O. Gonzalez-Benito

Chapter 6 Measuring Sustainable Culture among Construction Stakeholders in Hong Kong 159
Robin C. P. Yip, C. S. Poon and James M. W. Wong

Chapter 7 The Pros and Cons of ISO 14000 Environmental Management Systems (EMS) for Turkish Construction Firms 179
Ahmet Murat Turk

Chapter 8 Users' Preferences and Choices in Argentinean Beaches 197
A. Faggi, N. Madanes, M. Rodriguez, J. Solanas, A. Saenz and I. Espejel

Chapter 9	Soil Carbon Sequestration through the Use of Biosolids in Soils of the Pampas Region, Argentina *Silvana Irene Torri and Raúl Silvio Lavado*	**221**
Chapter 10	Developing an Ecosystem-Based Habitat Conservation Planning Protocol in Saline Wetland *Jennifer Hitchcock and Zhenghong Tang*	**237**
Chapter 11	A Process to Determine One Organization's Environmental Management System *Lindsay Thompson and Shirley Thompson*	**253**
Chapter 12	Are Ecosystem Models an Improvement on Single-Species Models for Fisheries Management? The Case of Upper Gulf of California, Mexico *Alejandro Espinoza-Tenorio, Matthias Wolff and Ileana Espejel*	**269**
Chapter 13	A Study about the Adoption of the Practice of Cleaner Production in Industrial Enterprises Certified ISO 14001 in Brazil *José Augusto de Oliveira, Otávio José de Oliveira and Sílvia Renata de Oliveira Santos*	**283**
Chapter 14	Towards a Watershed Approach in Non-Point Source Pollution Control in the Lake Tai Basin, China *Xiaoying Yang, Zheng Zheng and Xingzhang Luo*	**299**
Chapter 15	Socio-Environmental Marketing as an Environmental Management System to Protect Endangered Species *Juan José Mier-Terán, María José Montero-Simó, and Rafael A. Araque-Padilla*	**313**
Index		**331**

PREFACE

Environmental management systems are designed to improve environmental performance, which can reduce costs by lowering compliance costs, reducing waste, and improving efficiency and productivity. In this book, the authors present current research from across the globe in the study of the sustainability and current issues of environmental management systems. Topics discussed include an assessment of wildfire-related losses; the European Eco Management and Audit Scheme (EMAS) regulation; sustainable resources for water treatment; ISO 14001 research standards and developing an ecosystem-based habitat conservation planning protocol in saline wetlands.

Chapter 1 - Interactions between social and ecological systems can pose threats to humans and the natural environment. One example of this phenomenon is the increasing threat of losses from wildfire, which is influenced by both social processes (i.e., expanding human settlement, residential development patterns, forest management, and fire suppression) and ecological conditions (i.e., high surface and canopy fuel loadings, forest type, topography, and climate change). Methodological frameworks for evaluating multifaceted threats include various conceptualizations, assessments, and simulations of the interacting factors that determine human exposure to risk or that make humans less vulnerable to potential hazards. One such approach is agent-based modeling or agent based models (ABM), which simulate complex system behavior from the bottom up using simple decision rules for the behavior of different agents.This chapter presents an ABM framework for simulating the dynamics of a coupled natural-human system for wildfire in Flathead County, Montana, USA. The ABM has three interacting agents (i.e., homeowners, community and regional planners, and land and wildfire management agencies) that influence or are influenced by potential losses from wildfire. The ABM approach is part of the *wildfire climate* (FIRECLIM) model that simulates future wildfire risk in Flathead County using different assumptions about climate change, economic growth, residential development, and forest management.The proposed ABM framework incorporates methodologies for integrating monetary and non-monetary attributes that influence human behavior and decisions with respect to wildfire. The methodologies include collaborative approaches that can be used to evaluate the characteristics and behaviors of agents in a geographical location, providing an experientially driven way to capture the impact of social diversity on the dynamics of a coupled natural-human system for wildfire. The proposed ABM framework includes preliminary decision rules for agents that reflect how they interact with and modify the decisions of other agents and/or other processes that operate in a coupled natural-human system for wildfire.

Chapter 2 - In recent years, based on the voluntary commitment and the pro-active approach of organizations, the certification framework has gained a crucial role amongst the instruments of the European environmental policy. Voluntary instruments (such as EMAS and the "twin" regulation Eco-label) were designed by introducing concepts and mechanisms that, for the time being, led to a radical change in the environmental policies of the European Commission. The application of these patterns, in fact, originated a highly innovative policy trend, based on voluntary certification as a marketing tool providing a competitive edge. The Commission's purpose was clearly to attract the interest of companies and convince them to spontaneously mobilize their financial, technical and management resources towards a path of continuous improvement in environmental performance.

Chapter 3 - Primary agricultural production is often managed on a rather crude level (e.q. Sørensen et al., 2010). Plant nutrients and pesticides are applied equally not only within-fields, but also to all fields growing the same crop type. Fields and crops are treated according to standards defined by simple, easily observable parameters. Likewise, society's regulation of agricultural production is on a crude level as, for example, nitrogen quotas are regulated on a farm level, because only the farm's purchase and sales are available for the required supervision. Both farm management and environmental protection could be improved significantly by a more detailed management and regulation of various inputs.

Chapter 4 - The adsorption of contaminants is one of the most popular processes in water treatment. The feasibility of a long variety of materials, such as clays or active carbons and multiple waste products such as biomass, tires or other kinds of industrial residues, makes the production of new adsorbents a very interesting researching subject.This chapter is focused on a new type of natural adsorbent which is based on tannin extracts. Gelation of tannins is a recent chemical procedure that is often performed in order to obtain high quality adhesives. However, not many works have been carried out studying the ability of these gels for removing cationic pollutants from aqueous matrix. This is the scope of the current contribution.Wastewater techniques should be improved in order to guarantee their applicability in a wide scenario: from developed countries, where no economical problems are detected; to less developed zones, where technology is usually a natural barrier. Global sustainability is evidently depending on the correct care of natural resources all over the world, not only where socioeconomic conditions allow to keep the environment clean. Tannins from different trees, such as *Quebracho, Acacia, Pinus* or *Cypress*, have been gelified according to different gelation processes. Parameters such as aldehyde or NaOH concentration were varied and optimal conditions for gelification were applied in order to obtain efficient adsorbents for the removal of dyes, surfactants and heavy metals. Reliability of the results was technically confirmed by statistical methodologies.

Chapter 5 - An Environmental Management Systems (EMS) can be described by means of its primary measures; unfortunately, in contrast to the literature on quality management systems in which the most relevant aspects of the system are perfectly defined and compared, there is still lacking a consensus concerning the critical factors of an EMS. This has led to several different approaches in research on EMS, and has made it more difficult to compare scientific results. In this chapter, the authors aim to deepen the theoretical bases that underpin the study of EMS. To this end, the authors first review the history of environmental management in firms and the appearance of formal EMS. Next, the authors examine nature of the standards and how they are employed within various organizations. Finally, the authors reconsider the literature on and the structure of the standard represented by ISO 14001:2004, and propose a model to study EMS that involves eighteen critical factors under four headings: Top Management Support, HR Management, Information System and Externals Factors. This model enables us to develop a tool that allows for a more systematic approach to empirical analyses of EMS, especially those based on ISO 14001.

Chapter 6 - The construction industry is a leading contributor in improving the quality of the built environment, but concurrently it is a main producer of solid waste and greenhouse gas that damage the environment. Stakeholders of the construction industry thus have a decisive role to play in enhancement of sustainability and suppression of environmental damages. In the process of performing sustainable construction, stakeholders changed subconsciously their attitudes and behaviours towards a more sustainable culture. This paper aims to examine the extent of these attitudinal and behavioral changes by caterizing these changes in four sustainable cultural components. The attitudinal changes are classified into *awareness* and *concern*, while the behavioral changes are classified into *motivation* and *implementation*. The investigation was carried out by means of two surveys conducted in years 2004 and 2006 among different stakeholder groups of various disciplines including the Government, Developer, Consultant, Contractor and the frontline construction supervisors embracing site agents, site supervisors and foremen. The findings indicated that different stakeholder group carries different influential power to contribute sustainability. The consultant group and the frontline participants group demonstrated readiness in compliance by their willingness to adopt new practices favorable to sustainable construction. On the other hand, although embracing high influential power, the developer group had yet a relatively lower apprehension on sustainability, particularly in motivation and implementation aspects. Holding the highest influential power, the government group had a remarkable awareness and motivation on sustainable construction but inadequate in implementation when compared with other industry stakeholders. Although the contractor group exhibited an overall improvement in sustainable culture, but the improvement of various cultural components are relatively low. The results of investigation that reflect such a social phenomenon is an important reference for decision-makers in the government and in private sectors to formulate policies that couple with universal demands for sustainable development. The means of measurement so developed may also serve as a valuable reference for other industries.

Chapter 7 - Recently, the ISO 14000 environmental management system (EMS) has been widely utilized by all sectors throughout the world prepared by ISO (International Standard Organization). With the purpose of keep up and develop the environmental performance within all sectors, including the construction sector, some methods exist for the protection of sustainable development and the environment worldwide. ISO 14000 EMS originated from such necessity. Lately, an increased awareness has emerged related to the use of this system in the construction industry. Despite such interest, the research related to the implementation practice of the ISO 14000 EMS by the construction firms has not reached the desired level. In this study, the major motives for seeking ISO 14001 certification for Turkish construction firms is being examined by using the questionnaire survey method. Questionnaire survey was conducted with 68 individual construction firms, which represent the top firms in Turkey and operate in national and international markets as they are members of the Turkish Contractors Association (TCA). Descriptive and factor analyses were used with the obtained data from questionnaires of advantages and disadvantages related to ISO 14000 EMS. In the study, factor analysis was used to summarize many variables with a few factors. Each of the factors acquired from these analyses presents advantages and disadvantages of ISO 14000 EMS factors. As the advantages of ISO 14000 EMS, mainly two factor dimensions are found in the analysis. First dimension is entitled as "related to environment" and second one is called as dimension "related to company". As the disadvantage of ISO 14000 EMS, mainly three factor dimensions have found in the analysis. First one is named as "related to lack of the knowledge and personnel". Second one is called as "related to the cost and implementation" and third dimension is entitled as "related to no apparent benefits". This study shows that there is a positive approach to the ISO 14000 EMS within the construction sector in Turkey as well as indicating that the utilization of the ISO 14000 EMS is not yet at the ideal level. In particular, the problems of lack of information and qualified personnel revealed in the results of the analysis should be overcome. Personnel should be qualified in the concept of EMS and on the technical details. In the global construction market, an increase in the number of firms having EMS will both reduce environmental impact and develop the potential of awarding contracts to the construction firms from underdeveloped and/or developing countries.

Chapter 8 - This study analysed the profile and perception – composed of opinions and attitudes – of beach users in Argentina, using data from nine sandy beaches, each with unique environmental and socioeconomic features, located in two coastal municipalities. The authors distributed 329 surveys composed of 42 questions to Argentinean residents and tourists visiting the beach. Data on the profile were analysed by cluster analysis, and data on perception were explored by Principal Components Analysis. These results allowed grouping all beaches according to variables such as cleanliness, accommodation, infrastructure and services. Users' age and marital status were found to be associated only with certain beaches; married people visited urban and rural beaches, preferring those without infrastructure, while single and young people chose urban beaches with facilities. Contrasting answers regarding environmental beach features were recognised between both municipalities, indicating the success of awareness programs that enhance the beaches' natural values.

Chapter 9 - Carbon sequestration in agricultural soils through the increase of the soil organic carbon (SOC) pool has generated broad interest to mitigate the effects of climate change. Increases in soil carbon storage in agricultural soils may be accomplished by the production of more biomass, originating a net transfer of atmospheric CO_2 into the soil C pool through the humification of crop residues, resulting in carbon sequestration. This Chapter addresses the potential of carbon storage of representative soils of the Pampas region amended with different doses of biosolids, and their soil carbon sequestration potential. Increase in biomass or yields of crops cultivated in sludge amended soils compared to unamended control soils are also discussed. The crops considered in present chapter are a forage, rye.grass (*Lolium perenne*), an annual crop, maize (*Zea mays*) and two trees, pine (*Pinus elliottii*) and eucalyptus (*Eucalyptus dunnii*).

Chapter 10 - The ecosystem approach to management is in its integrative form a management method that considers the system as a whole instead of as individual components. This holistic approach focuses on habitats and system integrity, and on an objective aimed at the health and integrity of the ecosystem (Currie 2010). Ecosystem-Based Management is a unique management style that is intended to overcome the shortfalls of single-sector management and contains the following characteristics.

Chapter 11 - Implementing an Environmental Management System (EMS) can reduce operational costs for an organization as well as increase employee morale. An EMS is a component of an organization's overall management system that commonly includes an environmental policy, environmental aspects, objectives, targets, actions, significant aspects, training and auditing. Reasons for different approaches to environmental management include managerial interpretations of environmental issues as threats or opportunities, top management attitudes toward the environment, organizational champions pushing for a more proactive environmental approach, stakeholder pressures and regulations. An EMS provides a means to benefit business and environmental goals simultaneously, if it can integrate the two into day-to-day decisions. But how is this done? This paper provides a case study of integrating sustainability in one medium sized organization through both quantitative and qualitative methods to develop an EMS.

Chapter 12 - The authors review the recent applications of ecosystem models (EMs) as tools for fisheries management in the Upper Gulf of California (UGC), Mexico. EMs are compared with single-species model applications in the UGC, as a basis for assessing the benefits of each ecosystem model as a tool for evaluating management alternatives capable of diminishing impacts on marine ecosystems. The strengths and weaknesses of different types of EMs and their ability to evaluate the systemic mechanisms underlying observed shifts in resource production are also examined with respect to Ecosystem-Based Fisheries Management (EBFM) general goals. Findings showed that ecosystem modeling has increasingly resulted in support for EBFM in the UGC. However, outputs also proved evidence on that EMs are facing a most complicated situation than single-species models regarding the lack of data. Thus, the step from single-species models to EMs is a stage of the management of the area that does not require the elimination of the first approach, but rather the use of both approaches in a complimentary manner. The challenge is the integration of current ecosystem information to detect the gaps in the collective knowledge on the UGC. Insights from this study are valuable in defining a planning model scheme that supports ecosystem-based management policies in local fisheries.

Chapter 13 - The growth of developing countries has significantly driven the industrial activity, which has generated employment and development. However, this scenario also has its downside, in other words, due to lax laws and consumers still little conscious of the importance of effective environmental protection. The industry has been the main degrading environment in these nations. However, some technical programs and management tools have been used to minimize this serious problem, especially if the environmental management system ISO 14001:2004 and Cleaner Production (CP). For these reasons, this book chapter has as main objective to present good practices of cleaner production adopted by four industrial companies operating in Brazil with high profile and certified according to ISO 14001:2004. The companies act in the area of cellulose and paper and chemical. Their management systems ISO 14001:2004 will be characterized and the major practices of Cleaner Production will be reported. In general terms, it was observed that the EMS has greater coverage of these companies that ISO 14001:2004 due to demands from the processes related to cleaner production. They use specific methodologies and specific classifications to the elements of CP and do not follow exactly the elements proposed by UNIDO (United Nations Industrial Development Organization). It was observed that the companies have achieved significant results by adopting the cleaner production achieved environmental and financial gains from the reuse of production inputs, internal recycling in the production process, cost reduction with treatment and disposal of waste, reducing spending with environmental liabilities promoted by prevention of the pollution, among others, which are the main benefits expected by the practice of CP.

Chapter 14 - While accounting for 0.4% of its land area and 2.9% of its population, the Lake Tai basin generates more than 14% of China's Gross Domestic Production. Accompanied with its fast economic development is serious water environment deterioration in the Lake Tai basin. The lake is becoming increasingly eutrophied and has frequently suffered from cyanobacterial blooms in recent years. Although tremendous investment has been made to control pollutant discharge to improve its water quality, the Lake Tai's eutrophication trend has not been reversed due to the past emphasis on point pollution sources and the lack of effective measures for non-point source pollution control in the region. A watershed approach is proposed to deal with the serious non-point source pollution issues in the region with four guiding principles: (1) Control from the source; (2) Reduction along transport; (3) Emphasis on waste reuse and nutrient recycling; and (4) Intensive treatment at key locations. Research as well as field applications that have been conducted to implement this watershed approach are introduced. Existing problems and their implications for future research needs as well as policy making are also discussed.

Chapter 15 - The environmental management systems are normally used to obtain certain environmental objectives. That is why they are established as either business or institutional obligations aimed at identifying objectives that improve the environment and incorporate procedures in management to attain these objectives. Moreover, given that prevention is always better than cure, they ensure that the aforementioned is complied with. A common objective in these systems is to change behaviour that is thought to be negative with regard to the environment, whether it is that of those working for the aforesaid organisations or that of the final users of the services. Social marketing has proven to be a very suitable technology to change behaviour, and with regard to environmental issues, the specialisation of Environmental Marketing is a good example of how it is applied to a specific subject matter. In this case its basic principles and the techniques that are applied are aimed at modifying negative behaviour towards environmental problems. In this context, the basis and foundations of the Socio-environmental Marketing are shown as a management tool used to change behaviour in the sustainable management of protected areas. Furthermore a specific case is presented whereby this management system is used to protect an endangered species, the Iberian Lynx.

In: Environmental Management
Editor: Henry C. Dupont

ISBN: 978-1-61324-733-4
© 2012 Nova Science Publishers, Inc.

Chapter 1

INTEGRATING DYNAMIC SOCIAL SYSTEMS INTO ASSESSMENTS OF FUTURE WILDFIRE LOSSES: AN EXPERIENTIAL AGENT-BASED MODELING APPROACH

Travis B. Paveglio[1] and Tony Prato[2]

[1]College of Forestry and Conservation, The University of Montana, US
[2]Agricultural and Applied Economics and Center for Applied Research and Environmental Systems, University of Missouri, US

ABSTRACT

Interactions between social and ecological systems can pose threats to humans and the natural environment. One example of this phenomenon is the increasing threat of losses from wildfire, which is influenced by both social processes (i.e., expanding human settlement, residential development patterns, forest management, and fire suppression) and ecological conditions (i.e., high surface and canopy fuel loadings, forest type, topography, and climate change). Methodological frameworks for evaluating multifaceted threats include various conceptualizations, assessments, and simulations of the interacting factors that determine human exposure to risk or that make humans less vulnerable to potential hazards. One such approach is agent-based modeling or agent based models (ABM), which simulate complex system behavior from the bottom up using simple decision rules for the behavior of different agents.

This chapter presents an ABM framework for simulating the dynamics of a coupled natural-human system for wildfire in Flathead County, Montana, USA. The ABM has three interacting agents (i.e., homeowners, community and regional planners, and land and wildfire management agencies) that influence or are influenced by potential losses from wildfire. The ABM approach is part of the *wildfire climate* (FIRECLIM) model that simulates future wildfire risk in Flathead County using different assumptions about climate change, economic growth, residential development, and forest management.

The proposed ABM framework incorporates methodologies for integrating monetary and non-monetary attributes that influence human behavior and decisions with respect to wildfire. The methodologies include collaborative approaches that can be used to evaluate the characteristics and behaviors of agents in a geographical location, providing an experientially driven way to capture the impact of social diversity on the dynamics of a coupled natural-human system for wildfire. The proposed ABM framework includes preliminary decision rules for agents that reflect how they interact with and modify the decisions of other agents and/or other processes that operate in a coupled natural-human system for wildfire.

INTRODUCTION

The increasing risk of wildfire-related losses to residential properties and public lands is widely acknowledged to be the result of both human actions and biophysical processes. Human actions that contribute to wildfire risk for private and public property include: (1) the legacy of past forest management by land management agencies (e.g., fire exclusion or species composition changes); (2) the ever-expanding fringe of urban and rural development (e.g., the Wildland Urban Interface [WUI]) that influences the need for fire suppression; and (3) residents' action (or inaction) regarding fuel reduction and building materials for homes that influence the probability that homes burn [USDA 2006; Martin et al. 2007; Steelman and Burke 2007; Cohen 2008; Jensen and McPherson 2008]. Biophysical attributes that influence wildfire risk include climate change, which contributes to: (1) longer and more intense fire seasons; (2) changes in the composition of forest species; (3) increased availability of dry fuels; and (4) the invasion of non-native, exotic plant species [Arno and Allison Bunnell 2002; Brooks et al. 2004; Westerling et al. 2006; Spracklen et al. 2009].

The complex interactions between humans and the natural environment make it difficult to estimate or simulate current and future wildfire losses. Most previous research focuses on the site-specific ecological factors that influence fire severity, burn patterns, and consumption rates of wildfires or explore the influence of climate change on these processes [see Reinhardt and Dickinson [2010] and Weinstein and Woodbury [2010] for reviews]. Far less research has focused on how human decisions influence the threat of wildfire losses on private and public (i.e., timber) property. Meanwhile, a large and growing literature focuses on the factors underlying collaborative human actions for reducing wildfire risk, including controlling residential and commercial development, increasing community or homeowner fire mitigation, and targeting fuel reduction by land management agencies or private property owners to high risk areas. The growth of this literature in the past few decades (see Daniel et al. [2007] or Martin et al. [2008] for reviews) and significant U.S. policy targets focusing on collaborative human efforts to reduce wildfire risk (i.e., Healthy Forests Restoration Act, National Fire Plan) demonstrate the increasing need to model both biophysical and human dimensions of potential wildfire losses. Simulations of wildfire risk that integrate dynamic human and biophysical processes can inform land use planners, natural resource managers, and community members' decisions and policies for reducing future wildfire losses.

One promising methodological development for simulating the dynamic interactions between social and ecological systems is Agent-Based Modeling or Agent-Based Models (ABM). ABM allows researchers to dynamically simulate the interactions among individual, collective, or agency decision makers across time and space and the effects of those interactions on wildfire risk [Bonabeau 2002; Gilbert 2008; Heath et al. 2009]. ABM has been widely used, but is less prevalent in simulating the behavior of agents with respect to wildfire. The majority of wildfire research continues to model wildfire risk in a static fashion that may include elements of social and ecological systems, but rarely provide insights into how human actions interact and aggregate to the landscape level. In short, many existing studies of wildfire risk do not account for the diversity or variability of human decisions, their interactions with the natural environment, and how those interactions influence current and (likely) future wildfire risk. Such insights would increase the capacity of human communities and broader society to manage wildfire as both a necessary, natural process and a hazard that can result in losses of property and resources.

This chapter describes an ABM for simulating the dynamics of a coupled natural-human system for wildfire in Flathead County, Montana (referred to as the FIRECLIM ABM). The ABM incorporates three agents who make wildfire-related decisions: (1) land and wildfire management agencies (hereafter referred to as $Agent_a$); (2) homeowners (hereafter referred to as $Agent_h$); and (3) regional and community planners (hereafter referred to as $Agent_p$). The ABM assumes that members of each agent make decisions based on certain rules that differ across agents. The decisions of each agent, their mutual interactions, and their interactions with the natural environment (for wildfire) influence their exposure to or impacts from wildfire. This chapter does not discuss the computer software needed to implement the ABM.

The ABM for wildfire presented in this chapter is part of a larger FIRECLIM model, which is being developed as part of a National Science Foundation project. The project is simulating and demonstrating how communities can adaptively manage wildfire risk under alternative climate change and economic growth scenarios. Alternative climate change scenarios can influence the availability of fuels and composition of wildland vegetation and, hence, wildfire severity and future wildfire risk. Impacts of climate change scenarios are simulated using the Fire-BGCv2 model [Keane et al. 1996; 1999]. Impacts of economic growth scenarios are simulated using the RECID2 model [Prato et al. in press], which is an extension of the RECID1 model [Prato et al. 2007]. Both economic growth and climate change scenarios influence and are modified by agents' decisions. This chapter focuses on the conceptual framework of the ABM and not the structure of the Fire-BGCv2 or RECID2 models.

AGENT-BASED MODELING

ABM decomposes complex systems by simulating the decisions of individual agents. Agents often interact in a spatially explicit, simulated computer environment. Both agents and the decision environment in which they operate are designed to mimic the real-world system being evaluated. Each agent can have a knowledge base (e.g. memory about the outcomes of previous actions), communicate with other agents, interact with the environment, or have a learning capacity [An et al. 2006; Gilbert 2008; Yu et al. 2009]. Parameterized decision rules for agents integrate factors they consider before making a decision, including the outcomes of decisions made by other agents and the dynamics of the environment. Changes in simulation conditions or interactions among agents make the model dynamic as actors must reassess and act in response to new circumstances [Parker et al. 2003; Miller and Page 2007; Acosta-Michlik and Espaldon 2008].

The primary strength of ABM is that it can capture the functioning of a complex system across time and space from the "bottom up." The computer simulation of numerous interactions among actors and between actors and the natural environment is *process based*, which allows aggregate patterns to *emerge* rather than being determined by mathematical formulae or model design [Epstein and Axtell 1996; Wainwright 2008; Irwin 2010]. Because of the interactions among agents in the ABM, the outcomes do not necessarily conform to preconceived ideas or goals of individual agents and are not simply a consequence of model structure. In this regard, ABM and related approaches, such as multi-agent simulations and individual-based modeling, are superior to other modeling strategies for simulating individual parts of a system or that assume uniform behavior of agents or ecosystem processes [Helleboogh et al. 2007; Edmonds 2010].

Castle and Crooks [2006] and Bonabeau [2002] contend that ABM is more of a paradigm than a method because each ABM is designed to represent a particular system. As such, there is much variability and flexibility in the design or implementation of ABMs. Agents can represent individuals, groups of individuals, agencies, or organizations; they can be heterogeneous in terms of their demographic characteristics, knowledge, decision-making ability, and responses to other actors or the environment. The frequency of agent decisions, type of environmental processes, and/or length of the evaluation period can vary in ABMs depending on the specific research goals [Valbuena et al. 2008; Gilbert 2008]. ABMs are often spatially explicit in order to: (1) meaningfully characterize micro-site characteristics and/or interpret patterns at multiple scales; (2) provide further heterogeneity in terms of the factors that influence agent behavior; (3) improve simulations of the interactions between socio-economic and biophysical processes; and (4) enable simulation of land use change patterns at the parcel level [Gimblett 2002; Bousquet and Le Page 2004; Millington et al. 2008]. The spatially-explicit nature of ABMs is particularly important in simulating wildfire-related decisions, which are highly contingent upon site-specific patterns of fuel reduction or forest management activities, enforcement provisions of policies or regulations aimed at fuel reduction, residential development, and biophysical conditions such as fuel loading, topography, and weather patterns.

ABM researchers cite a number of additional benefits of using ABM to understand complex or coupled natural-human systems, including the ability to: (1) move beyond purely rational or economically driven agents by integrating non-market considerations, individual perceptions, or other social processes into decision rules that drive agent behavior; (2) dynamically test the efficacy of multiple policy scenarios; (3) discern incremental changes in phenomenon of interest (i.e., expected losses or risk from wildland fire) in situations where stakeholders cannot agree on the best way to proceed; and (4) more accurately portray linkages between social and environmental processes [Lempert 2002; Ramanath and Gilbert 2004; Janssen and Ostrom 2006; Chu et al. 2009].

Despite wide variability in the concepts used in developing of ABMs, they are all based on a conceptual model [Gilbert 2008; Heath et al. 2009]. The conceptual model for any particular ABM combines known system theories and data about the agents or their decision environment, drives model development and the choice of model assumptions, and specifies the rationale behind agent decision rules and simulations [Robinson 2008; Heath et al. 2009]. The purpose of this chapter is to present the preliminary conceptual model for the FIRECLIM ABM. The FIRECLIM ABM presented here has several advantages. First, decision rules for $Agent_h$ and $Agent_a$ include non-market values or other social factors because both empirical research and observation suggest that these agents respond to other factors besides economic values. Second, multiple economic growth and climate change scenarios are simulated to determine how agents' decisions are influenced by uncertainty regarding economic growth and climate change. Third, outcomes of agent decisions are paired with metrics of wildfire risk, allowing comparison of these metrics across different scenarios. Finally, by incorporating existing simulation models of socioeconomic and natural (wildfire) processes, the FIRECLIM ABM produces a flexible decision-making structure that is responsive to those processes.

Although ABM approaches have great promise for simulating complex coupled natural-human systems, they are often criticized for oversimplifying the way those systems are represented or for assuming that a modeling approach accounts for all variables in the system. The consequence, according to critics, is that results are difficult to validate and/or are based on incomplete or inaccurate information regarding agent reasoning [Galán et al. 2009; Bone and Dragiecivic 2010; Edmonds 2010]. Many of these difficulties stem from the complexity of the systems that are modeled using ABMs and the diversity of human actors operating in those systems. Often, there is little, incomplete, or inconclusive research on the factors that drive agent behavior. Assumptions, which are ubiquitous in any modeling effort, may work in tandem to confound research results. Additionally, the structure of the ABM, in terms of both the number of agents simulated and the size/complexity of system, makes them cumbersome to use.

In response to these criticisms, it is imperative to ground ABM in existing understandings of the processes, agents, and system(s) of interest. Often, this means collecting additional data about potential agents and their decision environment, including the use of empirical case studies, interviews, and surveys. Secondary datasets representing the decision environment or characteristics of agents are another source of information for improving ABM. Many authors suggest using a combination of both qualitative and quantitative data to achieve both conceptual and empirical validity for agent behaviors or decision processes [Janssen and Ostrom 2006; Berger and Schreinemachers 2006; Pohill et al. 2010]. The ABM described here incorporates many of these strategies as described in the following sections.

Researchers employing ABMs and other modeling approaches are increasingly reaching out to the populations being studied in an effort to better inform the conceptual basis of their models and/or collaboratively define the goals, outcomes, and procedures embedded within them. Such efforts, often termed *participatory modeling* or *collaborative modeling,* are not new; some argue that they are useful approaches when dealing with complex systems and ABM [Ramanath and Gilbert 2004; Matthews et al. 2007; Edmonds 2010]

In participatory modeling, the structure of the model is an outcome of the research. Stakeholders help define the parameters used in the ABM, provide feedback on the suitability of decision rules or interactions among agents, and contribute local knowledge about observed agent behavior. The FIRECLIM project established stakeholder panels for each of the three agents in the ABM. Panels include local informants with specific knowledge of Flathead County and the people who live there. Participatory modeling uses stakeholder panels to help guide development and provide verification of decision rules, specify parameters of management alternatives, and critique the ABM process as a whole. Specific instances where stakeholder panels are consulted during the development and implementation of the FIRECLIM ABM are identified below.

Greater computer processing power, the development of various software toolkits that streamline the implementation of ABM (e.g., NetLogo, RePast, MASON, SWARM), and the positive benefits of the approach described above have increased the utilization of ABM in simulating complex problems and systems [Railsback et al. 2006; Nikolai and Madey 2009]. ABMs have been used to study processes ranging from natural resource management to organizational effectiveness and influenza outbreaks [Jager et al. 2002; Chu et al. 2009]. Beyond the relatively few applications of ABM to wildfire (described in the next section), ABM applications relevant to this chapter include: (1) simulating forest management or conversion of forestland to agricultural uses [Hoffman et al. 2002; Bone and Dragiecivic 2010]; (2) modeling evacuation, emergency response, or mitigation of hazards [Chen et al. 2006; Yu et al. 2009]; (3) simulating land use change and resultant changes in property values and/or ecosystems [Ligtenberg et al. 2004; Matthews et al. 2007]; and (4) exploring the efficacy of alternative policies for solving environmental problems, such as water use or habitat fragmentation [An et al. 2006; Chu et al. 2009].

ABM AND WILDFIRE RESEARCH

Efforts to model the spread and severity of wildfire and its potential impacts on human settlement have a long history. Recent efforts to model potential losses from wildland fire integrate human actions by simulating the effects of different forest treatments on fire occurrence or the effect of human suppression tactics (dynamic or otherwise) on fire spread [Reinhardt et al. 2008; Kim et al. 2009; Reinhardt et al. 2010]. Other literature attempts to identify potential structure losses from wildfire by identifying the proximity of human development to wildland vegetation, historical wildfire occurrence, and areas with extreme topography and weather [Radeloff et al. 2005; Stewart et al. 2007; Platt 2010].

Despite these efforts, very little research has utilized ABM to dynamically model human decision makers as a component of the coupled natural-human system that includes wildfire occurrence, severity, or potential impacts. Thorp et al. [2006] paired ABM, geographic information system (GIS) data, and the fire modeling simulator FARSITE to simulate the evacuation potential of Santa Fe, New Mexico during different wildfires. Research by Hu and Ntaimo [2009] and Ntaimo and Hu [2009] used ABM to simulate the effect of different firefighting tactics on hypothetical fires. Yin [2010] used an ABM approach to simulate how Colorado homeowners' choice of where to build new homes and various land use policies influence wildfire risk. Niazi et al. [2009] utilized ABM (with non-human agents) to simulate wildfire spread and to validate a wireless sensor network for monitoring fires. Prato et al. [2008] proposed an ABM approach for simulating future wildfire risk that integrates the behavior of various agents that are assumed to act based on the expected losses and expected benefits of their actions. The FIRECLIM ABM presented here is an expansion of Prato et al.'s [2008] proposal.

Although other research on evacuation during fires has employed components of ABMs, researchers do not always identify with the ABM approach in an explicit way. For instance, Cova and Johnson's [2002] microsimulations of homeowner evacuation during wildfire combined dynamic traffic and fire models to determine whether additional egress would facilitate more effective evacuation. Pultar et al. [2009] paired models simulating wildfire spread and human mobility with a dynamic GIS to determine the most effective wildfire evacuation protocols. Finally, Korhonen et al. [2010] used ABM to simulate evacuation from structure fires.

The above applications of ABM to wildland fire do not address many of the social actors or human actions that influence wildland fire losses. Static simulations of wildfire-related decisions are insufficient because they exclude a variety of human factors that influence the potential impacts of wildfire. Among the important human factors currently neglected in many studies and identified in the social science literature on wildfire are: (1) the collective or individual actions of homeowners to reduce wildfire risk on their properties; (2) the importance of community networks in disseminating information about the ecological benefits of wildfires; (3) the use of land use policies to reduce vulnerability of private property to wildfire losses; and (4) the importance of local wildfire planning and mitigation backed by national or state policies (e.g., Community Wildfire Protection Planning described in the Healthy Forests Restoration Act) and/or local programs (e.g., Firewise, FireSafe Montana). Research that addresses aspects of items 1-4 include McCaffrey [2006], Grayzeck et al. [2009], Winter et al. [2009], and Paveglio et al. [2010]. Other research has documented the variety of characteristics that lead to wildfire mitigation by homeowners, agencies, or land use planners [Daniel et al. 2007; Martin et al. 2008].

Efforts to integrate the above social factors into dynamic models of wildfire spread, severity, or risk and possible wildfire losses are currently lacking. The majority of studies addressing wildfire risk treat social actors and their actions regarding preparation for, mitigation of, or recovery from wildfire in a static or uniform manner. Static or uniform modeling of human actions in regards to wildfire often extends to agency management of forests to balance ecosystem health and potential wildfire losses, private residents' decisions about mitigation actions that could reduce potential losses from wildfire, and community planners' decisions about residential growth in fire-prone areas. For instance, some studies combine the WUI (i.e., number and density of structures, and nearby fuel loadings) with dynamic simulation models of fire occurrence, severity, and spread (e.g., FlamMap and FARSITE) to improve prediction of where severe fires are likely to threaten life safety and property [Haight et al. 2004; Theobald and Romme 2007; Bar Massada et al. 2009]. Treating human agents in a uniform and static manner, as is done in these studies, is deficient because human actions often change in response to wildfire and are heterogeneous across agents. As with many other broad assessments of probable structure loss from wildfire, these authors assume that any structure within a fire perimeter is destroyed. Such assumptions are a weakness of large-scale wildfire risk assessments [Stockmann 2009; Platt 2010].

Another class of wildfire studies focuses on testing the effect of one or more characteristics (e.g., land use policy, climate change, and fuel reduction) on current or future wildfire risk to human settlements assuming other variables, including social actors, are static over time. Examples of this class of studies include simulating the effects of future climate change on wildfire risk [Westerling and Bryant 2008], possible WUI expansion [Gude et al. 2008; Hammer et al. 2009], and fuel reduction [Reinhardt et al. 2008; Schoennagel et al. 2009].

There are major advantages to developing a comprehensive model for simulating the effects of wildfire-related decisions on wildfire risk. Such a model needs to account for the interdependent behaviors of multiple social actors who make decisions that influence or are influenced by wildfire risk. In addition, the model needs to demonstrate how biophysical and social processes interact to influence the benefits, costs, and risks of wildfires, how those benefits, costs, and risks influence agent behavior with respect to wildfire, and how agent behavior influences wildfire losses. The ABM presented in this chapter contributes to the development of such a model.

ABM DECISION RULES FOR AGENTS

An agent's decisions influence and/or are influenced by their exposure to or potential impacts from wildfire. In addition, an agent's decisions can influence the behaviors of other agents. The ABM assumes that members of each agent make decisions based on certain rules that differ across agents. This section outlines the preliminary parameters, behavioral rules, and assessments that influence the decisions made by each of the three agents in the FIRCLIM ABM.

Actions and effects for the three agents are simulated at different levels (e.g., multiple agencies or individual parcel owners) and spatial scales (e.g., Flathead National Forest lands, private subdivisions, or individual parcels). The levels chosen for each agent depends on how members of that agent typically make decisions. Decision rules for each agent are influenced by two sets of factors: (1) endogenous factors that are controlled by members of the agent; and (2) exogenous factors that are controlled by other agents or are external to all agents. This approach reflects the way agents behave in a coupled human-natural system because it simulates how the interactions between agents and their environment continually modify both potential wildfire losses and the decisions of individual agents.

The following sections present the parameters, decision rules, and other factors that influence the actions of the three agents included in the FIRECLIM ABM. The section for each agent contains individual numbered steps that describe various procedures.

Definition of WUI and other Area Designations

This section defines the WUI and other area designations relevant to agents' decisions in the FIRECLIM ABM. The WUI is defined as the ever-expanding area where residential development is adjacent to or intermingled with wildland vegetation [USDA and USDI 1995; USDA and USDI 2001]. The cost of suppressing wildfires in the WUI is consistently cited as a primary driver of wildfire management decisions and costs. The WUI for Flathead County is defined based on the Federal Register [USDA and USDI 2001; Radeloff et al. 2005; Stewart et al. 2007] and additional criteria established during development of the Flathead County Community Wildfire Protection Plan (CWPP) [GCS Research 2005]. The Federal Register defines the WUI as an "…area where houses meet [interface WUI] or intermingle with [intermix WUI] undeveloped wildland vegetation." It is determined primarily by a density criterion for houses, which is more than one housing unit per 16 ha, and proximity to wildland vegetation. This definition excludes from the WUI areas containing commercial, institutional, or industrial (CI&I) facilities that meet or intermingle with wildland vegetation and parcels developed at a density less than or equal to one housing unit per 16 ha. The WUI is delineated at the end of each 10-year subperiod to account for simulated changes in residential development and wildland vegetation. Simulated future residential development places additional structures at risk from wildfire. A description of the complete procedure used to delineate the WUI during each subperiod is given in Appendix A of the FIRECLIM ABM description and can be found at: http://projects.cares.missouri.edu/fireclim-montana/Methods/ABM_Appendices.pdf

The Flathead County Community Wildfire Protection Plan process, which was encouraged by policy targets in the Healthy Forests Restoration Act, includes collaborative designation of county-wide and individual fire district *priority areas* for wildfire risk and fuels reduction. Priority areas reflect local professionals' opinions about where fuel reduction would be most beneficial in terms of reducing wildfire threat to resident and firefighter life safety or losses to private property [GCS Research 2005]. County-wide priority areas are broad regions that extend across fire districts while fire district priority areas are smaller, ranked areas determined independently by individual fire districts and restricted to acres within their borders. The FIRECLIM ABM includes CWPP priority as one of the areas where Agent$_a$ can choose to perform fuel reduction activities (i.e. forest harvest, forest thinning, prescribed burning). Some fire district priority areas are nested within county-wide priority areas and others are not. A map of these areas appears in Appendix B of the FIRECLIM ABM description and can be found at: http://projects.cares.missouri.edu/fireclim-montana/Methods/ABM_Appendices.pdf

Non-WUI areas are those for which the housing density criterion and/or the wildland vegetation criterion for the WUI are not satisfied. Such areas can be further distinguished as *developed non-WUI areas* and *remote non-WUI areas*. Developed non-WUI areas are those that do not meet the WUI criterion for vegetation and are near or in urban areas. Remote non-WUI areas are those that do not meet the WUI criterion for housing density and extend outward near or interspersed among wildland vegetation. The focus of the FIRECLIM model is WUI areas, although later versions of the model may be used to test the difference in wildfire risk for both developed non-WUI areas and remote non-WUI areas.

Homeowners (Agent$_h$)

The ABM simulates Agent$_h$ decisions at the level of the parcel owned and/or managed by each homeowner. Agent$_h$ makes three decisions: (1) whether or not to perform fuel reduction around residential structure(s) on their property; (2) if fuel reduction is performed, the level of fuel reduction conducted on their property and; (3) the roofing and wall materials selected for structures on new residential properties, which affects their flammability.

These three decisions have a direct bearing on the conditional probability that a structure burns given the parcel in which it is located burns. That probability is a primary determinant of expected residential property losses from wildfire, or E(RLW) in the full FIRECLIM model. E(RLW) influences the decisions made by all three agents and is used to define overall net wildfire risk for the study area.

Additional efforts by homeowners to reduce wildfire risk are a significant need in wildfire management and one way to alleviate the costly and unsustainable fire suppression tactics that characterize wildfire management in the United States [Steelman and Burke 2007; Daniel et al. 2008; Martin et al. 2008]. Specific definitions and types of fuel reduction or building materials simulated in the FIRECLIM model are described in step 2 of this section. Figure 1 provides an overview of the decision-making process for Agent$_h$.

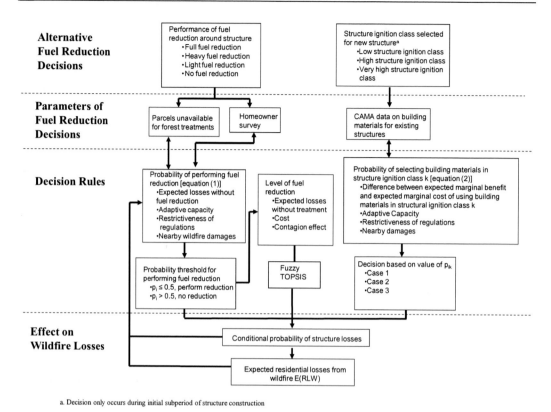

Figure 1. Schematic of parameters and decisions for Agent$_h$.

1. Nature of Agent$_h$

Agent$_h$ includes private citizens who own residential properties or parcels containing structures in or adjacent to the WUI, remote non-WUI, or CWPP priority areas. The RECID2 model simulates future residential development and the ABM simulates how homeowners manage properties with respect to wildfire. The parameters used in the simulation of homeowners' future management of properties with respect to wildfire are described in section 2 below.

Agent$_h$ excludes private citizens who own residential structures in *developed non-WUI areas* because: (1) such homeowners are much less likely to experience threats to life safety or property losses from wildfire; (2) given the first point, homeowners' personal actions are likely to have a much smaller effect on E(RLW), the primary metric for assessing wildfire impacts in the study area; and (3) areas outside the WUI are not a policy and program focus of much wildfire management in the United States. Additional restrictions on members of Agent$_h$ that influence their decisions about fuel reduction and building materials are described in the following steps.

2. Alternative Decisions for Agent$_h$

The FIRECLIM model simulates homeowner selection of one of four levels of fuel reduction areas around their homes: (1) full fuel reduction; (2) heavy fuel reduction; (3) light fuel reduction; and (4) no fuel reduction. Fuel reduction levels and their parameters (i.e., amount of vegetation removed, allowance of vegetation remaining) are based on recommendations by the Firewise Communities USA program [2011] and National Fire Protection Association standards [2007]. Successful performance of a fuel reduction level can reduce the conditional probability that structures burn given the parcel in which the structures are located burn, thus reducing structure losses from future wildfires. The reduction in the conditional probability that structures burn is based on research by Stockmann et al. [2010] and Cohen [2000, 2008]. Fuel reduction values are summarized in Appendix C of the FIRECLIM ABM description and can be found at: http://projects.cares.missouri.edu/fireclim-montana/Methods/ABM_Appendices.pdf

Members of Agent$_h$ whose properties contain vegetation that is not available for fuel reduction (e.g., grassland and shrubland) are removed from ABM simulation of fuel reduction at the beginning of each subperiod. LANDFIRE EVT and FIA vegetation classes maintained by the USDA Forest Service are used to determine wildland vegetation. Both data sets are widely used to simulate fire processes [Haight et al. 2004; Bar Massada et al. 2009]. Vegetation types subject to fuel reduction were determined by reviewing literature on the vegetation classifications included in the above data sets and consulting with ecologists at the USDA Forest Service Missoula Fire Sciences Laboratory. Major vegetation types considered and not considered available for fuel reduction treatment are identified in Table 1. The 80% threshold for the amount of vegetation in a given class and parcel is used: (1) to allow members of Agent$_h$ with property in mixed vegetation classes the option to perform fuel reduction around their properties; and (2) because fuel reduction on parcels with more than 80% of vegetation unavailable for fuel reduction treatments is unlikely to reduce the conditional probability that structures burn.

Because vegetation growth and change is dynamic, it is simulated using the Fire-BGCv2 model [Keane et al. 1996]. Hence, some homeowners unable to perform fuel reduction on their properties in early subperiods may be able to perform fuel reduction in later subperiods. Also, homeowners may be able to perform fuel reduction on their properties in early subperiods, but not in later subperiods if disturbance (i.e., wildfire, harvest, etc) causes changes in vegetation on the parcel.

The conditional probability that structures burn during a given subperiod is also influenced by the exterior wall and roofing materials used in that structure. The Montana Cadastral Mapping (CAMA) data [2010] is used to determine the exterior wall and roofing materials used in *existing residential structures* in Flathead County. The CAMA data includes 10 exterior wall material classifications and 11 roof material classifications, each of which are independently placed in three flammability categories: (1) low; (2) moderate; and (3) high. Various combinations of the exterior roof and wall classifications are then organized into three *structure ignition classes* that describe the overall flammability of building materials used in structures: (1) low; (2) high; and (3) very high. The basis for: (1) classification of exterior materials into flammability categories; (2) classification of structure ignition classes; and; (3) the effect of structure ignition classes on the conditional probabilities that structures burn are determined based on research by Stockmann et al. [2010], NFPA [2007] or Firewise [2011] recommendations for home construction, and other research on structural flammability [Cohen 1995]. The authors contacted structure ignition experts from the National Institute of Standards and Technology, various academic institutions, and the USDA Forest Service for feedback on how to evaluate flammability of building materials and structure ignition class. The final procedure for determining the conditional probability that a structure burns employs a decision-tree approach described in Appendix C of the FIRECLIM ABM description and can be found at: http://projects.cares.missouri.edu/fireclim-montana/Methods/ABM_Appendices.pdf.

3. *Specifying Initial Inputs for Management Alternatives*

The conditional probability that structures on *existing* or *future* residential properties in the study area burn are estimated based on the: (1) number of members of Agent$_h$ that perform fuel reduction; (2) level of fuel reduction by members that perform fuel reduction; and (3) building materials used in structures.

Table 1. Vegetation types that are or are not available for fuel reduction treatment in the FIRECLIM ABM

Vegetation type	Wildland vegetation[a]
Agriculture-General	No
Agriculture-Pasture/Hay	No
Agriculture-Cultivated Crops and Irrigated Agriculture	No
Rocky Mountain Aspen Forest and Woodland	Yes
Northern Rocky Mountain Dry-Mesic Montane Mixed Conifer Forest	Yes
Northern Rocky Mountain Subalpine Woodland and Parkland	Yes
Northern Rocky Mountain Mesic Montane Mixed Conifer Forest	Yes
Rocky Mountain Lodgepole Pine Forest	Yes
Northern Rocky Mountain Ponderosa Pine Woodland and Savanna	Yes
Rocky Mountain Subalpine Dry-Mesic Spruce-Fir Forest and Woodland	Yes
Rocky Mountain Subalpine Wet-Mesic Spruce-Fir Forest and Woodland	Yes
Inter-Mountain Basins Aspen-Mixed Conifer Forest and Woodland	Yes
Rocky Mountain Alpine Dwarf-Shrubland	No
Inter-Mountain Basins Big Sagebrush Shrubland	Yes
Northern Rocky Mountain Montane-Foothill Deciduous Shrubland	Yes
Columbia Plateau Low Sagebrush Steppe	No
Inter-Mountain Basins Big Sagebrush Steppe	Yes
Inter-Mountain Basins Montane Sagebrush Steppe	Yes
Northern Rocky Mountain Lower Montane-Foothill-Valley Grassland	No
Northern Rocky Mountain Subalpine-Upper Montane Grassland	No
Rocky Mountain Alpine Fell-Field	No
Rocky Mountain Subalpine-Montane Mesic Meadow	No
Rocky Mountain Montane Riparian Systems	Yes
Rocky Mountain Subalpine/Upper Montane Riparian Systems	Yes
Northern Rocky Mountain Conifer Swamp	Yes
Middle Rocky Mountain Montane Douglas-fir Forest and Woodland	Yes
Northern Rocky Mountain Subalpine Deciduous Shrubland	Yes
Introduced Upland Vegetation - Perennial Grassland and Forbland	Yes
Artemisia tridentata ssp. vaseyana Shrubland Alliance	Yes
Pseudotsuga menziesii Forest Alliance	Yes
Larix occidentalis Forest Alliance	Yes

Notes: Properties with more than 80% of the vegetation in one type or a combination of types marked yes are available for fuel reduction treatment.

The ABM procedure employs a representative survey of Flathead County homeowners to determine the current proportion of Agent$_h$ members performing various levels of fuel reduction around their structures. The survey includes a number of questions designed to better understand the factors (i.e., perceived fire risk, community initiatives, fire experience, income) that influence Agent$_h$'s performance or level of fuel reduction and other wildfire-risk averse activities (e.g., gutter clearing, establishment of water supply, evacuation planning, and installation of vent screens). The survey also includes questions comparing the importance of market (i.e., potential timber losses and cost of homeowner fuel reduction) and non-market values (i.e., aesthetic value of property, recreational opportunities, and fear of harm) that members of Agent$_h$ consider when making fuel reduction decisions. The simulation of existing and new homeowners' decisions regarding fuel reduction is determined by the endogenous decision rules outlined in step 4 below. Because individual members of Agent$_h$ make decisions about fuel reduction during each subperiod, fuel reduction treatments can vary over subperiods.

4. Decision Rules for Agent$_h$

The decisions about whether or not to perform fuel reduction and, if so, the level of fuel reduction are made in every subperiod, whereas the building materials selected for structures on future residential properties are made in the subperiod in which the property is developed. Each decision is influenced by different factors and is based on different decision rules for the agent, which is summarized in Figure 1. The decision rules for Agent$_h$ described below are only applicable when not superseded by the decisions made by Agent$_p$ (see Agent$_p$ section below). The outcome of the decision to perform or not perform fuel reduction around structures on a property is determined based on the value of the probability of performing fuel reduction. The latter is a function of expected property losses from wildfire without fuel reduction, adaptive capacity of the homeowner to perform fuel reduction, the restrictiveness of WUI regulations, and the recent impact of fires on nearby lands. The decision regarding the level of fuel reduction is relevant only when fuel reduction is performed on a property. Decisions regarding the level of fuel reduction are determined using a multiple attribute evaluation procedure known as the fuzzy Technique for Order Preference by Similarity of Ideal Solution—fuzzy TOPSIS for short [Hwang and Yoon 1981; Chen and Hwang 1992; Chen 2000; Berger 2006). The fuzzy TOPSIS procedure is described in section b below. Factors influencing the level fuel reduction decision include expected residential property losses from wildfire, cost of treatment, and a contagion effect, which are treated as attributes of treatments. Although other factors can influence fuel reduction decisions, the ABM for Agent$_h$ advances modeling research on wildfire by integrating additional complexity not currently found in other studies of wildfire risk.

a) Decision to Perform Fuel Reduction

The decision of whether or not a homeowner performs fuel reduction on a residential property is determined based on the value of the following probability:

$$p_i = a_1 E(LW_{iu}) + a_2 A_i + a_3 R_i + a_4 E(DN_{id}) \tag{1}$$

where:

p_i = probability a homeowner performs fuel reduction on property (or parcel) i;
a = weight for attribute i (i = 1, 2, 3, 4 and $\sum_{i=1}^{4} a_i = 1$);
$E(LW_{iu})$ = normalized expected losses from wildfire for property i without fuel reduction ($0 \leq E(LW_{iu}) \leq 1$);
A_i = normalized adaptive capacity of homeowner for property i to perform fuel reduction ($0 \leq A_i \leq 1$);
R_i = normalized restrictiveness of WUI regulations applicable to property i ($0 \leq R_i \leq 1$); and
$E(DN_{id})$ = 0 or fixed increases in p_i when simulated wildfire damages impact parcels within d_1 or d_2 distance of parcel.

Because the weights and normalized attributes fall in the zero-one interval, $0 \leq p_i \leq 1$ for all i. $R_i = 0$ for properties where Agent$_p$'s decisions do not influence Agent$_h$'s decisions (i.e., fuel reduction on residential properties outside the WUI in the case of the current and moderately restrictive subdivision regulation scenarios described in Agent$_p$ section below). Whether or not homeowners perform fuel reduction on their properties is determined using the following decision rule: (1) if $p_i \leq 0.5$, the homeowner for property i does not perform fuel reduction; and (2) if $p_i > 0.5$, the homeowner for property i performs fuel reduction. Final specification of equation (1) and selection of the attribute weights are determined in collaboration with the homeowners' stakeholder panel.

$E(LW_{iu})$ depends on the probability that property i burns, which is influenced by the decisions of Agent$_a$, the conditional probability of wildfire losses to the residential structures on property i without fuel reduction, the conditional probability of loss in aesthetic value of property i without fuel reduction, the value of residential structures on property i, and the total value of residential property i. p_i increases (or decreases) as $E(LW_{iu})$, A_i, R_i, and $E(DN_{id})$ increase (or decrease).

Adaptive capacity is commonly referred to as the characteristics of a local social system (i.e., local resources, knowledge and experience of population, and relationships between people) that allow for continual action in the face of disturbance or change [Nelson et al. 2007; Paveglio et al. 2010]. A_i is estimated based on information obtained from the stakeholder panels and two focus groups composed of Flathead County emergency professionals, firefighters, community foresters, and others with experience in reducing wildfire risk. Focus groups were conducted in September 2010. Fourteen participants attended the first focus group and 15 participants attended the second focus group. Each focus group lasted between 4 and 5 hours and was designed to explore meanings for the concept of adaptive capacity for wildfire and obtain assessments of its variable existence among populations in Flathead County. Focus group participants acknowledged that adaptive capacity is an indicator for, among other things, the willingness, likelihood, and ability of local populations to perform fuel reduction around their homes or build with fire resistant materials.

After discussing the parameters of adaptive capacity, focus group participants provided collective assessments for nine distinct areas or *functional communities* within the study area. The study area includes the majority of Flathead County with a focus on inhabited areas. A map of the study are can be viewed in Appendix D of the FIRECLIM ABM description at: http://projects.cares.missouri.edu/fireclim-montana/Methods/ABM_Appendices.pdf

Functional communities are described by Jakes et al. [1998] as areas that share similar resources, approaches, and problem conceptualizations regarding natural resource issues. Determining functional communities includes consideration of where social (e.g., demographic or local relationships), topographic (i.e., local aesthetics, forest boundaries, and roads) or common views of natural resource issues (i.e., conservation and forest harvest) lead to collections of individuals (i.e., communities) that act in similar ways regarding an issue such as wildfire risk. The authors determined initial geographic boundaries for the functional communities assessed during the focus groups and contacted key informants in the county to refine these areas. New structures are assigned the adaptive capacity score determined for the area in which the structures are located.

R_i is estimated based on information provided by the community-regional planners' panel (see Agent$_p$ section below).

Weights a_1, a_2, a_3 and a_4 are determined using the fixed point scoring method [Saaty 1987] that requires stakeholders to assign 100 points to the four attributes.

$E(DN_{id})$ is determined based on Fire-BGCv2 and FSIM [Finney 2007; Calkin et al 2010] simulations of wildfires within d distance of parcel i during the previous subperiod. Two distances are simulated because the results of focus group meetings with the FIRECLIM stakeholder panel and interviews with local stakeholders suggest that the probability of Flathead County residents' performing fuel reduction increases following wildfire events that either: (1) damage nearby properties; or (2) force evacuation of homeowners responsible for risk-averse actions. The fixed increases corresponding to d_1 and d_2 are separate values, with $d_1 > d_2$. Only one value of $E(DN_{id})$ is applied to each parcel in the study area during a subperiod. This procedure is illustrated in the decision tree in Figure 2.

Because there is little empirical data concerning how much more likely residents are to perform fuel reduction when a wildfire occurs on a nearby property or how close the wildfire needs to be to prompt fuel reduction action, preliminary values of d_1 and d_2 are determined in consultation with key informants in Flathead County and wildfire experts. Final values are determined in consultation with the FIRECLIM stakeholder panel.

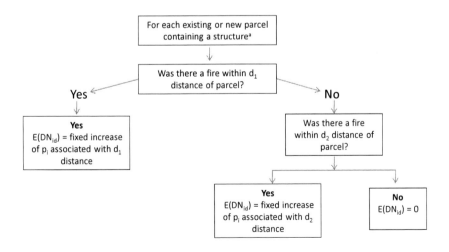

a. Excludes properties having more than 80% of their vegetation in classes not considered wildland vegetation.

Figure 2. Decision tree for determining the values of $E(DN_{id})$.

b) Decision about Level of Fuel Reduction Treatment

The best level of fuel reduction treatment (i.e., heavy, full, or light) for a property is evaluated and determined using a fuzzy TOPSIS method. TOPSIS is a variation of the ideal point method that ranks alternatives based on their closeness coefficients. A closeness coefficient measures how close the attributes for various decision alternatives, such as full, heavy, or light fuel reduction treatment, are to the attributes of the fuzzy positive-ideal solution and how far away they are from the attributes of the fuzzy negative-ideal solution. The positive-ideal solution has the most favorable and the negative-ideal solution has the least favorable attributes of treatments. The best fuel reduction treatment for a property is the one with the highest closeness coefficient. An advantage of the fuzzy TOPSIS method is that it does not assume utility independence of attributes and a risk neutral decision-maker[1], as does the more commonly-used utility function approach for ranking alternatives (e.g., Prato and Hajkowicz 2001; Prato 2003). Steps in the fuzzy TOPSIS method are as follows.

First, select the attributes of treatments. Three attributes are selected: $E(LW_{ij})$; cost; and contagion effect. $E(LW_{ij})$ is the expected loss from wildfire for property i with fuel reduction treatment j. It depends on the probability that property i burns, the probability of wildfire losses to residential structures on property i with fuel reduction treatment j given property i burns, the probability of loss in aesthetic value of property i with fuel reduction treatment j given property i burns, the value of the residential structure(s) on property i, and the total value of property i. Cost of treatment is the subperiod cost of conducting a treatment. A contagion effect measures the extent to which higher levels of fuel reduction on one property increase fuel reduction on nearby properties. The adaptive capacity score for a given area (see subsection a above for description) is used as an indicator of the strength of the contagion effect in that area.

Second, estimate the values of the three attributes of treatments for all residential properties in the study area using certain models/information. Models for estimating values include: (1) the Fire-BGCv2 [Keane et al. 1996] and FSIM [Finney 2007; Calkin et al 2010] landscape and fire simulation models; and (2) the IMPLAN model. The latter simulates economic output and employment for specified annual growth rates for eleven sectors of the Flathead County economy [Prato et al. 2007; Minnesota IMPLAN Group, Inc. 2011]. Information comes from: (1) the FIRECLIM stakeholder panel; (2) focus groups; and (3) homeowner survey.

Third, designate each attribute as positive or negative. The desirability of a fuel reduction treatment increases (or decreases) as a positive attribute increases (or decreases). In contrast, the desirability of a treatment decreases (or increases) as a negative attribute increases (or decreases). $E(LW_{ij})$ and cost of treatment are negative attributes and the contagion effect is a positive attribute. Therefore, a treatment with a higher (or lower) $E(LW_{ij})$ and/or a higher (or lower) cost is less (or more) desirable. A higher (or lower) contagion effect increases (or decreases) the desirability of all treatments.

Fourth, individual members of the homeowners' stakeholder panel: (1) rate the desirability of the fuzzy sets defined on the attributes by assigning each set a linguistic variable (e.g., very poor or very low, poor or low, medium poor or medium low, fair, medium

---

[1] A risk neutral decision-maker compares and ranks alternatives based solely on their expected values.

good or medium high, good or high, and very good or very high); and (2) rate the relative importance of each attribute by assigning it a linguistic variable (e.g., very low, low, medium low, medium, medium high, high, and very high).

Fifth, assign triangular fuzzy numbers to the linguistic variables used to rate the values and relative importance of attributes. Triangular fuzzy numbers corresponding to the ratings of attributes and their relative importance assigned by individual members of the homeowners' stakeholder panel are averaged to obtain triangular fuzzy numbers for the ratings and relative importance of attributes for the panel as a whole.

Sixth, calculate vertex distances between the values of attributes of treatments for individual residential properties and the values of attributes of treatments for the fuzzy positive- and the fuzzy negative-ideal solutions. Normalized attributes (i.e., raw values of attributes converted to the zero-one interval) are $E(LW_{ij}) = 0$, cost of treatments = 0, and contagion effect = 1 for the positive-ideal solution, and $E(LW_{ij}) = 1$, cost of treatments = 1, and contagion effect = 0 for the negative-ideal solution.

Seventh, calculate closeness coefficients for treatments using their vertex distances. Treatments with higher (or lower) closeness coefficients are more (or less) desirable because they have attributes that are closer to the positive-ideal solution and farther from the negative-ideal solution. The fuel reduction treatment with the highest closeness coefficient is the best fuel reduction treatment for a property.

c) Decisions about Building Materials

The ABM for Agent$_h$ assumes: (1) new residential structures added to parcels during subperiods incorporate one of three combinations of exterior wall and roofing materials (building materials for short); and (2) homeowners' preferences for building materials are revealed by their choice of residential structures. The combination of exterior wall and roofing materials is chosen by homeowners only once, during the initial subperiod of their construction (i.e., the FIRECLIM ABM does not simulate retrofitting). Building materials determine the structure ignition class for structures of which there are three: low (l); high (h); or very high (vh). Building materials for structures on new residential property i are determined by combining the following probabilities with several decision rules:

$$p_{ik} = a_1[E(MB_{ik}) - E(MC_{ik})] + a_2 A_{ik} + a_3 R_{ik} + a_4 E(DN_{id}) \quad (k = h, vh) \qquad (2)$$

where:

p_{ik} = probability that the homeowner for property i selects structure(s) with building materials in structure ignition class k;

a_j = weight for attribute j (j = 1, 2, 3, 4 and $\sum_{j=1}^{4} a_j = 1$);

$E(MB_{ik})$ = normalized expected marginal benefit of building materials in structure ignition class k relative to building materials in the low structure ignition class for property i ($0 \leq E(MB_{ik}) \leq 1$);

$E(MC_{ik})$ = normalized expected marginal cost of building materials for structure ignition class k relative to building materials for the low structure ignition class for property i ($0 \leq E(MC_{ik}) \leq 1$);

A_{ik} = normalized adaptive capacity of homeowner for property i to choose building materials corresponding to structure ignition class k ($0 \leq A_{ik} \leq 1$); and

R_{ik} = normalized restrictiveness of WUI regulations for property i with respect to the building materials required in structure ignition class k ($0 \leq R_{ik} \leq 1$).

$E(DN_{id}) = 0$ or fixed increases in p_{ik} when simulated wildfire impacts parcels within d_1 or d_2 distance of property i.

$E(MB_{ik})$ is the measured by the expected reduction in losses to buildings from wildfire between the low structure ignition class and structure ignition class k for property i during the first subperiod in which the property is developed. $E(MC_{ik})$ is measured by the increase in building cost between the low structure ignition class and structure ignition class k for property i. Because the weights and normalized attributes fall in the zero-one interval, $0 \leq p_{ik} \leq 1$ for all i and k. $R_{ik} = 0$ for properties for which Agent$_p$'s decisions do not influence Agent$_h$'s decisions (e.g., building material selected for residential properties outside the WUI in the case of the current and moderately restrictive subdivision regulation scenarios).

The decision rules for selecting building materials are:

Case 1: If $p_{ik} \leq 0.5$ for k = h, vh, the homeowner for property i selects building materials in the low structure ignition class;

Case 2: If $p_{ik} \leq 0.5$ and $p_{ik'} > 0.5$ for k ≠ k', the homeowner for property i does not select building materials in structure ignition class k because $p_{ik} \leq 0.5$, but does select building materials in structure ignition class k' because $p_{ik'} > 0.5$; and

Case 3: If $p_{ik} > 0.5$ for k = h, vh, the homeowner for property i selects building in the high structure ignition class if $p_{ih} > p_{ivh}$ or in the very high structure ignition class if $p_{ivh} > p_{ih}$.

$E(MB_{ik})$ is influenced by the decisions of land and wildfire management agencies (i.e., those decisions influence the probability that property i burns), the probability of wildfire losses to the residential structures on property i and the probability of loss in aesthetic value of property i with building materials in structure ignition class k and with or without fuel reduction given property i burns, the value of residential structures on property i, and the total value of residential property i. p_{ik} increases (or decreases) as $[E(MB_{ik}) - E(MC_{ik})]$, A_i, R_i, and $E(DN_{id})$ increase (or decrease). $E(MC_{ik})$ is determined in collaboration with building contractors and building supply companies. The remaining variables in equation (2) are defined in section b above.

d) Calibration of Equations (1) and (2)

There is a need to determine whether: (1) the outcomes of the decision rules for determining whether or not to perform fuel reduction based on equation (1) are consistent with the proportion of members of Agent$_h$ currently performing fuel reduction; and (2) the outcomes of the decision rules for selecting building materials based on equation (2) are consistent with historical use of building materials in the various structure ignition classes. Such calibration improves the accuracy of the decision rules used to simulate future decisions by Agent$_h$.

- Calibration of equation (1)

Proportions of Agent$_h$ currently performing fuel reduction is determined using a homeowner survey (see step 2 of Agent$_h$). Results from the survey are compared to the decision outcomes for subperiod 1 (2010-2019) based on equation (1). If necessary, equation

(1) is calibrated by varying the weights assigned to the attributes and the probability threshold, which is currently 0.50, until the proportions of properties doing fuel reduction are comparable to those obtained in the survey.

- Calibration of equation (2)

The CAMA parcel data contain the date of construction for existing structures in the study area. The structure ignition classes for structures built during the previous 10 years (1999-2009) are identified using the CAMA parcel data and compared to the classes obtained for subperiod 1 (2010-2019) using the initial version of equation (2). If necessary, equation (2) is calibrated by varying the attribute weights and probability thresholds until the proportions of residential properties in various structure ignition classes during the first subperiod are similar to the proportions for the period 1999-2009.

5. *Exogenous Factors for Agent$_h$*

Several exogenous factors influence the decisions made by Agent$_h$ (see Figure 5). First, decisions regarding land use policy can impact Agent$_h$'s decisions (see next section). Second, forest treatments selected by Agent$_a$ can alter the probabilities that parcels burn, which can influence $E(LW_{iu})$ for members of Agent$_h$. Third, climate change can influence $E(LW_{iu})$ and, hence, the probability that parcels owned by members of Agent$_h$ burn. Fourth, economic growth and associated residential development can alter WUI and non-WUI boundaries, which can alter the amount and types of fuel loads, forest treatments selected by Agent$_a$, the probabilities that parcels burn, and the building materials and level of fuel reduction treatments selected by Agent$_h$.

Community-Regional Planners (Agent$_p$)

1. *Nature of Agent$_p$*

Agent$_p$ includes the local governments of Flathead County and incorporated city governments (i.e., Whitefish, Kalispell, and Columbia Falls), most notably the planning and zoning departments that make decisions regarding residential and commercial development in the county. Members of Agent$_p$ make collective policy decisions about residential planning and development in each subperiod of the FIRECLIM simulation, including: (1) land use policies governing residential development; (2) requirements for fuel reduction around private residential structures located in subdivisions; and (3) building materials used in the construction of residential homes. These decisions influence the number and spatial pattern of residential structures in the study area that are vulnerable to wildfire damages. Inputs for considerations 2 and 3 are outlined in section 2 of the Agent$_h$ description. More restrictive land use policies have the potential to lower $E(RLW)$, which affects the conditional probability that private residential structures burn (see Agent$_h$ section). Less restrictive policies give Agent$_h$ more freedom in making decisions about fuel reduction around homes and choosing building materials for residential structures. Figure 3 provides an overview of the entire process for determining the ABM decisions for Agent$_p$.

2. Policy Alternatives and Decisions for Members of Agent$_p$

The ABM for Agent$_p$ assumes the members of Agent$_p$ choose one of nine possible policy alternatives for each subperiod at the beginning of the subperiod. Each policy alternative consists of unique combination of a land use policy scenario (i.e., current, moderately restrictive, or highly restrictive) designated ls and a subdivision regulation scenario (i.e., current, moderately restrictive, or highly restrictive) designated rs. The nine possible policy alternatives for each subperiod are:

1. Current ls and current rs
2. Moderately restrictive ls and moderately restrictive rs
3. Highly restrictive ls and highly restrictive rs
4. Current ls and moderately restrictive rs
5. Current ls and highly restrictive rs
6. Moderately restrictive ls and current rs
7. Moderately restrictive ls and highly restrictive rs
8. Highly restrictive ls and current rs
9. Highly restrictive ls and moderate rs

Not all the above policy alternatives are likely to be feasible or reasonable. Accordingly, the regional/community planners' stakeholder group is consulted to determine which, if any, of the nine policy alternatives should be omitted from consideration.

Effects of land use policy scenarios on residential development during subperiods are simulated using the RECID2 model [Prato et al., in press]. As the land use policy scenario becomes more restrictive (i.e., current to moderately restrictive to highly restrictive): (1) the percentage of the residential development at higher housing densities increases; (2) setbacks of residential developments from water bodies increase; (3) less residential development is

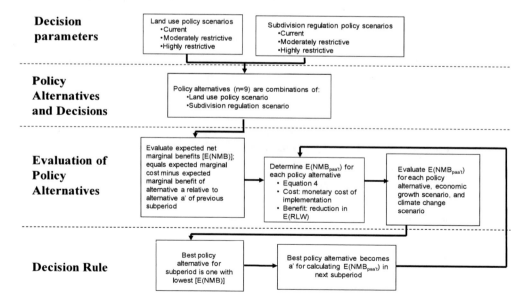

Figure 3. Schematic of parameters and decisions for Agent$_p$.

allowed in environmentally sensitive areas (e.g., state parks, national forests, national parks, and county parks); and (4) more limitations are placed on the types of residential development allowed on parcels outside of sewer accessible areas [Prato et al. 2007].

Subdivision regulation scenarios impose additional constraints on residential development in subdivisions located in CWPP priority areas, WUI areas, non-WUI areas. Constraints include: (1) whether homeowners are required to perform high levels of fuel reduction (full or heavy) around residential structures; (2) whether fuel reduction standards are enforced by $Agent_p$ after the final subdivision plat is approved; and (3) whether the structures on residential properties are required to use building materials in the low structure ignition class.

The preliminary parameters of the three subdivision regulation scenarios are:

a) The *current* subdivision regulation scenario is based on the existing Flathead County subdivision regulations regarding wildfire protection [Flathead County 2011]. Current subdivision regulations require that all parcels subdivided in the WUI after September 2007 must have full or heavy fuel reduction around residences and that residential structures use approved building materials in the low structure ignition class. There is no enforcement of fuel reduction following approval of the final plat. As such, the current subdivision regulation scenario assumes full or heavy fuel reduction in the areas occupied by new residential properties during the subperiod in which those properties are developed. The decision of whether or not to perform fuel reduction and the level of fuel reduction performed in subsequent subperiods is decided by the homeowner based on the decision process described for $Agent_h$ in section 3b.

b) The *moderately restrictive* subdivision regulation scenario requires that all new residential properties in WUI areas receive full or heavy fuel reduction around residential structures. Fuel reduction standards are enforced following approval of the final plat. Hence, all new residential properties in the WUI are assumed to receive full or heavy fuel reduction during each subperiod that the policy is in effect. The decision of whether or not to perform fuel reduction outside the WUI and the level of fuel reduction to perform is decided by the homeowner based on the decision process described for $Agent_h$ in section 3b. New residential properties added during or after the subperiod of the policy implementation are required to use building materials in the low structure ignition class. Building materials used in new residential structures outside the WUI are determined based on equation (2).

c) The *highly restrictive* subdivision regulation scenario requires that all new residential properties in WUI areas, CWPP priority areas, and non-WUI areas receive full or heavy fuel reduction around residential structures. Fuel reduction standards are enforced following approval of the final plat. Hence, all new residential properties in those areas are assumed to receive full or heavy fuel reduction during each subperiod that the policy is in effect. Residential properties added during or after the subperiod of the policy implementation are required to use building materials in the low structure ignition class. Equations (1) and (2) for $Agent_h$ do not apply to subperiods during which this scenario is in effect.

The community-regional planners' stakeholder panel can offer modifications to the preliminary parameters of the subdivision regulation scenarios described above.

3. *Decision Rules for Evaluation and Selection of Best Management Alternative for Agent$_p$*

The ABM for Agent$_p$ assumes that community-regional planners select the best policy pertaining to wildfire in the study area during each subperiod. Policies are selected for a number of reasons in addition to wildfire. For instance, planners may select more restrictive land use policies that promote clustered development, which has the benefit of reducing fragmentation of wildlife habitats and protecting environmentally sensitive areas [Howe et al 1997; Daniels 2001; Hansen et al. 2005]. The ABM makes the simplifying assumption that Agent$_p$ evaluates and selects policies based on their benefits and costs as they relate to wildfire. Specifically, the best policy for Agent$_p$ during each subperiod is the one that maximizes the expected net marginal benefits [E(NMB)], which is the difference between the expected marginal benefits of the policy in terms of reducing E(RLW) and the expected marginal cost of implementing the policy in the study area. This is a common decision rule in benefit-cost analysis [Prato 1998]. The procedures for estimating policy implementation costs for Agent$_p$ are described below.

E(NMB) of policy alternative a relative to policy alternative a' during subperiod t is defined as:

$$E(NMB_{paa't}) = E(MB_{aa't}) - E(MC_{aa't}) \quad (a, a' = 1, ..., 9) \tag{3}$$

where $E(MB_{aa't})$ is the expected marginal (or additional) benefit and $E(MC_{aa't})$ is the expected marginal (or additional) cost of implementing policy alternative a relative to policy alternative a' during subperiod t. $E(MB_{aa't})$ is defined as:

$$E(MB_{aa't}) = E_{at}(RLW) - E_{a't}(RLW) \quad (a, a' = 1, ..., 9) \tag{4}$$

where:

$E_{at}(RLW) = E(RLW)$ when policy alternative a is in effect during subperiod t; and
$E_{a't}(RLW) = E(RLW)$ when policy alternative a' is in effect during subperiod t.

$E_{at}(RLW)$ and $E_{a't}(RLW)$ depend on the probabilities that properties burn, the conditional probabilities of wildfire losses to the structures and aesthetic values of residential properties under policy alternative a and a', respectively, the values for structures and land on residential properties, and other variables.

$E(MC_{aa't})$ is defined as:
$$E(MC_{aa't}) = E(MP_{aa't}) + E(MN_{aa't}) \quad (a, a' = 1, ..., 9), \tag{5}$$

where:

$E(MP_{aa't})$ = expected marginal personnel cost of implementing policy alternative a relative to policy alternative a' during subperiod t; and

$E(MN_{aa't})$ = expected marginal non-personnel cost of implementing policy alternative a relative to policy alternative a' during subperiod t.

$E(MP_{aa't})$ is defined as:

$$E(MP_{aa't}) = [\sum_{g=1}^{n} E(H_{gaa't})E(W_{gt})] \qquad (a, a' = 1, \ldots, 9), \qquad (6)$$

where:

$E(H_{gaa't})$ = expected additional hours of employee type g required to implement policy alternative a relative to policy alternative a' during subperiod t;

$E(W_{gt})$ = expected hourly wage of employee type g during subperiod t; and

n = expected number of employee types involved in planning activities.

$E(H_{gaa't})$, $E(W_{gt})$, and n are determined in consultation with the community-regional planners' stakeholder panel. Combining equations (4) through (6) gives the following expression for the net marginal benefit of policy alternative a relative to policy alternative a':

$$E(NMB_{paa't}) = [E_{at}(RLW) - E_{a't}(RLW)] - [\sum_{g=1}^{n} E(H_{gaa't})E(W_{gt}) + E(MN_{aa't})] \qquad (7)$$

The best policy alternative for $Agent_p$ during subperiod t is the one with the highest $E(NMB_{paa't})$.

4. Exogenous Factors Influencing Agent$_p$'s Decisions

Several exogenous factors can influence $Agent_p$'s decisions (see Figure 5 for overview). First, forest treatments selected by $Agent_a$ and fuel reduction levels and building materials selected by $Agent_h$ can alter E(RLW).Changes in E(RLW) can change the best policy alternative for $Agent_p$. Second, climate change over subperiods can alter the burn probabilities for parcels containing residential properties, which can change E(RLW) and the best policy alternatives for $Agent_p$. Third, economic growth in Flathead County increases the number of residential properties added in each subperiod, which can increase the size of WUI and non-WUI areas. Both changes can alter E(RLW) and, hence, influence the best policy selected by $Agent_p$.

Land and Wildfire Management Agencies (Agent$_a$)

1. Nature of Agent$_a$

$Agent_a$ includes the following land and wildfire management agencies that perform forest treatments on lands they manage or own in the study area: (1) Flathead National Forest (FNF); (2) Plum Creek Timber Company (PC); (3) Glacier National Park (GNP); (4) Montana Department of Natural Resources and Conservation (DNRC); (5) Confederated Salish and Kootenai Tribes and other entities (i.e., county parks, Nature Conservancy) designated as Other State/Private Trust/Indian Lands (OL); and (6) lands receiving state or regional funds for fuel reduction (i.e., National Fire Plan or state funding), lands managed by other logging companies (i.e., Stoltze Lumber Co.), or residential properties managed by third party contractors (i.e., private logging operations) designated as Private Lands (PL). Members of Agent$_a$ influence wildfire primarily through the selection and placement of forest treatments in the study area. Types of forest treatments simulated in the FIRECLIM model are described below in steps 3 and 4.

Because of the diversity of members of Agent$_a$, the variables influencing decisions about forest treatments vary across members. These differences are accounted for by selecting different parameters for or imposing different constraints on forest treatment decisions of members of Agent$_a$, including the: (1) types of forest treatments applied to lands in the WUI, non-WUI or CWPP priority areas managed by different members; and (2) the amount of land managed by a member that conducts different forest treatments during each subperiod of the simulation. These parameters and constraints are based on current practices and policies of the six members of Agent$_a$ listed above and are determined in collaboration with the land and wildfire management agencies' stakeholder panel. The steps used to determine these parameters and constraints are described below. Figure 4 provides an overview of the process used to determine the decisions made by Agent$_a$.

2. Decision rules for Agent$_a$

At the beginning of each subperiod, each member of Agent$_a$ selects the best management alternative for the lands they own/manage from a set of management alternatives. The latter specify how many acres of forest treatments are allocated to lands owned/managed by each member and the location of treated acres. Forest treatments impact residential property losses by modifying fuel loads, the intensity of wildfires, and the severity of residential property losses from wildfires on private land or timber losses from wildfires on public land. The primary way forest treatments conducted by members of Agent$_a$ influence E(RLW) is by their

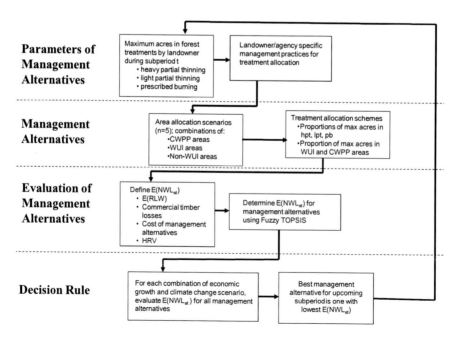

Figure 4. Schematic of parameters and decisions for Agent$_a$.

effects on the subperiod burn probabilities for parcels in the study area. Management alternatives for members of Agent$_a$ are described in steps 3 and 4 of this section. The best management alternative for a subperiod is the one that minimizes expected net wildfire losses for Agent$_a$ for that subperiod t [i.e., E(NWL$_{at}$)]. E(NWL$_{at}$) is defined in step 5.

3. Parameters of Management Alternatives for Agent$_a$

The FIRECLIM model simulates three forest treatments for members of Agent$_a$: (1) heavy partial thinning (hpt); (2) light partial thinning (lpt); and (3) prescribed burning (pb). Forest treatments are determined using focus groups and interviews with key informants on the land and wildfire management agencies' stakeholder panel. Individuals from each member of Agent$_a$ are consulted about landowner- or agency-specific management parameters needed to simulate forest treatments with the Fire-BGCv2 model, such as minimum and maximum DBH to harvest, minimum basal area to harvest, amount of slash left on stand, etc. Management parameters are used to create unique forest treatments for each member of Agent$_a$.

Information on current acreage in forest treatments conducted by members of Agent$_a$ is paired with information on the growth in the outputs of the wood products manufacturing and residential construction industries (taken from the RECID2 model) to calculate future acreage in forest treatments for members of Agent$_a$.

More specifically, for each member, forest treatment, and subperiod, maximum acres treated are calculated based on: (1) a triangular probability distribution of acres treated; (2) average acres per treatment during the 10-year period 2000-2009; (3) the percent by which maximum acres treated in future subperiods exceeds average acres treated during the 10–year period 2000-2009; (4) annual average growth rates for outputs of the wood products manufacturing and residential construction industries during the evaluation period for the economic growth scenario being simulated; and (5) annual growth in the size of the WUI. Members of Agent$_a$ use different forest treatments.

For instance, PC conducts hpt and lpt, but not pb. GNP conducts lpt and pb, but not hpt. These differences are summarized in Table 2. A complete description of the procedure used to calculate future maximum acres in forest treatments is given in Appendix E of the FIRECLIM ABM procedure found at: http://projects.cares.missouri.edu/fireclim-montana/Methods/ABM_Appendices.pdf

Table 2. Differences in parameters of forest treatments for members of Agent$_a$

Member of Agent$_a$	Forest treatments permitted	Increase in maximum acres over subperiods
Plum Creek (PC)	hpt, lpt	Yes
Glacier National Park (GNP)	lpt, pb	No
Flathead National Forest (FNF)	hpt, lpt, pb	Yes
Other State/Private Trust (OL)	hpt, lpt, pb	No
Private Lands (PL)	hpt, lpt	Yes
Dept. of Natural Resources (DNRC)	hpt, lpt	No

Two cases are specified and simulated for maximum acres by landownership, forest treatment, and subperiod: (1) annual maximum acres treated do not increase over subperiods; and (2) annual maximum acres treated increase over subperiods at the annual rates of growth in the outputs of the wood products manufacturing and residential construction industries for the economic growth scenario being simulated. Case 1 only applies to landowners whose forest management activities are linked to market values (i.e., US Forest Service, Plum Creek, private lands). Case 2 does not apply to Other State/Private Trust/Indian Lands, DNRC, and Glacier National Park because of these landowners' harvest policies.

4. Management Alternatives for Members of Agent$_a$

Modeling Agent$_a$'s decisions requires specifying management alternatives for each member and subperiod. A management alternative consists of an *area allocation scenario*, which designates the areas to which maximum acres are allocated. Area allocation scenarios are the primary management alternative for members of Agent$_a$ and are chosen at the beginning of each subperiod. Additional details for the area allocation scenarios are discussed below. Associated with each area allocation scenario are *treatment allocation schemes* that specify the proportions of maximum acres allocated to the three forest treatments. Additional details regarding treatment allocation schemes are described below.

Area allocation scenarios involve three areas: (1) Wildland Urban Interface (WUI), (2) Community Wildfire Protection Plan (CWPP) areas; and (3) areas outside of WUI areas (non-WUI areas) (see section on definition of WUI and other area designations).

Area allocation scenarios. Members of Agent$_a$ make a variety of decisions concerning the placement of forest treatments to reduce E(NWL$_{at}$) for the study area. The ABM for Agent$_a$ simulates the following area allocation scenarios and their effects on E(NWL$_{at}$):

1. CWPP priority areas only;
2. WUI areas (ignores CWPP priority areas);
3. WUI areas (includes CWPP priority areas);
4. WUI, CWPP and non-WUI areas (1); and
5. WUI, CWPP and non-WUI areas (2).

Area allocation scenarios are discussed with and possibly modified by the land and wildfire management agencies' stakeholder panel. Scenarios 3, 4 and 5 specify the proportion of maximum acres allocated to different areas and a treatment allocation scheme for each area. Scenarios 4 and 5 differ in terms of the proportions of acres allocated to each of the three areas. The procedure for determining the proportion of maximum acres allocated to different areas is described in the section below on treatment allocation schemes. It uses proportions collaboratively defined by stakeholder panel members, available policy targets, and the amount of land owned/managed by each member of Agent$_a$.

Treatment allocation schemes. Treatment allocation schemes for each member of Agent$_a$ are constant across area allocation scenarios. They can vary across members of Agent$_a$ based on the proportion of maximum acres allocated by each member to the three forest treatments during the 10-year period 2000-2009. These data are used to simulate maximum acres in future subperiods (see step 3) using the procedure described in Appendix E of the FIRECLIM ABM procedure found at:http://projects.cares.missouri.edu/fireclim-montana/Methods/ABM _Appendices.pdf.

A treatment allocation scheme specifies $\{a_{hpt}, a_{lpt}, a_{pb}\}$, where a_k is the number of acres of a landowners' subperiod maximum acres allocated to treatment k (k = hpt, lpt, pb), and $\sum_{hpt,lpt,pb} a_k = 1$. Note that $a_k = 0$ for landowners that do not use treatment k on their land. Alternatively, treatment allocation schemes for landowners can be specified in terms of the proportions (r) of maximum acres allocated to treatments (i.e., $\{r_{hpt}, r_{lpt}, r_{pb}\}$, where r_k is the proportion of maximum acres allocated to treatment k). The above procedure requires simulating the following five area allocation scenarios and associated treatment allocation schemes for Agent$_a$:

1. CWPP priority areas only;
   GNP $\{a_{hpt}, a_{lpt}, a_{pb}\}_{cw1}$
   PC $\{a_{hpt}, a_{lpt}, a_{pb}\}_{cw1}$
   FNF $\{a_{hpt}, a_{lpt}, a_{pb}\}_{cw1}$
   DNRC $\{a_{hpt}, a_{lpt}, a_{pb}\}_{cw1}$
   PL $\{a_{hpt}, a_{lpt}, a_{pb}\}_{cw1}$
   CT $\{a_{hmt}, a_{lpt}, a_{pb}\}_{cw1}$

2. WUI areas (ignores CWPP priority areas)
   GNP $\{a_{hpt}, a_{lpt}, a_{pb}\}_{wu1}$
   PC $\{a_{hpt}, a_{lpt}, a_{pb}\}_{wu1}$
   FNF $\{a_{hpt}, a_{lpt}, a_{pb}\}_{wu1}$
   DNRC $\{a_{hpt}, a_{lpt}, a_{pb}\}_{wu1}$
   PL $\{a_{hpt}, a_{lpt}, a_{pb}\}_{wu1}$
   CT $\{a_{hmt}, a_{lpt}, a_{pb}\}_{wu1}$

3. WUI areas (including CWPP priority areas)
   GNP $\{a_{hpt}, a_{lpt}, a_{pb}\}_{cw2}$
   PC $\{a_{hpt}, a_{lpt}, a_{pb}\}_{cw2}$
   FNF $\{a_{hpt}, a_{lpt}, a_{pb}\}_{cw2}$
   DNRC $\{a_{hpt}, a_{lpt}, a_{pb}\}_{cw2}$
   PL $\{a_{hpt}, a_{lpt}, a_{pb}\}_{cw2}$
   CT $\{a_{hmt}, a_{lpt}, a_{pb}\}_{cw2}$

   GNP $\{a_{hpt}, a_{lpt}, a_{pb}\}_{wu2}$
   PC $\{a_{hpt}, a_{lpt}, a_{pb}\}_{wu2}$
   FNF $\{a_{hpt}, a_{lpt}, a_{pb}\}_{wu2}$
   DNRC $\{a_{hpt}, a_{lpt}, a_{pb}\}_{wu2}$
   PL $\{a_{hpt}, a_{lpt}, a_{pb}\}_{wu2}$
   CT $\{a_{hmt}, a_{lpt}, a_{pb}\}_{wu2}$

4. WUI, CWPP and non-WUI areas (1)
   GNP $\{a_{hpt}, a_{lpt}, a_{pb}\}_{cw3}$
   PC $\{a_{hpt}, a_{lpt}, a_{pb}\}_{cw3}$
   FNF $\{a_{hpt}, a_{lpt}, a_{pb}\}_{cw3}$
   DNRC $\{a_{hpt}, a_{lpt}, a_{pb}\}_{cw3}$
   PL $\{a_{hpt}, a_{lpt}, a_{pb}\}_{cw3}$
   CT $\{a_{hmt}, a_{lpt}, a_{pb}\}_{cw3}$

5. GNP $\{a_{hpt}, a_{lpt}, a_{pb}\}_{wu3}$
   PC $\{a_{hpt}, a_{lpt}, a_{pb}\}_{wu3}$
   FNF $\{a_{hpt}, a_{lpt}, a_{pb}\}_{wu3}$
   DNRC $\{a_{hpt}, a_{lpt}, a_{pb}\}_{wu3}$
   PL $\{a_{hpt}, a_{lpt}, a_{pb}\}_{wu3}$
   CT $\{a_{hmt}, a_{lpt}, a_{pb}\}_{wu3}$

   GNP $\{a_{hpt}, a_{lpt}, a_{pb}\}_{nw3}$
   PC $\{a_{hpt}, a_{lpt}, a_{pb}\}_{nw3}$
   FNF $\{a_{hpt}, a_{lpt}, a_{pb}\}_{nw3}$
   DNRC $\{a_{hpt}, a_{lpt}, a_{pb}\}_{nw3}$
   PL $\{a_{hpt}, a_{lpt}, a_{pb}\}_{nw3}$
   CT $\{a_{hpt}, a_{lpt}, a_{pb}\}_{nw3}$

6. WUI, CWPP and non-WUI areas (2)
   GNP $\{a_{hpt}, a_{lpt}, a_{pb}\}_{cw4}$
   PC $\{a_{hpt}, a_{lpt}, a_{pb}\}_{cw4}$
   FNF $\{a_{hpt}, a_{lpt}, a_{pb}\}_{cw4}$
   DNRC $\{a_{hpt}, a_{lpt}, a_{pb}\}_{cw4}$
   PL $\{a_{hpt}, a_{lpt}, a_{pb}\}_{cw4}$
   CT $\{a_{hmt}, a_{lpt}, a_{pb}\}_{cw4}$

   GNP $\{a_{hpt}, a_{lpt}, a_{pb}\}_{wu4}$
   PC $\{a_{hpt}, a_{lpt}, a_{pb}\}_{wu4}$
   FNF $\{a_{hpt}, a_{lpt}, a_{pb}\}_{wu4}$
   DNRC $\{a_{hpt}, a_{lpt}, a_{pb}\}_{wu4}$
   PL $\{a_{hpt}, a_{lpt}, a_{pb}\}_{wu4}$
   CT $\{a_{hmt}, a_{lpt}, a_{pb}\}_{wu4}$

   GNP $\{a_{hpt}, a_{lpt}, a_{pb}\}_{nw4}$
   PC $\{a_{hpt}, a_{lpt}, a_{pb}\}_{nw4}$
   FNF $\{a_{hpt}, a_{lpt}, a_{pb}\}_{nw4}$
   DNRC $\{a_{hpt}, a_{lpt}, a_{pb}\}_{nw4}$
   PL $\{a_{hpt}, a_{lpt}, a_{pb}\}_{nw4}$
   CT $\{a_{hpt}, a_{lpt}, a_{pb}\}_{nw4}$

In the above list, cw designates CWPP areas, wu designates WUI areas, and nw designates non-WUI areas. Each numbered instance of cw, nw, and wu is a distinct area allocation. Also, for each landowner, there are unique treatment allocation schemes for each area allocation scenario (see next section). Treatment allocation schemes are determined for each subperiod.

Three factors differentiate treatment allocation scenarios: (1) the types of forest treatments simulated on lands owned/managed by each member of Agent$_a$ (see step 3); (2) whether maximum acres for members of Agent$_a$ increase over time due to changes in the wood products or residential construction industries (see step 3) and; (3) the proportion of Agent$_a$'s maximum acres allocated to each of the three forest treatments.

Specifying treatment allocation schemes. Specification of treatment allocation schemes requires additional information about forest treatments by members of Agent$_a$, including current practices or policies that influence the proportion of their efforts devoted to fuels reduction in CWPP, WUI, and non-WUI areas. In particular, treatment allocation schemes are specified by combining this additional information with the existing simulation results for maximum acres in forest treatments for each landowner (see section 3). The following steps describe how the additional information about forest treatments is obtained and used to specify treatment allocation schemes.

a. Determine the proportions of each landowner's maximum acres in the three forest treatments (i.e., r_{hpt}, r_{lpt}, r_{pb}, where r stands for proportion of acres). Values of r_k are used instead of actual acres to make it easier for stakeholders to specify future treatment allocation schemes. If a landowner does not perform a certain forest treatment, then the proportion for that treatment is zero. Values of r_{hpt}, r_{lpt}, r_{pb} specified for the WUI, CWPP priority, and non-WUI areas for each member of Agent$_a$ are listed in Table 3.

Consider the following example. Area allocation scenario 1 stipulates that landowners allocate their entire maximum acres in forest treatments to CWPP priority areas. Maximum acres treated by the PL member of Agent$_a$ during the first subperiod (2010-2019) is 51,337. Based on the proportions in Table 3, the acres in the three forest treatments are: a_{hpt} = (51,337)(.092) = 4,723 acres; a_{lpt}=(51,337)(.907) = 46,563 acres; and a_{pb}=(51,337)(0) = 0 acres.

Table 3. Values of r_{hpt}, r_{lpt}, r_{pb} specified for members of Agent$_a$

Member	r_{hpt}	r_{lpt}	r_{pb}
Plum Creek (PC)	0.44	0.56	0
Glacier National Park (GNP)	0	0.56	0.44
Flathead National Forest (FNF)	0.42	0.1	0.48
Other Lands (OL)	0.69	0.15	0.16
Private Lands (PL)	0.093	0.907	0
MT Dept. of Natural Resources (DNRC)	1	0	0

b. Determine the proportion of maximum acres that each landowner treats in the WUI. Values of maximum acres in each forest treatment (hpt, lpt, pb) for each landowner and subperiod vary across the three economic growth scenarios. The preliminary proportions are determined by: (1) using ArcGIS to calculate the proportion of each landowner's property that is within the WUI; and (2) existing policy targets or agency data on proportion of historical forest treatments conducted in the WUI. The preliminary proportion is the higher of the two values determined in (1) and (2) above. Preliminary proportions are discussed with area professionals and the stakeholder panel. The panel for Agent$_a$ is asked to modify the proportions as needed.

Consider the following example. Area allocation scenario 4 stipulates that members of Agent$_a$ conduct forest treatments in WUI, non-WUI and CWPP priority areas. Among other things, this means determining the proportions of maximum acres treated in each area. The Flathead National Forest manages approximately 700,000 acres in the study area. Of this amount, approximately 150,000 acres are in the WUI, which implies the proportion for the WUI is 0.21. The Forest Service allocates approximately one-half of its fuel reduction effort in WUI areas with a recognized CWPP. One-half is the preliminary proportion of acres allocated to the WUI because it is the larger of the two values. That proportion can be modified by the stakeholder panel.

The maximum acres treated by the Flathead National Forest during the first subperiod (2010-2019) is 46,116 acres. Assuming that the preliminary proportion of 0.5 for the WUI is not modified by the stakeholder panel, approximately 23,058 acres of Forest Service lands in the WUI are treated during the 2010-2019 subperiod. This acreage can be further modified after consideration of the proportion of acres that is within CWPP priority areas (see step c below). This modification is necessary because most of the CWPP priority areas are nested within WUI areas.

For this example, Forest Service areas outside the WUI (23,058 acres) allocated to the three forest treatments are determined using the procedure described in step a. Placement of treatments is determined using the Fire-BGCv2 model that simulates vegetation growth and includes parameters for determining where forest treatments are conducted.

c. For allocation scenarios that include both WUI and CWPP priority areas, determine the preliminary proportions of acres treated in the WUI that are in CWPP priority areas for each landownership and subperiod. These proportions reflect landowners' level of responsibility or effort toward reducing wildfire risk in areas delineated in the Flathead County CWPP. Table 4 lists the preliminary proportions of acres treated in the WUI that are in CWPP priority areas for each of the six landowners. Preliminary proportions can be modified by the stakeholder panel and other local experts.

Table 4. Proportion of acres treated in WUI that are in CWPP priority areas by landowner

Landowner	PL	USFS	GNP	DNRC	OL	PC
Proportion	0.70	0.50	0.70	0.30	0.10	0.0[a]

Notes: WUI treatments on Plum Creek lands are determined using the Fire-BGCv2 model and are not focused on protection of residential properties in the WUI. For that reason, the proportion is variable and based on Fire-BGCv2 simulations.

Consider the following example. Area allocation scenario 5 stipulates that members of $Agent_a$ conduct forest treatments on WUI, CWPP, and non-WUI areas. The maximum acres treated by the PL member of $Agent_a$ during the first subperiod (2010-2019) is 51,337. Using the logic outlined in step b, a total of 40,000 acres of forest treatments on private lands occur in the WUI during the subperiod 2010-2019 and the proportion of acres treated in the WUI that are in CWPP priority areas is 0.70 for member PL. Therefore, (40,000) (0.70) = 28,000 acres of forest treatments for member PL are allocated to CWPP priority areas. Those acres are distributed to parcels in the CWPP priority areas using the procedure outlined in step d.

Private landowners conduct either heavy (hpt) or light (lpt) mechanical thinning on their land. Thus, the proportions outlined in step a are used to determine the acreage allocated to forest treatments within CWPP priority areas. For instance, the ratio of hpt acres to total forest treatment acres for member PL during the 2010-2019 subperiod is 0.093 (i.e., r_{hpt} = 0.093; see Table 3). Therefore, acres of CWPP priority areas allocated to hpt for member PL during the subperiod 2010-2019 is (28,000)(0.093) = 2,604 acres. Similarly, the ratio of lpt acres to total forest treatment acres for member PL during the 2010-2019 subperiod is 0.907 (i.e., r_{lpt} = 0.907; see Table 3). Therefore, acres of CWPP priority areas allocated to lpt for member PL during the subperiod 2010-2019 is (28,000) (0.907) = 25,396 acres. No acres are allocated to prescribed burning for member PL. A similar process is used to allocate acres in each forest treatment to lands: (1) in the WUI, but outside CWPP priority areas (12,000 acres); and (2) outside the WUI (11,337 acres).

d. Specify the treatment order of county-wide and individually designated fire district priority areas outlined in the CWPP. The FIRECLIM model disproportionately allocates acreage among CWPP priority areas based on: (1) a ranked order for area treatments; and (2) a random allocation of treatments. The ranked order for area treatments is determined using a nested procedure that combines: (1) independent priority rankings of fire district areas within a given fire district; and (2) collective rankings of county-wide areas that cross fire district boundaries. The result is one priority list for all CWPP priority areas. A full description of the procedure used to create the ranked CWPP priority list is provided in Appendix B of the FIRECLIM ABM procedure found at: http://projects.cares.missouri.edu/fireclim-montana/Methods/ABM_Appendices.pdf The following preliminary proportions are used to allocate CWPP acreage: (1) 0.33 of allocated acreage is based on the ranked order of CWPP treatments; and (2) 0.66 of allocated acreage is based on a random allocation among CWPP treatments.

Table 5. Steps for calculating treatment allocation schemes for different area allocation scenarios

Area allocation scenario	Order
CWPP areas only	Steps a and d
WUI areas (ignores CWPP areas)	Step a
WUI areas (includes CWPP areas)	Steps a, c, and d
WUI, CWPP and non-WUI areas (1)	Steps a, b , c, and d
WUI, CWPP and non-WUI areas (2)	Steps a, b, c, and d

Progression through the ranked order for CWPP treatments occurs only once. Use of the ranked order for CWPP treatments is discontinued after every area in the ranking is treated. CWPP priority areas treated through random allocation and before their designated ranking will be removed from the ranked list. This reduces the chances that a given CWPP priority area is treated twice in rapid succession. Treated CWPP areas become available for retreatment (through random selection) during the third subperiod following initial treatment. For instance, if a CWPP area on Forest Service land is treated in subperiod 1 (2010-2019), it is not available for retreatment until the third subperiod (2030-2039). Therefore, it is possible for CWPP areas to be treated more than once during the 50-year evaluation period.

e. Determine which of the steps described above apply to area allocation scenarios simulated with Fire-BGCv2. As mentioned above, not all steps described in this document apply to the calculation of landowner treatment area schemes for each area allocation scenario. The primary driver of which steps are necessary for calculation of various treatment allocation schemes is summarized in Table 5.

5. Defining Expected Net Wildfire Losses for Agent$_a$

Expected net wildfire losses for land and wildfire management agencies E(NWL$_{at}$) depends on three negative attributes (i.e., expected residential property losses from wildfire [(E(RLW)], expected net commercial timber losses, and expected costs of management practices) and one positive attribute (i.e., expected ecological benefits of wildfire determined by the extent to which the wildfire restores vegetative conditions in the area being treated to the Historical Range of Variability or HRV). E(NWL$_{at}$) decreases (or increases) as the negative attributes decrease (or increase) and/or the positive attribute increases (or decreases).

E(RLW) is the sum of the present value of expected wildfire losses for existing residential properties (i.e., those that existed at the beginning of 2010) and the present value of expected wildfire losses for future residential properties (i.e., those added during the 50-year evaluation period). Present values are calculated for the 50-year evaluation period using a real (inflation-adjusted) discount rate of 4%. E(RLW) is based on the conditional probability that structures on residential properties in the study area burn, the value of those structures, and the expected losses in value (property or aesthetics) resulting from wildfires.

Net commercial timber losses and cost of forest treatments are based on subperiod changes in vegetation simulated by Fire-BGCv2 (Keane et al. 1996). Costs are estimated using the University of Montana Bureau of Business and Economic Research Harvest Cost Model (Keegan et al. 2002).

HRV is based on spatial and tabular output from a 1,000-year run of Fire-BGCv2, parameterized with *historical* fire regime and weather data. It is compared to the simulations of *future* landscapes under alternative forest management alternatives and future climate scenarios to determine how close to or departed from the HRV is the vegetation for future simulated landscapes.

This approach uses the HRV as a reference point for measuring how ecological conditions in landscapes are influenced by alternative forest management alternatives and future climate change scenarios, not as a target for management. The HRV concept has been criticized in terms of the utility of using historical conditions when climate is changing. However, the high uncertainty in predictions from highly complex general circulation models limits the utility of using future climate scenarios to determine target environments. The past is known and provides sufficient data to quantify historical vegetation and disturbance dynamics over large variations in historical climate with much greater certainty than climate models. Thus, departure from the HRV provides the best proxy of the effects of forest management alternatives and future climate change scenarios on the ecological condition of landscapes (Keane et al. 2009).

6. Evaluation and Selection of Best Management Alternative for Agent$_a$

Members of Agent$_a$ evaluate E(NWL$_{at}$) for the management alternatives at the beginning of each subperiod. Because the index of proximity of vegetative conditions to the HRV is a non–monetary attribute, it is not possible to evaluate E(NWL$_{at}$) in monetary terms. For this reason the values of E(NWL$_{at}$) for forest management alternatives and the best forest management alternative for members of Agent$_a$ are determined using the fuzzy TOPSIS method (Hwang and Yoon 1981; Chen and Hwang 1992; Chen 2000; Berger 2006), which entails the following six steps.

First, individual members of the panel for Agent$_a$ are asked to rate the simulated values of the attributes of E(NWL$_{at}$) using seven linguistic variables: very poor; poor; medium poor; fair; medium good; good; and very good. Ratings are based on the simulated values of the attributes for all management alternatives, subperiods, economic growth scenarios, and land use policy scenarios. A unique triangular fuzzy number is assigned to each linguistic variable based on Chen [2000].

Second, individual panel members are asked to rate the relative importance of the four attributes of E(NWL$_{at}$) using seven linguistic variables: very low; low; medium low; medium; medium high; high; and very high. Once again, a unique triangular fuzzy number is assigned to each linguistic variable based on Chen [2000].

Third, the triangular fuzzy numbers for the linguistic variables assigned to the ratings of attributes and the ratings of the relative importance of attributes by individual members of the panel are averaged to obtain triangular fuzzy numbers for the ratings of attributes and the relative importance of attributes for the panel as a whole.

Fourth, the vertex distances between the simulated attributes of management alternatives and the attributes for the fuzzy positive-ideal and the fuzzy negative-ideal solutions are calculated. The fuzzy positive-ideal solution has the most desirable values of the attributes (i.e., attributes that result in the lowest possible E(NWL$_{at}$) and the fuzzy negative-ideal solution has the least desirable values of the attributes (i.e., attributes that result in the highest possible E(NWL$_{at}$)).

Fifth, the resulting vertex distances are used to calculate the annual closeness coefficients for all forest management alternatives. A closeness coefficient measures how close the attributes for a particular forest management alternative are to the attributes for the fuzzy positive-ideal solution and far away they are from the attributes for the fuzzy negative-ideal solution. As the closeness coefficient approaches one (or zero), the forest management alternative becomes more (or less) desirable.

Sixth, the forest management alternatives are ranked from most to least preferred based on their annual closeness coefficients. The best forest management alternative for a subperiod is the alternative whose closeness coefficient is nearest to one.

The above six-step procedure is used to determine $Agent_a$'s best forest management alternatives during subperiods for each of the nine combinations of climate change and economic growth scenarios.

7. Exogenous Factors for $Agent_a$

Several exogenous factors influence the decisions made by certain members of $Agent_a$, including: (1) subperiod changes in maximum acres in forest treatments that occur when those acres are indexed to growth in the wood products manufacturing and residential construction industries; (2) climate change; and (3) economic growth and associated residential development (see Figure 5). The first factor is not relevant for Glacier National Park because it does not conduct commercial timber harvesting. The second and third factors influence $E(RLW)$, which can affect $E(NWL_{at})$ and, hence, the best forest management alternative for $Agent_a$.

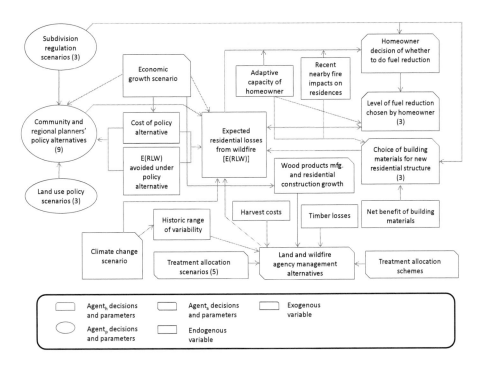

Figure 5. Relationships between agent decisions and endogenous and exogenous inputs simulated in the FIRECLIM ABM.

CONCLUSION

This chapter describes an ABM for simulating the dynamics of a coupled natural-human system for wildfire in Flathead County, Montana. The FIRECLIM ABM includes three agents: homeowners; land and wildfire management agencies; and regional-community planners. The framework assumes these agents make decisions at different spatial and temporal scales based on various decision criteria. Agents' decisions are linked to the outcomes of models that simulate economic growth and associated residential development, wildfire behavior, vegetation growth, and climate change.

The ABM framework presented here advances wildfire research by providing an integrated approach to simulating the actions of individual agents operating in a dynamic coupled natural-human system for wildfire. Rather than treating agents and their actions in a static manner, the FIRECLIM ABM assumes the decisions of multiple agents, including the interactions among agents and between agents and the natural environment, interact to influence wildfires and agents. In contrast, most previous wildfire research does not employ such an integrated approach. Moreover, studies that apply ABM to wildfire are scarce and those that do employ ABM typically focus on response (i.e., fire suppression or evacuation).

The FIRECLIM ABM provides a more comprehensive framework for simulating decisions and biophysical processes that influence the impacts of and responses to wildfire. In contrast, most past wildfire research addressing the potential impacts of wildfire concentrates on isolated elements of human behavior or features of the natural environment (e.g., forest management and climate change). Regardless of their spatial and temporal scales, results of studies that address the dynamic interactions among various human agents and how those interactions influence potential wildfire losses provide community-regional planners, land and wildfire management agencies, and homeowners with a solid foundation for devising and evaluating strategies to reduce future adverse impacts while retaining the benefits of wildfire. For this reason, it is worthwhile to conduct additional ABM studies of wildfire that build on or refute the ABM presented here.

Wildfire and human response to wildfire are complex because they are influenced by numerous biophysical and social processes. While the parameters, factors, and behavioral rules for agents used in the ABM described in this chapter are a good starting point, they do not account for all the factors that influence human decisions pertaining to wildfire. The FIRECLIM ABM makes certain assumptions about human behaviors based on existing research or data for the study area. Some of the data that are still being collected may be useful in further refining the decision rules for agents.

An important feature of the FIRECLIM ABM is the use of stakeholder panels comprised of local residents, leaders, land managers, and wildfire-related professionals who represent agents in the model. Data and information provided by the panels are used to develop the assumptions of the ABM, select key parameters of the model, and/or to refine model structure. The framework presented here will continue to evolve, change, and expand as a result of these interactions. In particular, additional refinements of the FIRECLIM ABM will no doubt occur in the process of: (1) developing linkages between the outputs of the ABM and outputs of other models used in the FIRECLIM model (i.e., Fire-BGCv2, FSIM, and RECID2); (2) modifying or parameterizing computer software used in agent simulation; and (3) considering how best to manage the volume of outputs produced by the FIRECLIM ABM.

Use of the FIRECLIM ABM is not limited to Flathead County. The FIRECLIM ABM can be adapted to other fire-prone regions. Such adaptations would require researchers to determine: (1) whether it is necessary to alter the assumptions about human behavior made in the FIRECLIM ABM to better fit agent behavior in the region of interest; and (2) how best to acquire, for the region of interest, the significant amount of data required to parameterize the models used in the framework.

Ongoing social science research regarding wildfire holds great promise for improving the conceptual and experiential basis for agent decision rules used in the ABM presented here. One common theme in such research is the diversity of perspectives that homeowners, agency professionals, and land use planners bring to bear on wildfire and its consequences. ABM approaches provide a tool for building such diversity into studies of human responses to potential or actual wildfire losses and constitute a powerful analytical tool for future studies of decision-making in a coupled natural-human system for wildfire.

ACKNOWLEDGMENTS

This research is funded by the Dynamics of Coupled Natural and Human Systems program of the U.S. National Science Foundation. The NSF project number is 0903562.

REFERENCES

Acosta-Michlik, L., and Espaldon, V. (2008). Assessing vulnerability of selected farming communities in the Philippines based on a behavioral model of agent's adaptation to global environmental change. *Global Environmental Change, 18*, 554-563.

An, L., Linderman, L.M., Qi, J., Shortridge, A., and Liu, J. (2006). Exploring complexity in a human environment system: an agent-based spatial model for multidisciplinary and multiscale integration. *Annals of the Association of American Geographers, 95*, 54-79.

Arno, S.F., and Allison-Bunnell, S. (2002). *Flames in our forest: Disaster or renewal?* Washington, DC: Island Press.

Bar Massada, A., Radeloff, V.C., Stewart, S.I., and Hawbaker, T.J. (2009). Wildfire risk in the wildland-urban interface: A simulation study in northwestern Wisconsin. *Forest Ecology and Management, 258*, 1990-1999.

Berger, P.A. (2006). Generating agricultural landscapes for alternative futures analysis: a multiple attribute decision-making model. *Transactions in GIS, 10*, 103-120.

Berger, T., and Schreinemachers, P. (2006). Creating agents and landscapes for multiagent systems from random samples. *Ecology and Society, 11*(2). http://www.ecologyandsociety.org/vol11/iss2/

Bonabeau, E. (2002). Agent-based modeling: Methods and techniques for simulating human systems. *Proceedings of the National Academy of Sciences, 99(3)*, 7280-7287.

Bone, C., and Dragićević, S. (2010). Simulation and validation of a reinforcement learning agent-based model for multi-stakeholder forest management. *Computers, Environment and Urban Systems, 34*, 162-174.

Bousquet, F., and Le Page, C. (2004). Multi-agent simulations and ecosystem management: A review. *Ecological Modeling, 176*, 313-332.

Brooks, M.L., et al. (2004). Effects of invasive alien plants on fire regimes. *BioScience, 54(7)*, 677-688.

Calkin, D.E., Ager, A.A., and Gilbertson-Day, J. (eds.). (2010). Wildfire risk and hazard: procedures for the first approximation. Gen. Tech. Rep. RMRS-GTR-235. Fort Collins, CO: U.S. Department of Agriculture, Forest Service, Rocky Mountain Research Station. 62p.

Castle, C.J.E., and Crooks, A.T. (2006). UCL working paper series paper 110: Principles and concepts of agent-based modelling for developing geospatial simulations. Centre for Advanced Spatial Analysis, University of College London.

Chen, C-T. 2000. Extensions to the TOPSIS for group decision-making under fuzzy environment. *Fuzzy Sets and Systems, 114*, 1-9.

Chen, S.J., and Hwang, C.L. (1992). *Fuzzy multiple attribute decision making*. Lecture notes in economics and mathematical systems, No. 375. Berlin: Spring-Verlag.

Chu, J., Wang, C., Chen, J., and Wang, H. (2009). Agent-based residential water use behavior simulation and policy implications: A case-study in Beijing City. *Water Resource Management, 23*, 3267-3295.

Cohen, J. D. (2008). The wildland-urban interface fire problem: A consequence of the fire exclusion paradigm. *Forest History Today,* Fall, 20-26.

Cohen, J.D. 2000. Preventing disaster: Home ignitability in the wildland-urban interface. *Journal of Forestry, 98(3)*, 15-21.

Cohen, J. D. (1995). Structure ignition assessment model (SIAM). In D.R. Weise, R.E. Martin (eds.) The Biswell symposium: fire issues and solutions in urban interface and wildland ecosystems (pp. 85-92). February 1-17, 1994; Walnut Creek, California. Gen. Tech. Rep. PSW-GTR-158. Albany, CA: Pacific Southwest Research Station, Forest Service, U.S. Department of Agriculture.

Cova, T.J., and Johnson, J.P. (2002). Microsimulation of neighborhood evacuations in the urban-wildland interface. *Environment and Planning A, 34*, 2211-2229.

Daniels, T. (2001). Smart growth: a new American approach to regional planning. *Planning Practice and Research, 16*, 271-279.

Daniel, T.C., Carroll, M.S., Moseley, C., and Raish, C. (eds.). (2007). *People, fire and forests: A synthesis of wildfire social science*. Corvallis, OR: Oregon State University Press.

Edmonds, B. (2010). Bootstrapping knowledge about social phenomena using simulation models. *Journal of Artificial Societies and Social Simulation, 13(1)8*. http://jasss.soci.surrey.ac.uk/13/1/8.html

Epstein, J.M., and Axtell, R. (1996). *Growing artificial societies: Social science from the bottom up*. Cambridge, Mass: MIT Press.

Finney, M.A. (2007). A prototype simulation system for large fire planning in FPA. Missoula, MT. http://www.fpa.nifc.gov/Library/Docs/FPA_SimulationPrototype_0705.pdf

Firewise Communities USA. (2011). Resources-For the Homeowner. http://www.firewise.org/resources/homeowner.htm

Flathead County Office of Planning and Zoning. (2011). Flathead County Growth Policy (adopted March 19, 2007, resolution #2015A). http://flathead.mt.gov/planning_zoning/growth_resolution2015a.php

Galán, J.M., et al. (2009). Errors and artifacts in agent-based modelling. *Journal of Artificial Societies and Social Simulation, 12(1)1*. http://jasss.surrey.ac.uk/12/1/1.html

GCS Research. (2005). Flathead County community wildfire fuels reductions/mitigation plan. Missoula, MT.

Gilbert, N. (2008). *Agent-based models*. Series: Quantitative applications in the social sciences. Thousand Oaks, Calf.: Sage Publications.

Gimblett, H.R. (ed). (2002). *Integrating geographic information systems and agent-based Modeling techniques for simulating social and ecological processes*. New York: Oxford University Press.

Grayzeck, S. A., Nelson, K. C., Brummel, R. F., Jakes, P., and Williams, D. (2009). Interpreting federal policy at the local level; the wildland-urban interface concept in wildfire protection planning in the eastern United States. *International Journal of Wildland Fire, 18*, 278-289.

Gude, P., Rasker, R., and Noort, J,vd. (2008). Potential for future development on fire-prone lands. *Journal of Forestry, 106*, 198-205.

Haight, R.C., Cleland, D.T., Hammer, R.B., Radeloff, V.C. and Rupp, T.S. (2004). Assessing fire risk in the wildland-urban interface. *Journal of Forestry, 102*, 41-48.

Hammer, R.B., Stewart, S.I., and Radeloff, V.C. (2009). Demographic trends, the wildland-urban interface, and wildfire management. *Society and Natural Resources, 22*, 777-782.

Hansen, A., et al. (2005). Effects of exurban development on biodiversity: Patterns, mechanisms, and research needs. *Ecological Applications, 15(6)*, 1893-1905.

Heath, B., Hill, R., and Ciarallo, F. (2009). A survey of agent-based modeling practices (January 1998 to July 2008). *Journal of Artificial Societies and Social Simulation, 12(4)9*. http://jasss.soc.surrey.ac.uk/12/4/9.html

Helleboogh, A., Vizzari, G., Uhrmacher, A., and Michel, F. (2007). Modeling dynamic environments in multi-agent simulation. *Autonomous Agents and Multi-Agent Systems, 14(1)*, 87-116.

Hoffmann, M., Kelley, H., and Evans, T. (2002). Simulating land-cover change in South-Central Indiana: An agent-based model of deforestation and afforestation. In M.A. Jansen (ed.), *Complexity and ecosystem management: The theory and practice of multi-agent approaches* (pp. 218-247). Northampton, MA: Edward Elgar Publishers.

Howe, J., McMahon, E., and Propst, L. (1997). *Balancing nature and commerce in gateway communities*. Washington, DC: Island Press.

Hu, X., and Ntaimo, L. (2009). Integrated simulation and optimization for wildfire containment. *ACM Transactions on Modeling and Computer Simulation, 19(4)*, 1-29.

Hwang, C.L., and Yoon, K. (1981). *Multiple attribute decision making*. Lecture notes in economics and mathematical systems No. 186. Berlin: Springer-Verlag.

Irwin, E.G. (2010). New directions for urban economic models of land use change: incorporating spatial dynamics and heterogeneity. *Journal of Regional Science, 50(1)*, 65-91.

Jager, W., Janssen, M.A., and Viek, C.A.J. (2002). How uncertainty stimulates overharvesting in a resource dilemma: three process explanations. *Journal of Environmental Psychology, 22*, 247-263.

Janssen, M.A., and Ostrom, E. (2006). Empirically based, agent-based models. *Ecology and Society, 11(2)*, 37-51.

Jakes, P., Fish, T., Carr, D., and Blahna, D. (1998). Functional communities: a tool for national forest planning. *Journal of Forestry, 96*(3): 33.

Jensen, S.E., and McPherson, G.R. (2008). *Living with fire: Fire ecology and policy for the twenty-first century*. Berkeley, CA; University of California Press.

Keane, R.E., Hessburg, P.F., Landres, P.B., and Swanson, F.J. (2009). The use of historical range and variability (HRV) in landscape management. *Forest Ecology and Management, 258*, 1025-1037.

Keane, R.E., Ryan, K., and Running, S.W. (1996). Simulating the effect of fires on northern Rocky Mountain landscapes using the ecological process model FIRE-BGCv2. *Tree Physiology, 16*, 319-331.

Keane, R.E., Morgan, P., and White, J.D. (1999). Temporal pattern of ecosystem processes on simulated landscapes of Glacier National Park, USA. *Landscape Ecology 14*, 311-329.

Keegan, C.E., Niccolucci, M.J., Fiedler, C.E., Jones, G.J., and Regel, R. (2002). Harvest cost collection approaches and associated equations for restoration treatments on national forests. *Forest Products Journal, 52(7/8)*, 96-100.

Kim, Y., Bettinger, P., and Finney, M. (2009). Spatial optimization of the pattern of fuel management activities and subsequent effects on simulated wildfires. *European Journal of Operational Research, 197*, 253-265.

Korhonen, T., Hostikka, S., Heliövaara, S., and Ehtamo, H. (2010). FDS+Evac: An agent based fire evacuation model. In W.F. Klingsch, C. Rogsch, M. Schadschneider and M. Schreckenberg (eds.). *Pedestrian and Evacuation Dynamics* 2008 (pp. 109-120). New York; Springer.

Lempert, R. (2002). Agent-based modeling as organizational and public policy simulators. *Proceedings of the National Academy of Sciences, 99*: 7195-7196.

Ligtenberg, A., Wachowicz, M., Bregt, A.K., Beulens, A., and Kettenis, D.L. (2004). A design and application of a multi-agent system for simulation of multi-actor spatial planning. *Journal of Environmental Management, 72*, 43-55.

Martin, W.E., Raish, C., and Kent, B. (eds.). (2008). *Wildfire risk: Human perceptions and management implications*. Washington, D.C.: Resources for the Future.

Matthews, R.B., Gilbert, N.G., Roach, A., Pohill, G., and Gotts, N.M. (2007). Agent-based land-use models: A review of applications. *Landscape Ecology, 22*, 1447-1459.

Martin, I.M., Bender, H., and Raish, C. (2007). What motivates individuals to protect themselves from risks: The case of wildland fires. *Risk Analysis, 27(4)*, 887-900.

McCaffrey, S.M. (ed.) (2006). The public and wildland fire management: social science findings for managers. General Technical Report. NRS-1. Newton Square, PA: U.S. Department of Agriculture, Forest Service, Northern Research Station. 202 p.

Miller, J.H., and Page, S.E. (2007). *Complex adaptive systems: an introduction to computational models of social life*. Princeton, New Jersey: Princeton University Press.

Millington, J., Romero-Calcerrada, R., Wainwright, J., and Perry, G. (2008). An agent based model of Mediterranean agricultural land-use/cover change for examining wildfire risk. *Journal of Artificial Societies and Social Simulation, 11(4)4*. http://jasss.soc.surrey.ac.uk/11/4/4.html

Minnesota IMPLAN Group, Inc. (2011). IMPLAN Economic Modeling. Stillwater, MN 55082. http://www.implan.com Montana Cadastral Mapping. (2010). http://gis.mt.gov/.

Nelson, D.R., Adger, W.N., and Brown, K. (2007). Adaptation to environmental change: contributions of a resilience framework. *Annual Review of Environment and Resources, 32*, 395-419.

National Fire Protection Agency (NFPA). (2007). NFPA 1144: standard for reducing structure ignition hazards from wildland fire (2008 edition). National Fire Protection Association: Quincy, MA.

Niazi, M.A., Hussain, A., Siddique, Q., and Kolberg, M. (2010). Verification and validation of an agent-based forest fire simulation model. Proceedings of the Agent Directed Simulation Symposium 2010, as part of the ACM SCS Spring Simulation Multiconference (pp. 142-149), Orlando, FL, April 11-15, 2010.

Nikolai, C., and Madey, G. (2009). Tools of the trade: A survey of various agent based modeling platforms. *Journal of Artificial Societies and Social Simulation, 12(2)2*. http://jasss.soc.surrey.ac.uk/12/2/2.html

Ntaimo, L., and Hu, X. (2009). DEVS-FIRE: Toward an integrated simulation environment for surface wildfire spread and containment. *Simulation, 84(4)*, 137-155.

Parker, D.C., Manson, S.M., Janssen, M.A., Hoffmann, M.J., and Deadman, P. (2003). Multi-agent systems for the simulation of land-use and land-cover change: A review. *Annals of the Association of American Geographers, 93(2)*, 314-337.

Paveglio T.B., Carroll, M.S., and Jakes, P.J. (2010). Alternatives to evacuation during wildland fire: Exploring adaptive capacity in one Idaho community. *Environmental Hazards: Human and Policy Dimensions, 9(4)*, 379-394.

Platt, R.V. (2010). The wildland-urban interface: Evaluating the definition effect. *Journal of Forestry, 108(1)*, 9-15.

Pohill, J.G., Sutherland, L., and Gotts, N. (2010). Using qualitative evidence to enhance an agent-based modelling system for studying land use change. *Journal of Artificial Societies and Social Simulation, 13(2)10*. http://jasss.soci.surrey.ac.uk/13/2/10.html

Prato, T., Barnett, Y., Clark, A., and Paveglio, T. (in press). Improving simulation of land use change in a region of the Rocky Mountain West. In: XX and XX (Eds.), *Planning, Regulations, and Environment*. Hauppauge, NY: Nova Science Publishers, Inc.

Prato, T. (1998). *Natural Resource and Environmental Economics*. Ames, IA: Blackwell Publishing.

Prato, T. (2003). Multiple attribute evaluation of ecosystem management for the Missouri River. *Ecological Economics, 45*, 297-309.

Prato, T., and Hajkowicz, S. (2001). Comparison of profit maximization and multiple criteria models for selecting farming systems. *Journal of Soil and Water Conservation, 56*, 52-55.

Prato, T., A.S. Clark, K. Dolle, and Y. Barnett. (2007). Evaluating alternative economic growth rates and land use policies for Flathead County, Montana. *Landscape and Urban Planning, 83*:327–339.

Prato, T., Keane, R., and Fagre, D. (2008). Assessing and managing wildfire risk in the wildland-urban interface. In J. Chen and C. Guô, *Ecosystem Ecology Research Trends*. (pp. 275-297). Hauppauge, NY: Nova Science Publishers.

Pultar, E., Raubal, M., Cova, T.J., and Goodchild, M.F. (2009). Dynamic GIS case studies: Wildfire evacuation and volunteered geographic information. *Transactions in GIS 13(s1)*, 85-104.

Radeloff, V.C., et al. (2005). The wildland-urban interface in the United States. *Ecological Applications, 15(3)*, 799-805.

Raisback, S., Lytinen, S., and Jackson, S. (2006). Agent-based simulation platforms: review and development recommendations. *Simulation,82(9)*, 609-623.

Ramanath, A.N., and Gilbert, N. (2004). The design of participatory agent-based social simulations. *Journal of Artificial Societies and social simulation, 7(4)*. http://jasss.soc.surrey.ac.uk/7/4/1.html

Reinhardt, E.D., and Dickinson, M.B. (2010). First-order fire effects models for land management: overview and issues. *Fire Ecology, 6(1)*, 131-142.

Reinhardt, E.D., Keane, R.E., Calkin, D.E., and Cohen, J.D. (2008). Objectives and considerations for wildland fuel treatment in forested ecosystems of the interior western United States. *Forest Ecology and Management, 256*, 1997-2006.

Robinson, S. (2008). Conceptual modeling for simulation part I: Definition and requirements. *Journal of Operational Research Society, 59*, 278-290.

Saaty, R.W. (1987). The analytic hierarchy process-what is it and how is it used. *Mathematical Modeling, 9*, 161-176.

Schoennagel, T., Nelson, C.R., Theobald, D.M., Carnwath, G., and Chapman, T.B. (2009) Implementation of National Fire Plan fuel treatments near the wildland-urban interface in the western U.W. *Proceedings of the National Academy of Sciences, 106(26)*, 10706-10711.

Spracklen, D.V., et al. (2009). Impacts of climate change from 2000 to 2050 on wildfire activity and carbonaceous aerosol concentrations in the western United States. *Journal of Geophysical Research, 114*, 1-17.

Steelman, T.A., and Burke, C. A. (2007). Is wildfire policy in the United States sustainable? *Journal of Forestry, 105(2)*, 67-72.

Stewart, S.I., Radeloff, V.C., Hammer, R.B. and Hawbaker, T.J. (2007). Defining the wildland-Urban interface. *Journal of Forestry, 105(4)*, 201-207.

Stockmann, K., Burchfield, J., Calkin, D. and Venn, T. (2010). Guiding preventative wildland fire mitigation policy and decisions with an economic modeling system. *Forest Policy and Economics, 12(2)*, 147-154.

Theobald, D.M., and Romme, W. (2007). Expansion of the wildland-urban interface and wildfire mitigation in the United States. *Landscape and Urban Planning, 83*:340-354.

Thorp, J., et al. (2006). Santa Fe on fire: agent-based modeling of wildfire evacuation dynamics.

Santa Fe, NM: RedfishGroup. http://www.redfish.com/wildfire/RedfishGroup_Thorp_ABMofWildfireEvacuations_v002.pdf

U.S. Department of Agriculture and U.S. Department of the Interior. (1995). Federal wildland fire management policy and program review. Washington, D.C.: USDI and USDA, 40 p.

USDA and USDI. (2001). Urban wildland interface communities within vicinity of Federal lands that are at high risk from wildfire. *Federal Register, 66*,751–777.

U.S. Department of Agriculture. (2006). Audit report: Forest service large fire suppression costs. Washington, DC: USDA Office of Inspector General Western Region. 143p.

Valbuena. D., Berburg, P.H., and Bregt, A.K. (2008). A method to define a typology for agent-based analysis in regional land-use research. *Agriculture, Ecosystems and the Environment, 128*, 27-36.

Wainwright, J. (2008). Can modeling enable us to understand the role of humans in landscape evolution? *Geoforum, 39(2)*: 659-674.

Westerling, A.L., and Bryant, B.P. (2008). Climate change and wildfire in California. *Climatic Change, 87(1)*, S231-S249.

Westerling, A.L. Hidalgo, H.G., Cayan, D.R., and Swetnam, T.W. (2006). Warming and earlier spring increase western U.S. forest wildfire activity. *Science, 313*, 940-943.

Weinstein, D.A., and Woodbury, P.B. (2010) Review of methods for developing probabilistic risk assessments. Part 1: Modeling fire. In J.M. Pye, H.M. Rauscher, Y. Sands, D.C. Lee, and J.S. Beatty (eds.) *Advances in threat assessment and their application to forest and rangeland management*. USDA Forest Service, Pacific Northwest Research Station, Portland, Ore. PNW-GTR-802.

Winter, G., McCaffrey, S., and Vogt, C.A. (2009). The role of community policies in defensible space compliance. *Forest Policy and Economics, 11*, 570-578.

Yin, L. (2010). Modeling cumulative effects of wildfire hazard policy and exurban household location choices: An application of agent based-simulations. *Planning Theory and Practice, 11(3)*, 375-396.

Yu, C., MacEachren, A.M., Peuqet, D.J., and Yarnal, B. (2009). Integrating scientific modeling and supporting dynamic hazard management with a GeoAgent-based representation of human-environment interactions: A drought example in Central Pennsylvania, USA. *Environmental Modeling and Software, 24(12)*, 1501-1512.

Chapter 2

THE EUROPEAN EXPERIENCE ON ENVIRONMENTAL MANAGEMENT SYSTEMS AND THE THIRD REVISION OF THE "ECO MANAGEMENT AND AUDIT SCHEME" (EMAS)

Fabio Iraldo[a,b], Francesco Testa[a], Tiberio Daddi[a] and Marco Frey[a,b]

[a]Sant'Anna School of Advanced Studies, Piazza Martiri della Libertà, Pisa, Italy
[b]IEFE – Institute for Environmental and Energy Policy and Economics, Milano, Italy

1. INTRODUCTION

In recent years, based on the voluntary commitment and the pro-active approach of organizations, the certification framework has gained a crucial role amongst the instruments of the European environmental policy.

Voluntary instruments (such as EMAS and the "twin" regulation Eco-label) were designed by introducing concepts and mechanisms that, for the time being, led to a radical change in the environmental policies of the European Commission. The application of these patterns, in fact, originated a highly innovative policy trend, based on voluntary certification as a marketing tool providing a competitive edge.

The Commission's purpose was clearly to attract the interest of companies and convince them to spontaneously mobilize their financial, technical and management resources towards a path of continuous improvement in environmental performance.

The inspiring criterion was the belief, stated in the European Commission Fifth Environmental Action Programme, that manufacturing sectors and, more generally, all private (and public) actors whose activities had an environmental impact, could not only be seen as a part of the "problem" but also as a crucial part of the "solution", and therefore it was necessary to encourage them to participate and cooperate in building sustainable development paths.

The guiding principle behind the definition and the implementation of EMAS and Ecolabel was very simple: if the most active players on environmental improvement had been granted official recognition as a marketable value or in social relationships as a guarantee of their credibility, then two ambitious goals were achieved: first, the increase of their competitive edge and, secondly, the improvement of the environmental performance in the economic and productive industry.

Participation in EMAS is entirely voluntary, determined as it is by competitive and social pressures perceived by organizations, rather than by binding regulatory requirements. For this reason, the framework does not set quantitative limits, technology standards or emission thresholds, but it outlines the characteristics that a system of environmental management of an organization must have to be granted a public recognition of its correctness and efficiency. The basic steps an organization must take to participate in EMAS are the following: adopt an environmental policy; carry out an initial review of the existing environmental aspects; set the objectives and targets and establish a program to improve its performance; adopt a management system aimed at achieving the program objectives; perform audits to verify the functioning and the effectiveness of the system, and then draft an environmental statement to prove its commitment to the community.

It may be interesting here to briefly describe the stages of development and dissemination of EMAS, to which the competitive pressures described above have given rise.

The framework, issued by EC Regulation 1836 of 1993, required several years of preparation, due to the need to establish supervising national bodies and define the appropriate set of rules as provided by the EU: the first registrations in Europe date back to August 1995. After a very rapid development, particularly in Germany and Austria, the spread of EMAS suffered a slight slowdown in the early 2000s, partly for the implementation of the new guidelines, issued according to EC Regulation 761/2001.

Once at full capacity, the new version of EMAS Regulation clearly produced a further acceleration in the participation of small and medium-sized businesses, public sector organizations and services, especially in tourism. To encourage the involvement of these organizations, the Regulation introduced some major changes intended to correct some critical issues resulting from the previous EMAS experience and enhanced the opportunities for development and dissemination of the framework.

This chapter aims to describe the European evolutionary path about EMAS and how European institutions attempted to evaluate hindrances, starting from the main evidences emerging from literature, and identify favoring factors and efficient solutions to overcome them. A particular focus will be devoted to the description of new requirement of the EMAS III Regulation and which are the main differences from another formal scheme, such as ISO 14001.

2. Evaluation and Perspectives of EMAS Regulation

In 2006 the European Commission formally started the review of EMAS, which ended in late 2009. To fulfill this important stage in the evolution of the EU voluntary instruments, the Commission started a debate on the effectiveness of the framework, especially on the main expected results, such as increased competitiveness and improved environmental performance of registered organizations and, consequently, of the entire economic and productive system.

The revision carried out by the Commission focused on the preliminary assessment of a framework aimed at providing a set of clear guidelines for the sensitive decision-making processes. This study was commissioned to a team of international consultants and research institutions coordinated by IEFE Bocconi, such as SPRU University of Sussex, the IOEW of Heidelberg, Adelphi Consult and Valor and Tinge. The study, called Ever (Evaluation of EMAS / Ecolabel for their Revision) had a twofold objective:

1. To perform an in-depth assessment of EMAS effectiveness specifically in terms of environmental improvement and marketing opportunities and, more generally, as a growing factor for the community.
2. On the basis of the previous phase, to develop and advance ideas and concrete proposals for the review of EMAS, to outline scenarios the Commission may pursue in the expected review process.

The following sections introduce the main findings on literature about the factors influencing the adoption of Community regulations on eco-management, as well as the barriers and the obstacles that involved organizations must deal with.

2.1. The Main "Driver" for EMAS Adoption

The literature that since the mid-nineties has analyzed the reasons that drive organizations to obtain EMAS registration, produced a wealth of information about drivers that prove most effective in encouraging the adoption of the Regulation. A first indicator is the extreme heterogeneity of factors "driving" companies towards Environmental Management Systems - EMSs (and, specifically, towards EMAS). These vary significantly in connection with different aspects, like the size of the organization (SMEs vs large companies), its sector (e.g: manufacture vs Public Administration), national or regional contexts, and so on.

For instance, drivers can be either economic/strategic or "environment-led"; they can deal with the internal sphere of an organization (e.g: optimization of organizational activities), or be "external", such as the desire to gain a competitive advantage or benefit from fiscal/normative incentives and facilitations.

According to the outcome of a German UBA research (Clausen et al, 2002): economic and competitive motivations (such as energy/resources savings, better image, etc.) are very important. However, the lack of homogeneity of literature data makes it difficult to establish a well defined ranking in terms of potential efficacy or to prioritize internal economic variables, as external economic and strategic drivers seem to play an equally important role. An example refers to a pilot project carried out in Saxony-Anhalt. According to this study, companies were equally motivated by the expected improvement of competitiveness and corporate image, as well as by the reduction of energy or water consumption, and by the significant decrease of waste production (Schmittel, et.al. 1999). Moreover, it is noteworthy to mention that drivers can play a strategic and economic role even if they do not necessarily translate into quantifiable benefits or "monetization".

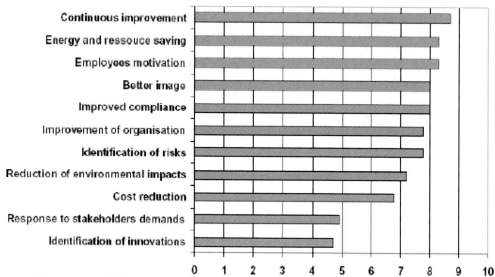

Source: Clausen et al., 2002.

Figure 1. Reasons leading to EMAS registration.

For instance, Perkins and Neumayer (2004) agree that cost-reductions, benefits and profitability of EMAS are major drivers, but they are unlikely to be the only ones, as companies often adopt organisational innovations based on management quests for external legitimacy, and specifically, for the need to conform to widely held beliefs of rational and efficient business practice. Hence, the participation in EMAS is likely to be shaped by two sets of factors: those influencing financial costs, benefits and profitability of the scheme, and "ideational forces" such as the requirements of external stakeholders.

A peculiar and very important "external" driver is represented by the communicational dimension of EMAS.

As reported by the relevant literature on environmental reporting and EMAS statements (e.g.: Gorla et al. 2001, Imperial College, ISO 14001 solutions and IEFE 1999, Grafé 1996, Jones et al 2000), the willingness to communicate with the stakeholders can be a powerful driver for EMAS participation. Some of the analyzed studies put an emphasis on the fact that, in some cases, EMAS has been preferred over ISO 14001 thanks to the possibility to use and disseminate credibly validated environmental information (Gorla et al. 2001).

Other studies place greater emphasis on environment-related drivers. As shown by a research conducted on registered French organizations, (Schucht, 2000), the improvement of environmental performance is the main justification supporting EMAS participation, relatively more important than strategic factors such as corporate image, cost reduction, staff motivation, etc.

Starting from a wide and diverse literature scenario, the EVER study aimed at collecting specific and updated empirical evidence, by means of a field investigation carried out through direct interviews with EMAS participating and non-participating organization and with involved stakeholders. The field survey partially confirmed the literature, whereas in many cases it revealed significant discrepancies.

Table 1. The main driver of Emas adoption

The main driver of Emas adoption	Relevance (da 1 a 5)
better management and guarantee of legal compliance	4,0
Improvement of our environmental performance	3,9
better risk management and environmental liability prevention	3,7
Improvement of our organisational and managerial capabilities in the environmental area	3,6
improvement of the relations with our stakeholders and the local community	3,5
improvement of competitive capabilities or satisfaction of a specific request by customers	3,4
keeping up with our main competitors/members of our trade association	3,2
satisfaction of a request by our corporate headquarters	3,1
benefits from regulatory relief	2,9
increase of our rating in having access to public funding or procurement procedures	2,3

Source: Ever study.

A first important aspect to point out is that interviewees seemed to attach great importance to "compliance" and "environmental" issues as drivers for EMAS registration. Indeed, "better management of legal compliance" and "improvement of environmental performance" are singled out as the most effective drivers, with 4 and 3,9 scores.

Considering, for example, the environment-related driver, more than 37% of interviewed organizations identified it as "very important", and an extra 33% rated it as rather or somewhat "important", while the figures depicting a scarce importance of the environmental issue are statistically not relevant.

According to EVER interviewees, together with "compliance" and "environmental improvement", other key drivers seem to be more of "internal" nature, dealing with better organisation and overall level of the activities. Contrary to the literature review findings, competitive variables lag behind (the improvement of competitive capabilities is indicated only as the seventh driver in terms of importance, and the willingness to keep up with competitors as the eighth).

It is worth noting that these strategic/economic drivers, even if they lag behind other motivations, have nevertheless achieved fair "overall" scores: indeed, all drivers seem to have a "positive" motivational effect on companies (with scores higher than 3), exception done for those drivers that are closely linked to the public sector and the environmental regulation (regulatory relief, public funding, green public procurement, etc.), since these potential benefits are currently rather scarce and perceived as such by the interviewees.

2.2. The Barriers to the Dissemination of EMAS

Barriers to EMS adoption are generally categorized into external and internal to an organization (Milieu Ltd and Risk and Policy Analysis Ltd, 2009). Different "keys of interpretation" apply to such a broad issue: indeed, barriers are heterogeneous in nature and form: they can be further classified into smaller groups according to different criteria, as hindrances can be either internal or external, organizational or economic, general or category-specific (e.g: SMEs), and so on.

The literature review shows, firstly, that the cost of implementation is a considerable drawback for smaller organizations, which suffer greater shortages of financial resources (Hillary, 1999, 2004; Biondi et al. 2000). Several studies agree in highlighting the crucial role played by this barrier: a survey of 2005, for example, shows how the lack of financial resources (33%) and the cost of certification (23%) are among the main hindrances to 'implementation of an environmental management system (ISO, 2005).

On the other hand, studies on EMS costs (Hamschmidt and Dyllick 2001, Milieu Ltd and Risk and Policy Analysis Ltd, 2009) suggest that the above mentioned figures might be underestimated. The discrepancies in the outcome of different investigations are due to many factors, not least the fact that most organizations do not have a system for the accounting of environmental costs. The table below collected evidence from previous studies on the costs of EMAS implementation in different countries.

The evidence gathered (Biondi et al. 2000, Cesqa and Sincert, 2002) suggests that external consulting and verification costs have a stronger impact on organizations, and are felt like a heavier burden compared to other costs such as those related, for instance, to the necessary modifications regarding production processes, or linked to product innovations.

The costs relating to EMAS registration, for example, are generally low, although they depend on each national Competent Body. In some countries the cost is related to the site dimension and turnover, in a positive attempt to knock down a financial barrier for SMEs. For example, in Italy the cost varies from 50 €, for small firms, to 1500€, for larger firms.

On the one hand, to give an idea of the financial resources required, it is worth mentioning the "EMAS toolkit" (European Commission, 2000), which provides figures and average expenditures for different size-categories of organisations:

€ 10,000 for very small companies (< 10 employees)
€ 20,000 for small companies (< 50 employees)
€ 35,000 for medium companies (50 <250 employees)
€ 50,000 for large companies (> 250 employees)

On the other hand, studies on EMS costs (Hamschmidt and Dyllick 2001, Milieu Ltd and Risk and Policy Analysis Ltd, 2009) suggest that the above mentioned figures might be underestimated. The discrepancies in the outcome of different investigations are due to many factors, not least the fact that most organizations do not have a system for the accounting of environmental costs. The table below collected evidence from previous studies on the costs of EMAS implementation in different countries.

Moreover, the previously mentioned Cesqa Sincert study shows how the average annual investment for the implementation of an EMS amount to about 1,9% of sales revenue for SMEs, and 5,2% for larger organisations. The problem rises from the coupling of two factors,

Table 2. Studies on the costs of EMAS implementation

Size Country	Small < 100 emp	Medium < 500 emp.	Large >500 emp.	Average
Austria (BMUJF 1999)	109.000€	225.000€	153.000€	
Denmark (Kvistgaard, 2001)				62.000€
Germany (UBA 1999)	37.000€	84.000€	85.000€	59.000€
Switzerland (Dyllik and Hamschmidt, 2000)	56.000€	93.000€	322.000€	172.000€
Hungary (INEM 2001)	3.200€-6.2.00€	5.800€-11.000€	>11.000€	
EU member States (Ec, 2009)[1]	21.000€-38.000€	17.000€-40.000€	38.000€-66.000€	26.000€-48.000€

like the relevance of the costs for a business activity and the uncertainty of their precise entity. This is consistent with the evidence emerging from the EVER study, which argues that one of the main problems faced by SMEs when considering the possibility of registering in EMAS is the existence of "a priori" undefined costs, mostly related to the implementation phase (IEFE et al. 2006).

One of the few variables that are indirectly "linked" to the evaluation of the costs of registration, that can be gathered from literature, concerns the time-length organizations take to implement or to maintain an EMS.

In a recent study on the costs and benefits of EMAS (Milieu Ltd and Risk and Policy Analysis Ltd, 2009), registered organizations were asked to indicate the number of person-days (of either their own staff or outside contractors) required to first implement EMAS. The range of responses was quite varied. External consultancy was used by most respondents to implement EMAS (59%). There may be a trade-off between the complexity of the EMAS system (lower in smaller organizations) and the expertise available (also likely to be lower in smaller organizations). The most time-consuming tasks for internal staff are the environmental review, EMS development and internal audit.

The lack of public recognition and interest affecting EMAS (and its logo) is well known, and most studies and surveys are in line with such assumption (Ends surveyed that only 6% of respondents admit EMSs being the main environmental factor orientating purchasing habits). Obviously, scarce awareness means scarce market response.

This goes for all kinds of organizations, but is probably more tackling for SMEs, which have to put a greater effort to implement the scheme, due to their limited resources. Participants of a workshop on SMEs and EMAS arranged during the EVER project argued

---

[1] The second amount refers the first year cost; the first amount refers the yearly cost after the first year.

Table 3. External barriers according to "in-field" research

External barriers	Relevance (from 1 to 5)
Lack of competitive rewards and advantages	3,2
Lack of recognition by the public institutions (including regulatory relief)	3,2
Lack of economic incentives (including funding)	3,1
Lack of recognition by the stakeholders	2,9
Lack of recognition at the international level (outside the EU)	2,9
Too expensive (including costs of verification and registration)	2,7
Difficulties in communicating EMAS to stakeholders and customer	2,7
Too difficult to maintain the EMS under the organisational and managerial point of view	2,6
Difficulties linked to the role of the CB	2,2
Difficulties linked to the role of the verifier	2,1

that "*an important proportion of SMEs who have invested the effort and resources to register in EMAS do not receive any relevant benefits or appreciation... and finally drop out with a negative impression of the scheme*".

Brouhle (2000) goes a step forward analyzing as well the scarce level of EMAS knowledge that characterizes firms themselves. He mentions a research study by UNI/ASU, establishing that over one quarter of executive managers did not know about EMAS (Freimann and Walther, 2001), and another study by the Institute for Research in Social Choices, which identified 33% who had no knowledge of EMAS and another one third who claimed to know it only partly.

Finally, the EVER study revealed that a major disadvantage is the lack of competitive advantage and recognition by public institutions, whereas costs seem to be marginally important.

Internal barriers can be defined as obstacles preventing or impeding EMSs adoption/ implementation (Hillary, 2004). They are a vast category, including factors such as lack of resources (time and human capital), difficulties in understanding and perceiving the EMS framework, drawbacks in its implementation, organizational internal culture, and so on.

For instance, according to the relevant literature (Biondi et al. 2000) a first substantial hindrance on the way to EMAS registration, is the difficulty in effectively understanding the framework and its requirements and in identifying relevant environmental aspects. Indeed, it appears that many organizations are unable to accurately understand EMAS, especially as far as the Initial Environmental Review and the EMS are concerned. Those difficulties are highlighted by many studies (Hillary et al 1999, Hillary 2004). Zackrisson et al. (2000) shows that 49% of companies find it challenging to identify relevant environmental aspects, and more than 1 out of 4 fail to identify some significant environmental aspects. Moreover, it has been assessed by some studies that many companies evaluate the relevance of environmental aspects by the so-called "rule of thumb", and not by an objective and reproducible method (IEFE at al. 2006). For many companies, the drafting and the diffusion of the EMAS

Table 4. Internal barriers according to "in-field" research

Internal barriers	Non participants	Stakeholders	Participants
Difficulties originating from the set up and functioning of the EMAS scheme	2,5	3,1	2,7
Difficulties in implementing the requirements	2,3	3,2	2,6
Difficulties related to disclosure through the Environmental Statement	2,2	3	2,3
Difficulties in involving, motivating or obtaining the commitment of personnel	2,2	2,6	2,8
Lack of human resources and competence	2	3,5	2,9

Source: Fonte: Iraldo et al. (2006).

statement represent other difficult requirements in the implementation process. This is often due, especially within SMEs, to a lack of competences and knowledge within the organization (Biondi et al., 2000).

However, other studies assert how this is not merely a matter of lack of competences. The problem can assume a different connotation: MacLean (2004) defines it a matter of "harmony" within an organization (e.g: interaction between business executives and EHS managers) on business priorities. No surprise if, given such situation, it is very difficult to set performance objectives and to hence recognize relevant aspects within EMAS to be dealt with.

3. THE THIRD VERSION OF EMAS REGULATION

By publishing the Regulation of the European Parliament and of the Council 1221/2009/CE in the Official Journal on 22nd December 2009, the Community institutions have complied, in extremis, to the public commitment taken on many occasions to complete the second review of EMAS by the year 2009. The new Regulation expressly repeals the earlier 761/2001/EC (EMAS II), but also the Commission Decision 2001/681/EC, which contains guidelines for its implementation; the Decision 2006/193/EC laying down rules on the use of the logo, as well as two accompanying Recommendations (2001/680/EC and 2003/532/EC), thus summarizing the official text of all the requirements for its implementation.

The Regulation, called EMAS III, entered into force on 11th January 2010, becoming immediately binding in its entirety and directly applicable in the Member States with a transitional period, during which the organizations registered according to the 2001 Regulations continued to figure in the EMAS register.

For a correct interpretation of the innovations introduced by the new Regulation, it should first be understood that it meets targets for significant expansion of the numbers of EMAS, on the one hand, and to strengthen the credibility and the guarantees offered by the registration, on the other.

Preliminary studies (first of all the EVER study), outlined the failure to achieve the framework potential especially in terms of its circulation, and the difficulties of SME to participate as well as the lack of advantages and benefits arising from EMAS application.

3.1. Simplification for Small Organizations

As anticipated, one of the main objectives of the review process dealt with the enlargement of the number of registered organizations. To achieve this, the changes introduced were designed primarily to break down the barriers to registration for small organizations which, notably, represent a majority target than larger enterprises (SMEs account for 99% of European companies and generate 57% of value added products).

A major change concerns the duration of certificates and the frequency of audits for SMEs. Article 7 provides that a small organization may require the competent body to extend the maximum period of three years of registering up to four years, and the annual frequency of surveillance for up to two years provided that the verifier confirms that they have complied with the following conditions:

- There are no significant environmental risks;
- The organization does not plan significant changes;
- The organization does not contribute to significant environmental problems at local level.

Small organizations could thus reduce the regular audit by the accredited verifier from 4 (3 "monitoring annual" checks and 1 renewal) to 2 (1 'annual monitoring' verification and 1 for renewal), with consequent and significant cost savings.

Nonetheless, small organizations receiving the extension must prepare and submit annually their updated environmental statement, although not validated, to the competent body.

The new Regulation also proposes specific recommendations for the monitoring of small organizations, providing, under art. 26, that the verifier should accept exemptions and exceptions to the conventional structure of an environmental management system based on written procedures and formalized organizational procedures, enhancing rather typical aspects of smaller businesses, such as: direct communication and informal multifunctional staff (who covers more functions, environmental and otherwise), training provided through coaching in the workplace and, above all, limited documentation.

Other simplifications for SMEs are encompassed as support and incentives, as treated below.

3.2. Environmental Management System Requirements

The new Regulation considers EMAS registration as the culmination of a "journey towards excellence" in the field of environmental management, against which other forms of certification may represent only "intermediate steps". There are many innovations that are aimed at realizing this vision.

First, EMAS III continues to be based on the environmental management system introduced by ISO 14001, but complements specifically a distinctive set of requirements, starting by strengthening the mechanism to ensure compliance with environmental legislation.

The attention towards this aspect emerges promptly in many aspects of the new Regulation. Article 2, for example, defines for the first time compliance with regulatory obligations, such as full implementation of the obligations applicable to the organization being certified, including the requirements contained in permits. Furthermore, it clarifies that the initial environmental review has to provide not only a comprehensive framework of obligations under applicable law, but also describe how the organization works to ensure compliance. The Regulation states that organizations submit material or documents certifying compliance with all applicable legal requirements in environmental matters.

The focus on regulatory compliance is also apparent from the requirements for the Internal Auditor, among which it is particularly emphasized the need to assess the management system for compliance, and also compared to the policy and the organization's environmental program in relation to the applicable legal requirements. It then explicitly states that the internal audit must be designed to also respect the laws.

The continuous emphasis on ensuring regulatory compliance of the EMAS applicant organizations has led the author of the new Regulation to include all over again the "legal requirements and permit limit" even in the non-exhaustive list of environmental aspects to be considered in the EMAS process. See Annex I, Section 2 (in addition to the use of additives and processing aids, as well as semi-finished). It is clear that such integration is dictated by the Commission's desire to emphasize the importance of compliance itself, rather than the idea that this really represents an environmental aspect, an aim which is methodologically misleading with respect to the same definition of the feature ("element of an organisation's activities, products or services that has or can have an impact on the environment").

To counterbalance the considerable effort required to organizations in terms of concrete security and sustaining regulatory compliance, art. 32 of the new Regulation introduces the request to Member States to offer assistance in fulfilling their regulatory obligations, in terms of ease of access to information related to these obligations, and activation of communication channels (e.g. to obtain clarification) among the organizations interested in EMAS, and the authorities responsible for such obligations. This role can be played directly by the competent organizations or other entities of support appropriately identified by Member States.

In this respect, there is an immediate connection with the Environmental Compliance Assistance Program for SMEs (ECAP) of the European Commission, that as a curious sleight of hand, indicates precisely in the EMAS one of the most effective tools to support small organizations in keeping up to date on (and fulfilling) legal requirements.

With regard to new management system, there should also be noted that the new Regulation combines in a single annex (Annex II) system requirements derived from ISO 14001 and the additional information which the organizations implementing EMAS should take into account (previously included in an annex), thus improving the effective integration and, at the same time, highlighting the distinctive characteristics of the EMAS process.

In addition to the role played by the initial environmental review, and to importance for regulatory compliance, for continuous improvement and widespread communication and transparency (as hereinafter specified), Annex II gives special attention to training and to the involvement of the organization's personnel, whose active participation continues to be a prerequisite and a vital resource, both to the functioning of the system and to improving environmental performance.

By this logic, as well as extend the provisions contained in the former Annex IB, Section 4, the new text makes it, in fact, mandatory part of the guidelines related to participation of employees as part of EMAS, as already suggested by the Recommendation 2001/680/CE .

The innovations quoted above are accompanied by brief clarifications on the environmental management system, that is in its practical implementation by many verifiers throughout the EU were well established by experience. Just think of the need in view of the first registration, to plan and launch, but not to complete, an audit program (at least about the most significant environmental impacts).

3.3. Reference Documents

An important innovation in introducing EMAS III regards the "reference documents". These are documents that describe best practices for environmental management, i.e. the most effective means by which an organization may apply a management system able to produce the better environmental performance in specific economic and technical conditions, besides those indicators that best measure these benefits in a given sector. The Commission will develop these reference documents , with the primary objective to promote the homogeneous implementation of best management practices.

The use of reference documents is not compulsory but, if available, organizations should at least take into account what they reported, both in the deployment of their management system, and in preparing the environmental statement. Besides, the verifiers are also required to use them as a benchmark to evaluate the effectiveness of a system, especially for the evaluation of the organization's environmental performance. These facts show that organizations may well justify a failure to properly align to what has been reported in the reference documents applicable to their business sector.

Originally, in the intention expressed by the European Commission in the "Explanatory Memorandum" (the strategic lines of the revised EMAS), the reference documents should be also "intersectorial" and refer to the methodological and operational aspects of the scheme under further consideration. This would fill some obvious gaps of the new Regulation, and to provide guidance that, although expected by many, it is in fact ignored.

Consider the issue of "indirect" environmental aspects, very complex for some sectors, which EMAS III offers only a confirmation about approaches already established in the practice implementation of many Member Countries. On the one hand, the regulation confirms the interpretation that the indirect aspect is what "results from the interaction of an organisation with third parties", and that it can "to a reasonable degree be influenced by an organisation". However, it also demands that the same organization assesses the significance of this aspect, by considering how much influence it can exercise on them .

On the other hand, it simply states that for those organizations that are not part of the industrial sector, as local governments or financial institutions, it is essential that they consider the indirect aspects related to their main activity and that, in this case, an environmental review and a management system limited to the "physical" structures (and the way they are managed) are absolutely insufficient.

Another aspect on which much was expected form EMAS III, especially after the enactment of Guidelines in 2005, regarded the integration of the EMAS management system with the product dimensione. and of services belonging to an organization. On this issue, innovations compared to EMAS II are almost untraceable: we find evidence about the size of the product among the skills that auditors should have, while it is reported verbatim the "life cycle" between the indirect aspects of Annex I and, finally, in the group of elements to consider when evaluating the significance of environmental aspects we have the following: design, development, manufacturing, distribution, maintenance, use, reuse, recycling and disposal of products of the organization.

3.4. Tools and Incentive Mechanisms

Innovations that relate more directly to implementing the requirements of EMAS by the organizations concerned, have been accompanied also by a set of important changes introduced by the new Regulation concerning the role and responsibilities of others actors involved in the scheme: the competent bodies, the Member States, environmental verifiers, etc.. From an in-depth reading about the innovations planned for these subjects, it emerges that some actions (under their responsibility) could have very positive implications on individual organizations. It is essentially a set of measures of support, encouragement and promotion of EMAS, aimed at increasing membership to the scheme to facilitate and make more "tangible" the benefits that are associated with registration. See Table 3 for a more detailed examination of these measures, in the following paragraph we simply highlight some of the main keys issues.

First, the review clearly shows its intention to "empower" the Member States concerning the initiatives to support EMAS: from the request to introduce incentives for certified organizations, such as access to funding or tax relief (it is advisable to link it to the ability to demonstrate a real improvement in environmental performance by the beneficiaries); to the obligation to develop and implement ways to simplify legislation for certified organizations, to the full enhancement of EMAS in terms of legal rules, control and management of tendered contracts and public procurement.

Second, a series of innovations designed to encourage and facilitate the completion of the EMAS process to achieve registration, relying on other forms of interim certification or feeding it through cooperation and networking.

On the one hand, the Regulation requires Member States to propose a staged approach to organizations, and initiates an interesting procedure for the recognition by the European Commission (on proposals of the Member States themselves), of "other" systems of environmental management in conformity, in whole or in part, to the requirements of EMAS. If the European Commission recognizes the equivalence between "another" management system-based certification scheme (national or regional) and the new EMAS Regulation, the organizations that already adhere to (and that are certified in accordance with) it, should not refer to the relevant requirements of further verification, because they will be automatically considered compliant in the first EMAS registration.

On the other hand, the new Regulation proposes the approach, also known as " EMAS Cluster", which was developed mainly in Italy, thanks to considerable supportive work by the Committee Ecoaudit-Ecolabel, of Apat (today Ispra) and by the Network

Descartes/CARTESIO (promoted by the Regions Emilia Romagna, Lazio, Lombardy, Liguria, Sardinia and Tuscany). Once more, however, there are positive and negative aspects of it: although there is a recognition of the effectiveness of the cluster approach and the request to Member States to encourage its development, it should be noted that it is not expected to be a real cluster registration, thus in the text are missing those useful, albeit meager, operating instructions introduced in the Decision 681/2001/EC that has been repealed.

As already noted, this type of methodological shortcomings may eventually be filled by specific "Reference documents".

4. BASICS STEPS TOWARDS AN ENVIRONMENTAL MANAGEMENT SYSTEM ACCORDING TO EMAS III REGULATION

4.1. Initial Environmental Review

The initial environmental review is a crucial phase in the implementation of EMAS, as the voluntary effort of self-assessment is not yet required by law (at least with regard to environmental issues), and EMS design is strictly bound to the outcomes of said analysis in terms of strategic choices, framework and management approach within the organization. Therefore, this review is the first and fundamental step to make when starting-up the process of complying with the EMAS Regulation. The initial environmental review is "an initial comprehensive analysis of environmental aspects, environmental impacts and environmental performance related to an organisation's activities, products and services" (art. 2). The main objectives of the preliminary environmental analysis are to:

- Identify, assess and document the key environmental issues associated with the activities performed by an organization;
- Study the interaction between these factors and the technical and organizational management of the activities
- Verify the compliance with laws and regulations;
- Draw up a preparatory assessment of environmental performance in the light of the environmental policy of the organization (if existing);
- On the basis of the above, provide information and advice to set priorities, objectives and define an environmental program;
- Set up a clear EMS framework to provide adequate ground and to detail the environmental policies of the organization in terms of compliance with the Regulation on the occasion of the first audit.

The initial review takes into account all environmental aspects of the organization (products, services, activities) to focus on those more relevant to the "assembly" of the different factors required by the Regulations (environmental policy, program, management system, auditing and environmental statement).

The environmental review should consists of basic activities such as:

- identify applicable laws and regulations and subsequently assess the organizational compliance;
- identify and analyze the overall environmental aspects of the organization associated with activities, products and services;
- evaluate and select the most environmentally significant aspects.

In order to correctly organize these activities, it is advisable for the organization to accurately review:

- Manufacturing processes and/or product and service portfolio;
- Raw materials, semi-finished products and purchased good and services;
- Product and service portfolio in terms of actual or expected environmental impact.

Moreover, to implement an integrated reference framework, it is also desirable to collect data and information on:

- the updated corporate structure (including any parent organization) and its evolution;
- plants and infrastructure (e.g. production facilities for an industrial plant, the network structure in for energy retail companies, transport fleet for a forwarding agent or a carrier, etc.).
- the context of reference, in full detail spatial planning, urban settlement, landscape, socio-economic and environmental aspects (geology, hydrography...). the organization surroundings - as detailed as possible - with reference to territorial, urban, and landscape planning, socio-economic and environmental aspects (geology, hydrography).

In summary, the environmental review is the tool used by the organization to define its position on environmental issues. Therefore, the analysis is like a "snapshot" of the environmental conditions of the organization at the time when it is made, thus becoming the "point zero", against which the organization will evaluate the evolution of its environmental performance over time. At this stage, it is useful to examine more closely the factors leading to the "development" of this "snapshot".

1) Identification of the Applicable Legal Requirements Relating to the Environment

Legal compliance is one of the fundamental elements of the Regulation and a prerequisite to obtain the registration. Before everything else, the environmental review should verify that the organization is familiar with all relevant legislation and complies with it.

The basic requirements to which the organization must comply, may have a different nature and come from different legal sources. Environmental laws (EU, national, regional or local), for example, may relate to the specific production activities of the organization and its impact on the environment (air, water, waste, soil, noise, transport, etc..), the products and services provided, the specific area in which the organization operates. Further to considering legislative references, the analysis can dwell on the internal regulations already in place, the organization's directives on the environment and all liabilities arising from voluntary agreements or on the participation to initiatives promoted by external parties (environmental groups, local associations, etc.). The Regulation, however, does not require the organization only to assess its state of compliance with the law by intervening in a timely manner to correct any deficiencies, but also to verify and ensure the maintenance over time of its legal compliance. This requirement can be satisfied not only by pursuing technological, engineering and production efficiency, but also by defining management methods and organizational useful to the continuous monitoring of the relevant legislative provisions and requirements. The implementation of the environmental review may represent an important element for the identification, completion and updating of the relevant legislation, and the set up of appropriate arrangements for its correct/proper "management".

2) Identification and Analysis of Environmental Aspects

The Regulation states that an organization should consider all aspects of its activities, of its products and services and decide on the basis of the criteria it defines, what aspects have a significant impact. The organization must therefore consider first of all the whole range of environmental aspects linked to its business, and only after they have been properly assessed, focus on what it considers the most significant ones (i.e.: those that have a significant environmental impact). The Regulation distinguishes between direct and indirect environmental aspects. The Regulation uses the concept of management control to distinguish between direct and indirect environmental aspects. We define as direct environmental aspects, in fact, those aspects under the management control of the organization and as indirect environmental aspects those on which it may not have full management control. Further to an in-depth analysis of the definitions offered by the Regulation, we can assume that the indirect aspects (i.e.: those aspects on which, according to the definition, the organization has only partial control) are also considered according to the contribution (whether conscious or not) of at least one other actor other than the organization - hereinafter referred to as intermediary - with whom it shares management control (Testa et al. 2010). Some examples will help clarify the above. The Regulation provides two lists (not exhaustive) of direct and indirect environmental aspects. With regard to the direct aspects, the Regulation requires that the organization takes into account at least the following:

1) legal requirements and permit limits;
2) emissions to air;
3) releases to water;
4) production, recycling, reuse, transportation and disposal of solid and other wastes, particularly hazardous wastes;
5) use and contamination of land;
6) use of natural resources and raw materials (including energy);
7) use of additives and auxiliaries as well as semi-manufactured goods;
8) local issues (noise, vibration, odour, dust, visual appearance, etc.);
9) transport issues (both for goods and services);
10) risks of environmental accidents and impacts arising, or likely to arise, as consequences of incidents, accidents and potential emergency situations;

11) effects on biodiversity.

The list, again not exhaustive, of the indirect aspects of the process include the following:

1) product life cycle related issues (design, development, packaging, transportation, use and waste recovery/disposal);
2) capital investments, granting loans and insurance services;
3) new markets;
4) choice and composition of services (e.g. transport or the catering trade);
5) administrative and planning decisions;
6) product range compositions;
7) the environmental performance and practices of contractors, subcontractors and supplier

As mentioned above, whereas the direct aspects are generated solely by activities and decision-making processes of an organization, those produced by indirect factors also depend on activities and decision-making powers of other entities, which are active parties in the interaction between the organization and the environment. Therefore, the analysis for the identification and assessment of the environmental aspects must include (Annex VI, paragraph 6.4):

- standard operating conditions (i.e. ordinary course of business and, for example, routine maintenance and extraordinary repairs of the facilities);
- non-routine operating conditions (including for example conditions of initiation and cessation of activities or cutback of facilities);
- incidents, accidents and predictable emergency situations (in this case the initial analysis should assess, together with the probability that the event happens, the possible consequences and preventive measures taken to prevent them);
- past, present and planned activities.

Special attention must be paid to this last issue. In fact, to faithfully describe the past activities of the organization, the initial environmental review must also include noncurrent operations or abandoned areas, still likely to exert environmental effects (i.e. an underground, disused and not reclaimed reservoir that may continue to leak polluting agents into the aquifer). As to the planning of new activities, products and services, their impact must undergo a prior evaluation in compliance with the national legislation in force on the subject of environmental impact assessment.

4. ENVIRONMENTAL EVALUATION AND IDENTIFICATION OF SIGNIFICANT ASPECTS

After the identification and the analysis of all the environmental factors revolving around its activities, the organization should focus on the most important ones, which should be pivotal in the outline of the environmental management system. The definition of the criteria for the assessment of environmental aspects is therefore left to the organization, and represents a crucial element of the environmental review. The Regulation, however, makes it clear that these criteria should be "comprehensive, capable of independent checking, reproducible and made publicly available " (Annex I, Section 3).

For example, an environmental aspect can be considered significant under one or more of the following circumstances (the list is not exhaustive but only indicative):

- upon verification, environmental parameters often or constantly show borderline values;
- the preliminary analysis highlighted a particularly critical environmental factor in relation to magnitude, frequency or degree of reversibility of its impact;
- the environmental analysis detected local, regional or global weakness factors related to a specific environmental aspect;
- the organization records frequent reports from local stakeholders (surrounding communities, employees, public administration/government) on the persistence of particular unpleasant facts deriving from the its operations (e.g. air emissions of dubious nature, fish-plague, etc.);
- the organization envisages upcoming restrictions in environmental laws on given subjects and decides to concentrate its efforts on said subjects to forestall the legislative evolution.

In these cases, or in any other case involving significant issues, the environmental management system should pay special care and support the organization in dealing with them. In other words, the system must be "tailored" on the characteristics of the organization to ensure highly effective control over sensitive environmental aspects.

It is also important for the management system to ensure the review of environmental aspects in the presence of changing circumstances and conditions (introduction of new systems, products, process and organizational changes, lay-out modifications, new parameters monitored by law...) to guarantee the constant adequacy of the whole environmental framework in relation to production, organization and management features associated to environmental aspects.

4.2. Policy and Programme

Upon completion, the environmental review will provide a broad and in-depth information framework on:
- type and extent of significant environmental aspects associated with the activities of the organization;
- strengths and weaknesses in its organizational, managerial, technological, and operational procedures in dealing with these aspects;
- potential for improvement and priorities for action.

This framework provides the organization with all relevant information to set up a tailor-made environmental policy and an appropriate "action plan".

The environmental policy defines the commitment that the organization's management intends to make on environmental protection and sets out the purposes and principles of action (Article 2) that will guide all its actions in the management of environmental issues connected with the conduct of its business.

Given its strategic value, the environmental policy must be endorsed by the top management, must be consistent and integrated with all the principles and objectives representing values and identity of the organization and must inspire decision-making processes and governance. Moreover, it must be translated into specific objectives and concrete actions to improve the environmental management and its performance. In the case of complex organizations (e.g., multinational organizations and/or multi-site corporations) the environmental policy must be shared with the parent organization(s). The policy must also be formalized in writing, signed by the top management, made operational and maintained over time: the substantial and unequivocal involvement of the top decisional level of the organization is critical to the process of implementation and maintenance of the management system, ensuring the adequate follow-up of environmental aspects.

The EMAS Regulation also calls for environmental policy to be communicated to all staff and made available to the public: it cannot therefore be regarded as a mere formal act, but must be an explicit statement of commitment that the organization takes against employees - called to share and participate in its implementation - and external stakeholders.

As for the content of the environmental policy, the definition given by the Regulation refers explicitly to the commitments to "comply with "all applicable legal requirements relating to the environment " and to "continuous improvement of environmental performance" (art. 2). Although not mentioned in the definition, there are other aspects to which the Regulation pays a particular importance and that it identifies as being the key features of EMAS: external communication and dialogue and the participation of employees. Having identified significant environmental aspects and formulated, through its environmental policy, a commitment to prevent, manage and control these aspects, the organization should establish specific targets for improvement and plan appropriate interventions for their pursuit. The environmental policy, the most significant environmental aspects and the targets for improvement must be consistent with each other. The transition from overall objectives to specific objectives translates the commitment made by the top management into performance objectives to meet productive, organizational, technological and financial goals in terms of environmental aspects.

The Regulation also requires that the objectives are quantified where possible, and translated into goals, or in detailed goals and/or intermediate steps aimed at achieving the same objectives: quantification and articulation in goals allow, first, to have measurable indicators of results and, secondly, to have references to cross-check the progress of the environmental program. Based on the problems emerged from the preliminary analysis, specific objectives can be pursued either by means of technical interventions, aimed at preventing or reducing the environmental impact, or by a rationalization of environmentally-related operations to be carried out by the organization management. To achieve a specific objective (e.g. reduction of a particular pollutant in waste water), preventive actions acting on the causes (e.g. the replacement of manufacturing materials resulting in polluting byproducts) are preferable to protective measures acting on the effects (e.g. the improvement of the purification system).

It is also important that the organization makes an assessment of cross-media effects arising from an intervention: it is not infrequent, in fact, that the improved performance on an environmental aspect can cause the deterioration of others (for example, the introduction of a new system for the breaking down of a specific pollutant can cause the increase of sludge out of the purification process).

Once set the improvement goals and identified the steps needed to achieve them, the organization's management drafts the environmental program. A key aspect is the involvement of staff in the plan for environmental improvement. This may result in a minimal commitment to get the employees acquainted about their direct involvement, while establishing the roles, responsibilities and tasks for their prosecution at a later stage. In summary, environmental programs must:

1) be consistent and closely linked with the environmental policy;
2) act on significant issues in accordance with the priorities identified by the preliminary analysis;
3) define the implementation of tools and methods to attain the objectives;
4) define responsibilities and powers and adequate financial resources;
5) quantify the expected results and set the means and instruments for a constant monitoring of their progress toward the achievement;
6) take into account comprehensive environmental improvement programs (plans and programs at local, regional or national level, district or area environmental programs, action plans in the Local Agenda 21, etc.).

4.3. The Environmental Management System

The next step in the "planning", namely the definition of policy, objectives and programs, is to create an organizational structure and a management system consistent with all the previous steps. To do so, the organization must define a specific management, organizational and technical structure, as well as the environmental management system (EMS), which represents the "heart" and the engine of the activities and processes aimed at managing the environmental aspects, to actually implement the strategy for environmental improvement.

After defining the organizational structure it is also necessary to determine the training requirements for each part of the environmental management system, and define the operational conditions for the proper management of the significant environmental aspects, and for a proper functioning of the system (Iraldo et al. 2009).

By analyzing the definition of EMS provided by ISO 14001 Standards [2] it is easy to understand that in many cases, the design and implementation of an EMS require a rationalization rather than a revolution within the organization, as well as a systematization of some existing processes (i.e. the operating mode for the management of waste or air pollutants generated by the production process).

---

[2] Part of the management system of an organization used to develop and implement its environmental policy and manage its environmental aspects.

Provided the autonomy of the organization to adapt its structure to the requirements of ISO 14001 or EMAS according to technical and management requirements, when defining the system it is necessary to underline some items. In particular the organization shall:

- adapt its organizational structure by defining (and properly describe) the "structure" of the environmental management system and the functions involved, indicating the respective tasks;
- involve employees by supplying appropriate awareness tools, training and skills enhancement procedures for the management of environmental aspects;
- develop and implement effective ways of working for a proper environmental management, and appropriate response to emergencies;
- monitor the internal environmental performance and the system functioning and guarantee responsiveness in case of criticalities;
- define communication processes, both top-down and bottom-up, between different business functions and as to third parties;
- documenting the system and record performances.

Here following are some examples of implementation:

1) To define clear roles and responsibilities as to environmental management. Establish a clear and well-balanced structure, consistent with issues and objectives the organization must pursue for an effective environmental management system. First, it is necessary to grant responsibilities and define roles amongst all the subjects involved in environmental management in any capacity or to those parties whose activities may directly or indirectly lead to environmental issues. To provide departments and employees with detailed references and information so as to support them in the correct performance of the tasks required by the EMS, the organization must clearly define:

- the roles of each subject of the organization as to environmental management;
- the responsibilities conferred on subjects with regard to environmental issues;
- the duties and tasks assigned within the environmental management system
- working methods required to fulfill these tasks and duties.

Moreover, the organization must take into special account the balance between powers and responsibilities so as to guarantee to each delegated subject the proper role in decision-making processes on environmental issues. In doing so, the organization may adopt many specific tools, such as organizational charts, matrices of responsibilities, job descriptions, function charts, etc.., which may provide an effective support to operations.

Secondly, the top management plays a key role in the proper functioning of the whole system. The direction must express and convey its commitment to environmental principles, enhance the awareness on these issues, and on the need to be consistent with these principles at all levels. To ensure a strong commitment, the top management must identify a representative endowed with adequate powers and decision making responsibilities. The representative of the top management (or representatives), irrespective of other responsibilities in the organization, must have well defined roles, responsibilities and authority order to:

a) ensure that the requirements of environmental management system are established, implemented and maintained in accordance with ISO 14001 or EMAS Regulation; reporting to the top management on the performance of the system, in terms of its continuous improvement. In addition to the Representative of the Top Management, the organization must identify a figure to be entrusted with an operational role for the coordination of environmental management activities. This figure, usually the environmental manager or the "Responsible of the management system", should not be seen as the sole specialist in charge of the whole environmental management, with a full responsibility and related burdens, but rather as a support and stimulus to the Top Management (and therefore to all employees) in managing environmental issues[3].

In order to maintain its EMS, it is essential that the organization ensures the availability of adequate resources, namely technical, financial and human resources aimed at achieving a proper and effective management of environmental issues. First, the organization must have the knowledge and technical equipment and technology to pursue the environmental improvement as its main objective. Second, the economic and financial planning, fundamental to the ordinary activities of the organization, should be extended to environmental management.

Finally, the *human resources* on which the organization relies for the implementation of the EMS, as well as all personnel involved in activities affecting the organization's environmental performance, must comply with the set objectives. Define who should do what (roles, responsibilities, duties and tasks), how to do it (working methods) and how to ensure the subject in question is able to do it (training, information and communication) are crucial factors for an effective and efficient Environmental Management.

2) Awareness-building, training and participation. An effective EMS is only possible if all the staff, regardless of tasks and functions, is adequately informed and trained. The primary objective of information and training procedures is to sensitize the personnel to actively engage in environmental management and to achieve improvement targets as well as to provide any person working for the organization or on its behalf, with the necessary skills to perform environmental-sensitive operations identified by the EMS.

The principles of the environmental policy, transferred to the employees at all levels, must be put into practice on a daily basis in their respective fields of operation. It is a process of "maturation" of the organizational culture affecting individual behaviour, and it calls for time and gradual development. Hence, it is reasonable, as well as appropriate, that training and awareness-bulding are designed and carried out in parallel with the implementation of environmental management system.

The EMAS Regulation (and ISO 14001 Standards) emphasize the need to spread environmental awareness to all workers, and in particular:

---

[3] As for EMAS, in particular, with the adoption of the User Guidelines, prepared by the relevant offices of the EC, the Responsible person of the system becomes practically mandatory: "A Responsible of the EMS must be appointed by the organization manager. The role of this person is to make sure That all the system requirements are in place and updated as well as to keep informed the team about the general management system Functioning, Strengths and weaknesses and improvements needed for future actions"

- on the importance of compliance with environmental policy, procedures and requirements of the EMS;
- on significant environmental impacts, actual or potential, consequence of their activities and on environmental benefits due to improved individual performance;
- on the roles and responsibilities in achieving compliance with the environmental policy, on the procedures and requirements of the environmental management system, including emergency preparedness and response requirements;
- on the potential consequences of discrepancies from specified operating procedures and instructions.

The training of workers must be primarily designed to promote behavioural change through learning processes asking them to "put issues forward", rather than "to help solve them." This is why training should meet the learning needs of workers in cognitive (knowledge), operational (the skills) and behaviour (knowing how to be) areas. Each area needs dedicated tools and training and assessment for learning and change. On the basis of the learning needs different ways of training and information can be designed. These three areas of learning and staff development are complementary.

In particular, to ensure proper awareness to all staff whose activities may have a significant impact on the environment, the organization can train workers on:

- objectives and contents of the EMAS Regulation;
- principles and commitments outlined in the environmental policy;
- organization's environmental program;
- direct and indirect environmental aspects identified by the environmental review;
- responsibilities, duties and tasks, related to environmental management (including emergency management);
- technical management of the EMS;
- metrics, and systematic monitoring of environmental performance;
- proper implementation of procedures, operative instructions and procedures for managing the production process and/or the activities of the organization;
- possible environmental impacts of each activity and consequences related to the role perception and subsequent behaviour of the employee (e.g. Negative consequences of non-compliance of procedures and positive consequences of a correct application);
- information channels and participatory tools adopted by the organization in terms of environmental management (e.g.: procedures for reporting non-compliance).

In addition to training, it is also very useful, that the environmental manager draws up a monitoring system for the outcomes of training and learning. Training can be considered adequate if it had a positive impact not only on the level of *knowledge*, but also - and especially - on *skills* and *behaviours*, pivotal for the improvement of performances. The Assessment must be repeated regularly over time to ensure that training activities are geared to the evolving training needs. This Assessment can be performed, for example, through the distribution and collection of questionnaires or through direct observation of the trained employees. The training seminars and the evaluation of this training must be properly documented and recorded. The head of environmental education may also be required to properly record and store all documentation, providing any personal files to reconstruct the basic curriculum for each professional staff.

The training workshops and the relevant assessment must be properly documented and recorded. The head of environmental training may also be required to properly record and store all documentation, providing any personal files to reconstruct the basic curriculum for each professional of its staff.

3) Define and implement correct working methods for environmental management and emergency responsiveness. One of the most important requirements of the environmental management system certainly concerns the ability of the organization to devise and implement methods for the *in field* application of the principles of the environmental policy, and the fulfilment of improvement objectives. The organization must therefore be based on the results of the environmental analysis to identify the activities associated with the most significant aspects (both direct and indirect) and define the most suitable actions and behaviours to minimize their environmental impact. Once the "mapping" of these activities is complete, correct working methods must be planned and applied in accordance with the following conditions:

- establishing and maintaining documented procedures to cover situations in which the absence of such procedures could lead to deviations from the environmental policy, the objectives and targets;
- stipulating the operating criteria of procedures;
- establishing and maintaining procedures related to significant (and identifiable) environmental aspects in relation to goods and services used by the organization, and communicating to suppliers and contractors the applicable procedures and requirements.

To ensure the effectiveness and application of procedures, it is important to share their content with the workers directly involved in the related activities by means, for example, of consulting tools made available by the management. Additionally, the implementation of a new procedure and/or operative[4] instruction needs a period of testing to verify its effectiveness, the degree of implementation with employees (also through appropriate forms of practice and training), the organizational and technical feasibility, and to identify any changes necessary to an effective implementation.

In general terms it is appropriate that the organization is equipped with:

- procedures and operational instructions regulating the activities carried out by both employees of the organization and by other subjects acting on its behalf;
- procedures for purchases and contracts to ensure that suppliers and those acting on behalf of the organization comply with its environmental policy;
- procedures for the control of the intrinsic characteristics of the process;

---

[4] According to the definiton of the EMAS User Guideline:"Working instructions must be clear and easy to understand. The content should contain: relevance of the activity, environmental risk associated to that activity, specific training for the staff in charge of it, and supervision of the activity.

- procedures for the approval of processes and for the maintenance of equipment used in operation;
- procedures to identify and respond to potential accidents and emergency situations and to prevent and mitigate perspective environmental impacts (the organization must review and revise, where necessary, these procedures, in particular after the occurrence of accidents or emergencies and should schedule regular drills, where possible);
- procedures aimed at communication and management of interactions with all stakeholders external to the organization from which can result in an indirect environmental aspect, namely the management of those activities that involve "intermediate" actors. procedures aimed at communication and management of interactions with all external stakeholders which may be involved in indirect environmental aspect, namely the management of activities related to intermediate actors.

An important aspect, emphasized by the EMAS Regulation (and by the standard ISO 14001), is the management and monitoring of suppliers and subcontractors. For example, the actions that an organization can adopt to manage relationships with its contractors may include:

- the introduction of rules and environmental performance requirements in tender specifications (as well as in subcontract clauses), and related contractual non-compliance clauses (e.g. exclusion of hazardous substances; use of machines with high environmental performance; contract provisions on the supervision of sub-contractors);
- perform audits on contractors and subcontractors to verify compliance with environmental requirements of the contract of service (e.g., audit services for the maintenance of thermal systems, document checks on compliance statements);
- the installation of internal facilities and/or temporary structures (and the setting up of related management procedures), to facilitate the proper conduct of on-site contractors and subcontractors;
- the definition of internal procedures for selection, qualification and monitoring of contractors and subcontractors (e.g., including the presence of internal personnel to supervise technical services);
- consequent actions against suppliers as outcome of audits, control and surveillance (e.g. corrective action: reports, official communications; punitive actions: decrease in the rating of the qualification of suppliers, payment of penalties);
- the definition of shared plans and procedures for emergency management to be adopted by onsite contractors and subcontractors (code of conducts in emergency situations; emergency management training).

The procedures for operational control, as defined by the organization, should be:

- present for all activities that require a clear definition of responsibilities, duties and tasks;
- properly disseminated and available in places of use;

- known by the staff involved;
- periodically reviewed and updated;
- documented in a systematic way.

It is a widespread belief that the definition in writing of the working methodology, and the detailing of operational criteria may be useful for the organization. The goal in the formalization of the EMS should be to achieve a balanced degree of documentation, especially in relation to the complexity and the problems of the organization in managing its environmental aspects. It should be noted that:

- an organization can choose the degree of formality that befits the most, being neither procedures or operating instructions subject to any form of constraint or requirement regarding content, length and degree of detail, except to reflect the real situation in this organization and to be functional.
- in particular, a small organization is entitled to rely on established practices for certain activities (implicitly or explicitly shared by all players although not formalized) that are, in fact, part of an environmental management system.

The procedures should be proportionate to the needs of the organization's management system.

4) Performance measurement, monitoring and improvement. The EMAS Regulation and the ISO 14001 Standards provide that the organization shall establish and maintain documented procedures to monitor and regularly measure the key characteristics of its activities and its operations that can have a significant impact on the environment. This includes recording information to track the development of environmental performance, relevant operational controls and conformity with the objectives and targets. The task of measuring and monitoring thus allows to collect quantitative data, processed in the form of summary indicators that provide important information for the evaluation of organizational performance, the efficiency of the EMS and its ability to achieve environmental objectives.

In addition to audits, the organization must implement two different level of monitoring system

- monitoring of "management"
- monitoring of "performance"

The first category includes the periodic monitoring of the implementation of the objectives, and the effectiveness of controls. An organization should establish operating procedures to effectively monitor over time the level of achievement of improvements. This verification should be carried out at least every six months (including once in connection with the Management Review), in order to promptly intervene in case of difficulties in respecting the deadlines. The second category concerns the measurement of environmental performance to verify the achievement of targets by means of performance indicators (e.g. m2 of removed asbestos roof, training hours per employee/task, etc.), the compliance with operational criteria, and to monitor incidents with potential environmental consequences. The tools and equipment used for monitoring and surveillance should be subject to calibration and maintenance, while records of this process shall be retained according to the organization's procedures. The monitoring of environmental performance has also the objective to cross-check (and possibly confirm or reduce) the significance of environmental aspects previously considered as significant and to identify new ones, consistently updating the relevant register. It should be noted that the measurement and monitoring of environmental aspects must be carried out on *a regular basis*, in accordance with the agenda set by the top management, or whenever significant changes in internal or external environmental conditions to the organization so require. Under the EMAS Regulation and the ISO 14001 Standards, the organization should define responsibility and authority for handling and investigating potential non-compliances (NC), taking action to mitigate any impacts and to amend the course of activities or the equipment/ machinery that gave rise to the NC and prevent them from happening in future.

"Non-conformity" thus refers to a failure to meet one or more of the requirements defined by the organization through its management system, by the reference standards (ISO 14001 or EMAS), or by applicable law and regulation, which is likely to affect the environmental performance. Examples of failure are, for instance, wrong instructions given to employees (resulting in incorrect management of the temporary waste storage), failure to achieve a goal, misapplication of a legal provision (e.g. non-control of the quality of emissions on a given time), non-compliance to a specific requirement of the standard of reference (e.g. failure to identify the documents of external origin).

Therefore, it is advisable for the organization to introduce a procedure defining the methodology for the identification, documentation, evaluation and treatment of NC to manage any corrective and preventive actions. These actions are taken to prevent the recurrence of NC due to systematic factors, eliminating the causes and enabling preventative measures.

Any corrective or preventive action taken to eliminate the causes of NC, real or potential, must be appropriate for the problems and environmental issues in question. Therefore, the organization could introduce a procedure similar to the following:

- reporting of NC to be carried out on appropriate forms by employees in charge of the process;
- registration of the CN to be carried out by the Environmental Manager;
- analysis of NC causes, possibly in cooperation with other functions of the organization affected by the same NC;
- management of the NC as agreed with the Heads of the Departments concerned, possibly with the assistance of the person in charge of the report;
- implementation of any corrective or preventive action (if necessary) to avoid a repetition of the NC;
- assessment of the outcome of the implemented action (in case of failure, the process must be repeated).

5) Communication processes within the EMS. The commitment to an effective environmental management also requires the activation of an appropriate and systematic communication on environmental issues. The flow of environmental communication must be addressed both internally and externally. As to environmental aspect and EMS, ISO 14001 Standards and EMAS Regulation require the organization to:

a) ensures internal communication among the various levels and functions of the organization;
b) Receive, document and meet the requirements of external stakeholders.

To facilitate *internal communication,* the organization should set up suitable channels and tools, and identifies management methods, possibly by establishing a specific procedure. The tools and information channels should be efficient and effective (first, to ensure that the requests reach the right people and, secondly, to guarantee adequate and timely feedback). The flow of information and communication must be bidirectional, so as to enable the employees not only to be informed but also to express requests and suggestions, and receive appropriate and timely feedback in order to be involved and participate in environmental management. The following charts describe tools and methods to disseminate environmental commitments and objectives within the organization. The establishment and maintenance of relationships and opportunities for an external interaction on environmental issues, if properly set, can result in a a profitable mechanism for exchanging information with the stakeholders. The external communication can take place through various channels and tools, depending on the target audience:

- *institutional communication*, i.e. intended for public bodies, ministries, public administrations and supervisory bodies, through, for example, meetings, conferences, and public awareness programs (e.g. the "open days");
- marketing communication i.e. participation in fairs/conferences, press releases in industry press, corporate brochures, environmental reports and sustainability reports
- *communication* to suppliers to introduce the organization, by editing descriptive brochures or brochures that summarize the characteristics of the management system implemented in the company, and the principles of its Environmental Policy;
- Communication to suppliers to introduce the organization, by means of descriptive brochures which summarize the characteristics of the management system implemented in the company, and the principles of its Environmental Policy;
- *communication to the stakeholders* such as associations, population and individual citizens, through direct meetings, attendance at public meetings, editing of brochures or leaflets, messages in the local press etc.

The organization, however, in addition to the flow of external information (external communication), should also provide tools and suitable channels to enable the implementation and management of all incoming information, useful for the functioning of the EMS. Only by enhancing this "bi-directional" value of communication and external relations, the organization can benefit from all the advantages of strategic information network. To achieve this result, the organization may implement a systematic complaint record in order to prevent any "side effects" (reporting, pressure on institutions, etc.), initiating, for example, a toll free number, issuing questionnaires designed to assess environmental perception and the degree of social consensus or, more simply, by drafting information sheets including blank fields for complaints, comments, suggestions, to distribute among local stakeholders (neighbouring communities, mayor, representatives of environmental associations, etc.). It is recommended that the organization willing to develop a strategy for internal/external communication, establishes (or identifies) a specific function to be entrusted with the collection, analysis, development, reporting and recording of said information while in charge of managing input, and dissemination of the official communication flow among interested stakeholders.

6) Document the system and record the performance: maintenance and document control. The ISO 14001 Standards and EMAS Regulation provide for the organization to establish and maintain information, in paper or electronic form, to:

- describe the core elements of the management system and their interactions, and issue guidelines on the relevant documentation
- record the activities relevant to management, monitoring and control of the environmental significant aspects

The required documentation focuses on two types of documents:

- the so-called "management documents" describing the activities of the EMS, which are used as a reference for the proper conduct of environmental activities
- documents relating to "registration", aimed at demonstrating the proper implementation of the EMS, which provide an updated picture of environmental performance (as well as a "reconstruction" against the past performances).

As to the first first category, the key objective of the formalization is to closely fit to the actual needs of the management system (and to be verifiable by external auditors). The overall criteria is to pursue the most possible coincidence between the description given in the documentation and the actual operational system of the organization. Furthermore, the degree of formalization and complexity of management documents must be in line with the real needs of the organization and the employees. So, small organizations in particular, should avoid overly elaborate documents or procedures as they should simply describe (even through a simple list of actions) the various operating modes that the different companies involved must perform to comply with the standard requirements. This approach is also increasingly appreciated and valued for the purposes of third party certification under ISO 14001 Standards. It is clear that in these case, said verification will be primarily done by observing the conduct of employees, and by ascertaining their knowledge of the proper operating procedures by means of interviews.

As regards the second type, it should be noted that the ISO 14001 Standards and the EMAS Regulation require the organization to establish and maintain procedures for the identification, preservation and disposal of environmental records. These records must include data relating to training, as well as the results of audits and reviews. The organization, as previously highlighted, is granted complete discretion and flexibility in defining its own documental system, provided it ensures the effective management of the EMS. A third group refers to "operational" documents or operational instructions detailing the operations for the management of particular environmental aspects. These documents must necessarily arise from actual and special management needs. In order to ensure their proper and effective application, it is desirable that the definition of their content is edited with the active involvement of employees. It is also important to note that the contents of the operating instructions should in line with the skills and know-how of the various recipients. To this end, it is possible to use a common terminology as well as industry specific language (including jargon, should the productive sector and the local situation requires it), as well as figures, diagrams, and explanatory images of various types to guarantee easier understanding and memorizing. This solution is particularly suitable if the organization employs foreign workers, who may be hindered by language barriers.

Whether they are system files, Systems files and registration files must be kept under control in order to ensure that: they are easily referable to the activity, product or service to which they relate. They are properly filed, and therefore can be easily located and retrieved, and protected against damage, deterioration and loss.

4.4. Audit

Once implemented, the EMS must be tested to assess its efficiency and effectiveness in ensuring the expected performances and in achieving the goals of the environmental program, in compliance with the system. Any organization willing to implement an EMS is therefore required to plan adequate procedures for the monitoring and internal control, to achieve basic environmental performances, and to effectively monitor the "virtuous circle" of a continuous improvement. The operational framework designed by the EMAS Regulation and by the ISO 14001 Standards attaches great importance to the role of auditing when it comes to the correct and complete implementation of appropriate environmental management systems. The EMAS Regulation III defines environmental audit as "a systematic, documented, periodic and objective evaluation of environmental performance of an organization, of its management system and processes devoted to environmental protection." Through auditing activities the organization aims at assessing:

- the "merits" of the environmental performances, the adherence to EMS criteria and principles, the adequacy of the productive, technological, administrative, organizational and managerial characteristics of the organization, as well as its ability to achieve objectives;
- "methods", i.e. the sound application of EMS and the compliance of conducts with the existing rules.

The assessment must be: systematic or based on certain and recognized methods; strict/objective, which means deriving from objective evidence, verifiable and reproducible in a systematic audit process; documented or based on the existing documents as a guarantee of the traceability of such evidence and consequently, of the its conclusions; periodical, or rather scheduled or performed regularly in order to set up and ensure over time the cycle of a continuous improvement.

Whereas the basic objective of implementing an EMS is the planned management of environmental aspects related to a task, a correct and developed audit program is an essential element for achieving this goal. From a management perspective, the ISO 14001 Standards and the EMAS require the organization to establish and maintain operating procedures for periodic audits, to be carried out starting from the results of previous audits and the environmental relevance of the activities involved.

In particular, these procedures should define the methods to adopt for:

- the training of internal auditors and/or the appointment of external auditors;
- the establishment of the group of auditors;
- the planning, the scheduling and implementation of activities
- the reporting and the use of outputs for the review of the EMS to be carried out by the management

The training of internal auditors or the choice of external one is a crucial step for the success of the auditing: the reliability of its process and the trustworthiness of its results are connected, to the independence and impartiality of the auditors and to their expertise. In general, larger companies with more units can support the auditor's team with internal EMS specialists usually operating in other departments and therefore not involved in the assessment of their own unit. This activity, also known as "cross-audit" (peer audit) can provide a degree of independence from the party carrying out the audit and, at the same time, competent support from EMS internal professionals. The auditing activity must also be properly prepared and planned, by identifying objectives and scope of each audit (or audit cycle). As to the audit frequency, the ISO 14001 Standards do not set restrictions of any kind, whereas the EMAS Regulation prescribes $a complete audit cycle every three years or every four years in the case of small organizations with no significant environmental impacts, subject to simplifications established in art. 7 of the Rules. The audit cycle is the period in which all areas/activities/elements of an organization undergo auditing.

The program is very important to properly schedule the steps of the auditing cycle so as to ensure the verification of all EMS areas in the reference period. In some cases, for example, for the "start-up" of the management system, it may be necessary to schedule more frequent audits (certainly more often than once a year) in order to underline the commitment of the organization and its willingness to properly and rapidly implement the environmental framework.

In complex organizations, in terms of organizational, managerial and/or operational aspects, the auditing activity may be structured in a number of specific audits to be carried out in a sequence, defined in accordance with the auditing scope. In this case, the organization, further to appropriate and prior assessments, may decide to subject some sensitive areas / activities / elements to more frequent audits. In summary, prior to carrying out the audit, the organization must have:

- trained the internal auditors or have-selected the external ones;
- planned the stages of the audit cycle;
- appointed the audit team, that will provide, together with the organization, all the necessary work tools (worksheets, checklists, protocols, questionnaires, etc.).

At this stage, it is possible to start up the planning and implementing phase of one or more audits, as provided in the program.

4.4. Management Review

As previously seen, the definition of responsibility is a critical step in the structuring of the EMS, especially for the senior management which is in charge of strategic planning and decision-making. The executive level should first define the policy principles, setting objectives and decide on programs, ensuring appropriate resources. Besides, it should oversee the management and ensure the functioning of the EMS through the promotion, supervision, monitoring and review of objectives, programs as well as the overall system. In this context, the top management is required to regularly review the management system to assess its capability and effectiveness in implementing policy and programs.

In a logic of continuous improvement, the task of review essentially aims at identifying seeks improvement areas. Through these activities, the top management intends to ensure the relevance of the commitments supporting the policy in a changing environment, to assess performances with respect to objectives and to verify the consistency of said objectives in terms of commitment, resources and timeframe. The review will highlight needs and opportunities as to policy, objectives, programs and system. These needs may arise:

a) from cases of non-compliance:

- of performance against quantified program targets;
- of obligations with respect to corresponding law or regulation;
- of the EMS with respect to the objectives of improvement (adequacy and effectiveness of the organizational and managerial structure; correctness in the implementing rules, effectiveness of awareness building measures, information and training, etc.).

b) from internal changes like the introduction of new manufacturing technology to amend environmental aspects; from the achievement of a a goal that changes priorities, from the launch of a new activity, product or service; or rather from a change in the organizational structure, etc.

- from changes due to external factors: for example, the introduction of a new legal requirement on the subject of environment; from the market availability of a new production technology reducing a specific environmental impact; or rather from special requests of various kinds coming from stakeholders, etc..

The review is carried out during regular inter-functional meetings, personally directed by the highest organization ranks (e.g. CEO). Frequency and methods of implementation, to be decided by the organization, should be defined in advance and described in a separate procedure. As to the frequency of meetings, in general, it is preferable that they are hold at least once a year and in case of any abnormal situation or emergency calling for immediate action. Moreover, the management review should follow the completion of the audit procedures carried out in the EMS: the findings and reports of these audits may in fact be an important base of information - detailed, objective and documented - used by the organization's management to evaluate the adequacy and completeness of the measures taken, and to the capacity of the entire EMS to achieve the defined policy and programs. As regards the review methodology applied, it is appropriate that the manager, in consultation with the leaders, sets the date for the meeting and, in due time, informs the participants in writing. The meeting should involve, as well as the top management (or its representative), the Environmental Manager, the main ranks of environmental management, the middle and upper management levels of the organization with specific tasks in the environment (e.g. the Production Manager), and, if necessary, HR Managers. The organization may implement the decisions taken at the meeting in many a way. Sometimes, the review is an opportunity to set new goals and new programs for improvement resulting from the meeting, and subsequently defined in detail. Once the previous "management cycle" is complete, and its results properly assessed, the management review serves also to provide impetus to a new management cycle by planning future activities.

4.6. The Environmental Statement

One of the main objectives pursued by the EC through the Regulation is the promotion of of relational processes between the organization and its stakeholders. These processes must be based on trust, dialogue and transparency. A fundamental element of any relationship of trust between two parties is the willingness to provide clear, comprehensive and above all truthful information. The drafting of an environmental declaration according to the requirements of the Regulation, the acceptance of the mechanism of verification and validation of the data entered, the commitment to continual updating and dissemination to the general public should not be perceived simply as a sequence of steps towards registration, but as the willingness of making the declaration a tool for social control.

The Environmental Statement is a collection of information concerning the organization and its activities, the impacts that these activities have on the environment, the rules adopted for the pursuit of improved environmental performance and results, and the statement of objectives and programs defined for the future. This information must be updated every year and its amendments validated by the environmental auditor.

One of the distinctive features of the environmental statement, as a genuine means of communication, concerns the communication objectives that the organization can achieve through its application. The Regulation grant this tool a wider role than the one-way information channel. The Regulation, in addition to state that the environmental declaration is to "provide the public and other interested parties with information on the impact and environmental performance, and the continual improvement of environmental performance," adds that "it is also a tool to respond to requests from the interested parties ...»». It is this bidirectionality, if properly pursued, which makes the document not only a channel of information but an opportunity for external dialogue.

A second issue concerns the possibility of using the environmental declaration as a communication tool to meet the information needs of various interest groups. The biggest issue in the definition of a communication strategy is the heterogeneity of the actors and their different know-how, scientific and environmental culture. This makes it relatively more difficult to define a common approach. It is evident, for example, that a basic technical information on a chemical process of an industrial plant is perfectly understandable for a "specialized" party, although it may not be so for the people living in the vicinity of that plant. Hence, there are different type of audiences, with well defined characteristics in terms of perception of the information contained in the declaration, each of whom requires a different linguistic approach and tailored content.

The basics and effective environmental communication shall include:

1. the organization and its activities;
2. the environmental aspects related to that activity, the nature of the impacts and environmental performance data;
3. the measures taken to avoid or mitigate those impacts, the commitments and intentions for the future improvement of its performance, and the means used for its pursue.

1) The Organization and its activity. First, the organization should provide a presentation in which to outline the business sector (NACE code), its history, the number of employees and the size of the business, its current corporate structure (including, in the case of multinational companies or multi-plants, the current relations with the parent company), and any perspective changes. The organization should also clearly express the field of application of the Regulation: this need has strongly increased further to the transition from the concept of site to that of organization, thus making the EMAS registration available to entities (organizations or parts thereof) that could be very complex and therefore in need of clear definition.

Whereas possible, it is important to maintain the local characteristics and apply the concept of site. The organization should provide information about its location, which includes the intended use of the area (residential, agricultural, industrial, commercial, recreational, environmental, etc.), and the surrounding areas, with particular reference to any constraints or requirements of natural interest (e.g. national parks), about the characteristics of the territory (flat, fitted, pre-mountain zone wet, dry, wooded, etc.., and if an area is subject to natural events of particular gravity such as earthquakes, floods, etc..), and some information on its geological as well as about any particular social and settlement patterns. If the organization is located in a sensitive environment (e.g. industrial areas, districts, companies subject to the laws of major accidents) it may be appropriate to mention the stakeholders of the nearby settlements, especially in relation to any agreements for emergency management.

The company should also describe its production in detail. The description of the type, nature and volume of products (and their possible uses as intermediate goods) or services offered, can raise the awareness of the interlocutor about their usefulness in everyday life, thus not to underestimate the role of a company when receiving information on environmental issues.

The description of the activities may include a simplified explanation of the manufacturing processes (and, if necessary, a description of the evolution of the manufacturing process). In addition to production facilities, the document should refer to auxiliary units (such as, for example, supply systems, sewage collection and treatment systems, emission control systems, utilities - steam, heating, electricity, etc.).

2) Direct and indirect environmental aspects connected with the activity and its environmental performance. By means of the environmental analysis the organization identifies the *most significant environmental aspects* and it is therefore in a condition to circulate them to the public and to describe the methods used to manage them and keep them under control. The Regulation requires the organization to consider both the *direct* (not very relevant for some non-industrial sectors) and *indirect aspects*. There is also an explicit reference, also reflected in the Recommendation, to the *correlation between significant aspects* (e.g., the emission of carbon dioxide from combustion) and its *environmental impact* (global warming).

It is important that the exposition is truly complete and balanced, and does not neglect to underline, with the necessary accuracy, any significant aspects resulting from the preliminary analysis.

The need for balance and transparency is also present in the exposition of the quantitative data relating to the significant aspects. The regulation focuses on the significance of the data to be reported (linking it to the significance of the aspects to which they refer to), and their correlation with the objectives and goals pursued. Provide greater relevance to data of remarkable environmental performance whose importance is negligible, may be a futile exercise in persuasion of well-informed consumers. Similarly, to minimize significant aspects and deprive them of numerical feedback may eventually be harmful to the company, should these issues come to the attention of the public in all their importance. In this case, it may be preferred to admit the lack of or the delay in dealing with these issues, paying wide space to the introduction of the improvement programs that the company intends to adopt face to the problem.

It is also required that the data allow an *inter-temporal comparison* (at least three years) to offer the reader evidence of the trend of environmental performance. In the case of a significant change of data on environmental parameters under consideration or of negative performance of major importance, it is appropriate that the organization illustrates the *reasons of this trend*.

The *units of measurement* must be clearly represented, as well as any applicable *legal limits* to the environmental aspects considered, with a precise indication of its legislative authority or authorization.

It may also be appropriate, in some respects particularly significant (for example, the quality of water discharges), to provide monitoring systems, sampling rates and the list of monitored substances, including those not currently considered to be particularly significant for which, therefore, the organization does not consider to include specific data. Particularly interesting and effective in terms of commonly reported in diagrams, where applicable, beyond the limits of the law, any limitations/goals more stringent target set internally.

The discretion in the choice of indicators, however, suffered a decline. The Regulation requires that both the three-year environmental declaration, and its yearly update, the organizations report at least on key indicators specifically described in Section C of Annex IV. The Regulation requires that both the three-year environmental declaration, and its yearly update, must include at least the key indicators specifically described in Section C of Annex IV.

With regard to the key indicators (*Core Indicators*) the organization is required to report against a list of key environmental issues (*energy efficiency, material efficiency, water, waste, biodiversity, emissions*) "insofar as they relate to the direct environmental aspects." The reference to the direct aspects could mean that explicit requests for indirect environmental aspects would require, for certain organizations, such as those in the industrial sector, a significant deployment of resources. The calculation of the key indicators, as expressed in section 2, paragraph 2 of Annex IV of the Rules, can be summarized in the table below.

Table 5. Key Performance Indicators[5]

Environmental Issues	Consumption/ Yearly Overall Impact (A)	Total annual production (B)	Indicator (A/B) – Examples
Energy efficiency	Total annual energy consumption expressed in MWh or GJ	The indication of the total annual production is the same for all sectors, but it is adapted to different types of organizations, depending on the type of activity: i) for organizations involved in the production sector (industry), indicating the total annual gross value added in million € (EUR million) or the total physical production in tons per year or, for smaller organizations, the turnover or the total annual number of employees; ii) for the organizations involved in the productive sector (administration / services), refers to the size of the organization in terms of number of employees.	MWh / € added value GJ / ton of finished product
	% Total annual energy consumption (electricity and heat) produced from renewable sources		
Efficiency of materials / goods;	Annual mass flow of different materials used - in tons		ton/€ of added value ton/ ton/ N ° of employees
Water	Total annual water consumption, in m^3;		m^3/€ of added value
Waste	Total annual production of waste, divided by type, in tons		waste ton / ton of finished product
	Total annual generation of hazardous waste in kilograms or tons;		tons of hazardous waste / € added value
Biodiversity	Land use, expressed in m^2 of built-up area;		m^2/ton of finished product
Emissions	the "total annual emissions of greenhouse gases", expressed in tons of CO2 eq		ton CO2 eq / € added value CO2eq/ton ton of finished product
	Total annual emissions including at least at least emissions of SO2, NOx and PM, expressed in kilograms or tons		kg NOx/ton of finished product

The Regulation also requires the organization not to neglect the description of "*other factors regarding environmental performance*" of particular interest to the activities and relationships with the company stakeholders. Among these, for example:

- relationships with suppliers and customers, with special reference to conditions or political negotiations for the purchase of goods of environmental value (consider, for example to green procurement in the institutional context);
- relationships with business partners: relationships with neighbours and the local community (e.g., claims or actions of cooperation, awareness, information)
- current expenditure and investments in environment and safety
- any incidents, with the details of the procedures to react to them and actions taken to prevent their recurrence;

---

[5] In addition to the listed indicators, organizations may also use other values to express the consumption/total annual impact in a given field.

- research and development that permit an improvement in environmental performance
- particular modes of internal awareness and involvement in the adoption of a safe and eco-friendly internal or external behaviour (e.g. incentive system) to the organization.

3) The measures taken by companies to avoid or minimize environmental impacts, the proposals for the future improvement of its performance and the means by which it intends to consolidate them. The presentation of the environmental management system consists in summarizing in a clear set the whole of practices, procedures, processes, responsibilities and resources that the company enshrines to achieve the set goals and implement the programs. The bulk of these requirements creates a great deal of flexibility, making it difficult for the organization to understand what information to provide and at what level of detail. For example, full information should include a brief description of the following aspects:

1) procedures enabling the introduction, the review and revision of the improvement programs;
2) the organization of staff, describing how the environmental liabilities have been allocated in line with the managerial needs of the company, that may be summarized in a chart;
3) initiatives set up to implement awareness and training of staff and to encourage its active participation in environmental management;
4) the measures taken by the assessment and registration of environmental aspects;
5) the summary of the practice and operational control procedures in use at the plant;
6) the methodologies used for the auditing of the plant and the audit programs;
7) The description of the potential environmental emergencies and procedures to manage them.

Furthermore, it is essential that, in addition to presenting environmental policy containing the principles of the corporate environmental strategy, the declaration sets out clear objectives (quantitative where possible) and the programs, with an indication of responsibilities, tools, resources and deadlines. The communication on the commitment of the organization is particularly effective if, on the one hand, it demonstrates the consistency and streamlining in front of real problems, needs and opportunities for improvement, and if it links the objectives and targets pursued to significant environmental aspects; and secondly, if it shows transparency and consistency, by offering the reader an account of the achievements (or the difficulties encountered) compared to the objectives set in previous years.

CONCLUSION

The third version of the EMAS Regulation comes at the end of a lengthy review process, which began in 2005 and developed from conducting a special evaluation study ("Study EVER - Evaluation of the EMAS and Ecolabel for Their Revision"), aimed at identifying and analyzing, on a European scale, the strengths and weaknesses of applying EMAS, and to provide the Commission a number of options and recommendations in support of its review. These studies have identified three possible scenarios:

- maintaining the "status quo", scenario does not envisage substantial amendments compared to the current goals and contents of the scheme, unless administrative or institutional;
- the gradual "closing" of the scheme, in the medium term, having as a goal the elimination of EMAS;
- the strengthening of the scheme through key changes.

The last option has been identified by the Commission as the only viable alternative. The evaluation process has highlighted, in fact, the timeliness and validity of the principles of EMAS and its effectiveness, as a tool to benefit the environmental management of the organizations. Certainly, the challenge of the European Commission is very ambitious, both for the difficulty in combining the two objectives of a high growth in the dissemination and rigor in its implementation: and for the difficulty to act on a real enhancement of the SMEs.

The EMAS scheme, however, is in a position of excellence among the existing international standards of environmental management, in which integration (in business management, in the supply chain) and accountability are two key elements. This was confirmed by the new version of the Regulations, which reinforces this holistic approach, especially through the enhancement of the role of indicators.

The coming years are a crucial period to ascertain the true effectiveness of the changes adopted for the scheme, and the actual achievement of the ambitious objectives that the European Commission has set itself.

More crucial elements to the new EMAS are the passage from an communication-based approach to one firmly based on real stakeholders engagement, or rather the full involvement of other actors of the life cycle through the effective implementation of its indirect environmental aspects.

REFERENCES

Biondi, V., Frey M. and Iraldo F., (2000). Environmental Management Systems and SMEs, *Greener Management International*, Spring, pp. 55–79.

Brouhle K. (2000). Information sharing devices in environmental policy: the EU Ecolabel and EMAS. Working paper series 721 European Union Center, University of Illinois.

Cesqa and Sincert (2002), Indagine sulla certificazione ambientale secondo la norma UNI EN ISO 14001; risultati indagine Triveneto.

Clausen, J., Keil, M., Jungwirth, M. (2002). The State of EMAS in the EU: Eco-Management as a Tool for Sustainable Development - Final Report for European Commission, European Community; Brussels.

Freimann, J. and Walther M. (2001). The impacts of corporate environmental management systems: a comparison of EMAS and ISO 14001, *Greener Management International* .36, pp.91-103

Gorla N., Iraldo F. (1998). La comunicazione ambientale d'impresa: uno studio sulle dichiarazioni EMAS. *Economia delle fonti di energia e dell'ambiente* 3, pp. 49-83

Grafé A. (1996), Study on Emas environmental statements, Final Report to European Commission DG XI, Bruxelles.

Hamschmidt J., Dyllick T., 2001. "ISO 14001: profitable? Yes! But is it eco-effective?", *Greener Management International*, 34, pp. 43-54.

Hillary R. (2004). Environmental management systems and the smaller enterprise, *Journal of Cleaner Production* 12, pp. 763-777.

Hillary, R. (1999)., Evaluation of study reports on the barriers, opportunities and drivers for small and medium sized enterprises – the adoption of environmental management systems Report for DTI Envirodoctorate 5th October, 1999, NEMA, London.

IEFE Bocconi, Adelphi Consult, IOEW, SPRU, Valor and Tinge, (2006). EVER: Evaluation of eco-label and EMAS for their Revision – Research findings, Final report to the European Commission – Part I-II, DG Environment European Community; Brussels. available from www.europa.eu.int/comm/environment/emas.

Imperial College of London, IEFE Bocconi, ISO14001 Solutions (1998), An Assessment of the Implementation Status of Council Regulation (No 1836/93) Eco-management and Audit Scheme in the Member States (AIMS-EMAS), Final Report Project No. 97/630/3040/DEB/E1, European Commissin Dg Environment, Brussels.

Iraldo F, Frey M. (2007). A cluster-based approach for the application of EMAS *Working Paper MandI* 03 (2007), MAIN Laboratory Sant'Anna School of Advanced Study.

Iraldo F, Testa F and Frey M. (2009) Is an environmental management system able to influence environmental and competitive performance? The case of the eco-management and audit scheme (EMAS) in the European union, *Journal of Cleaner Production* 17 , pp. 1444–1452.

ISO, (2005) ISO, The Global Use of Environmental Management System by Small and Medium Enterprises: Executive Report by ISO/TC207/SC1/Strategic SME Group, ISO, Geneva.

Jones K., Alabaster T., Hetherington K. (1999), "Internet-based environmental reporting: current trends", *Greener Management International*, n. 26, pp. 69-90.

Kvistgaard, M; Egelyng, H.; Frederiksen, B.S.; Johannesen, T. L. (2001): MilijÆstyring og MilijÆrevision i danske virksomheder. Kobenhagen.

MacLean R. (2004). Getting the most from your EMS, Manager's Notebook, Environment Proctecion March. Available from http://eponline.com/Articles/2004/03/01/Environmental-Management-Systems--Part-2.aspx?Page=1

Milieu Ltd and Risk and Policy Analysis Ltd, (2009). Study on the Costs and Benefits of EMAS to Registered Organisations. Final Report for DG Environment of the European Commission under Study Contract No. 07.0307/2008/517800/ETU/G.2.

Perkins, R. and Neumayer, E. (2004). Europeanisation and the uneven convergence of environmental policy: explaining the geography of EMAS. *Environment and Planning* (22), pp. 881-897.

Schmittel, W., Tempel, H., Bankert, K. and Johannes, M., 1999, Öko-Audit in Sachsen Anhalt.

Schucht S, (2000). The implementation of the Environmental Management and Eco-Audit Scheme (EMAS) Regulation in France', RP 2000-B-2, Centre d'Economie Industrielle, Ecole Nationale Superieure des Mines, Paris.

Testa, F., Iraldo, F. (2010) Shadows and lights of GSCM (Green Supply Chain Management): determinants and effects of these practices based on a multi-national study, *Journal of Cleaner Production*, 18, ππ 953 – 962.

Zackrisson, M., Enroth M. Widing A. (2000) Environmental management systems – paper tiger or powerful tool. Assessment of the environmental and economic effectiveness of ISO 14001 and EMAS. Industrial Research Institutes in Sweden IVF Research Publication 00828, Stockholm.

In: Environmental Management
Editor: Henry C. Dupont

ISBN: 978-1-61324-733-4
© 2012 Nova Science Publishers, Inc.

Chapter 3

MINIMIZING ENVIRONMENTAL IMPACT FROM APPLYING SELECTED INPUTS IN PLANT PRODUCTION

Claus G Sørensen[1], Dionysis D Bochtis[1], Thomas Bartzanas[3], Nikolaos Katsoulas[2] and Constantinos Kittas[2]

[1]University of Aarhus, Faculty of Agricultural Sciences,
Department of Biosystems Engineering, Tjele, Denmark
[2]University of Thessaly, School of Agricultural Sciences,
Department of Agriculture Crop Production and Rural Environment,
Laboratory of Agricultural Constructions and Environmental Control,
Fytokou Str., N. Ionia, Magnisia, Greece
[3]Center for Research and Technology of Thessaly,
Institute of Technology and Management of Agricultural Ecosystems,
Technology Park of Thessaly, Volos, Greece

INTRODUCTION

Primary agricultural production is often managed on a rather crude level (e.g. Sørensen et al., 2010). Plant nutrients and pesticides are applied equally not only within-fields, but also to all fields growing the same crop type. Fields and crops are treated according to standards defined by simple, easily observable parameters. Likewise, society's regulation of agricultural production is on a crude level as, for example, nitrogen quotas are regulated on a farm level, because only the farm's purchase and sales are available for the required supervision. Both farm management and environmental protection could be improved significantly by a more detailed management and regulation of various inputs.

In order to achieve an improved management in terms of increased profitability and reduced environmental impact in plant production, there is a need to develop and use decision support systems and other assessment tools to plan and control selected resource inputs. Such tools will cconsists of a wide range of techniques and technologies from information technology, sensor and application technologies aimed at farm and operations management and economics (Ohlmer et al., 1998). The majority of farm managers are not trained to use the vast amount of operations data efficiently and face many challenges on how to interpret these data as the basis for decision making on crop management (Auernhammer, 2001).

In this book chapter it is discussed how the application of different decision support systems and technologies affect the environmental impacts form inputs necessary for plant production systems. Specifically, the state-of-the-art and future perspectives concerning the following input factors are considered.

MACHINERY

One third of CO_2 emissions from the agriculture sector are attributed to on-farm fuel consumption and two thirds of this fuel is consumed during farm fieldwork (Dyer and Desjardins, 2003). Beyond the direct emissions due to fuel consumption, there are a number of reasons originated from machinery fieldwork that indirectly causes green house gas (GHG) emissions. For example, wheel traffic has damaging effects on greenhouse gas balances via different mechanisms due to the soil compaction which increases the fuel energy requirements of all soil-engaging operations, providing the major motivation for tillage, which encourage oxidation of soil organic matter and the release of CO_2, and has generally negative effects on nitrogen fertiliser efficiency and soil emissions (Tullberg 2010).

For the reduction of the environmental impact from farming activities, it is very important to consider, at one hand, environmentally friendly systems and, on the other hand, operations management tools for the optimisation, given the appropriate criteria, of the operations carried out by the machinery systems.

Environmental-Friendly Field Traffic Systems

Reduced Tillage

The practice of reduced tillage has been applied for many years in areas, where wind and water erosion, together with limited water resources, are major problems (Lindwall et al., 1998: Sandretto, 2001). In Europe, reduced tillage has been on the agenda with varying degrees of intensities and success (Davies, 1989). Specifically, there was a considerable interest in the practise of reduced tillage in the 1970s and the beginning of the 1980s, which then declined because of increasing weed infestation problems, damaged soil structure and, in some cases, reduced yields. However, a renewed interest has been experienced recently because of the need to adapt to declining market prices and potential new possibilities posed by the development of new management strategies and techniques. Reduced tillage is a concept, which requires a detailed specification of the actual methods and machinery components intended to be used in customised situations. Direct drilling or no-till without any kind of prior tillage constitutes the ultimate degree of reduction in tillage efforts (Philips et al., 1980). In-between this method and the traditional method of using a plough for primary soil tillage, there are various other options using combinations of different operations and implements in order to achieve the required amount of soil preparation. The labour input, energy input, capacity and costs are distinctly different for each of these systems. Typical, the systems employ strictly sequential operations, and thus require only a single operator, in contrast with traditional methods with parallel operations of ploughing and sowing.

Global initiatives as well as nationa regulations have targeted a reduction in the use of fossil fuels and the consequent carbon dioxide (CO_2) emissions (Energy Action Plan, 1990, Kyoto, 2008). The average consumption of diesel attributed to arable farming is 107 l ha-1, representing a CO_2 emission of 284 kg ha-1. A reduced number of machinery operations for soil preparation will imply a significant potential to reduce CO_2 emissions.

Numerous studies on reduced tillage and no-tillage (e.g. direct drilling without any prior tillage) are reported in the literature, although the amount of information given on methods, techniques, labour and energy has often been limited in terms of specific quantifications and evaluations. Most studies concentrate on biological factors such as yields, weeds, pests and soil structure, rather than include a complete system evaluation including the technology and energy demands of the different methods. Koeller (1989) indicated that it was possible to save as much as 70% of both the energy demand and the labour input without affecting the yield significantly for different combinations of ploughing, stubble cultivating and rotary cultivating. Ball (1989) determined that the adoption of reduced tillage and direct drilling were limited to specific soil types (e.g. stony soils) presenting problems for conventional soil tillage. Wiedemann et al. (1986) showed a 35% reduction in fuel consumption as well as a 27% reduction in the power requirement by adapting the tractor and implements and optimising the overall driving pattern in the field. Danish studies have shown a 27–44% reduction in fuel consumption and labour requirements for reduced tillage and a 72–78% reduction for direct (Nielsen and Luoma, 2000). Recent studies in the US reports a significant savings in labour, machinery and energy by changing from conventional tillage to no-tillage (CTIC, 2002).

The different resource inputs for the different tillage systems determines the potential economic advantages in terms of lower fuel costs, lower capital costs due to a reduced number of tillage equipment components, and lower labour costs (Weersink et al., 1992). Harper (1996) found a 22% reduction in operating costs with changing from convential tillage to reduced tillage. Additional changing to direct drilling reduced the operating costs by an additional 47%, giving a total 69% cost reduction by changing the tillage system from conventional tillage to direct drilling. However, new technology and methods have substantial changed in recent years in terms of functions and sizes and thus creating a demand updated operations data adhering to thise new technologies. Sørensen & Nielsen (2005) evaluated and designed a number of plausible scenarios depicting the energy input, labour input, machine performance, and costs of soil tillage, sowing and plant care for different machinery sets and tillage systems. A number of different machinery systems with varying degrees of soil tillage intensity were designed involving current and possible future configurations in terms of machinery types and field conditions. The baseline method consisted of the traditional stubble cultivating, ploughing, seedbed harrowing, seed drilling and spraying for plant care while at the other end of the range, direct drilling and spraying were the only operations. Within this range of systems, a varying number and types of machines were simulated. The results showed that the energy input associated with the range of operations directed for crop establishment ranges from 122 to 177 kWh ha^{-1} in the case of traditional soil tillage. This is reduced by 18–29% for reduced tillage with ploughing and by approximately 52-53% in the case of reduced tillage with no ploughing. The most significant reduction is achieved for the direct drilling (75-83%). The labour input and CO_2 emissions are reduced by approximately similar rates. The central obtainable savings from reduced tillage result from the elimination of the ploughing operation. The operations costs in the studied scenarios ranged from €78 to €150 ha^{-1}, depending on the methods used and assuming a 100% utilisation. The lowest costs were incurred for direct drilling, while reduced tillage methods involving a varying number of stubble cultivations but no ploughing increased the costs by 5-20%. Methods which involved ploughing increased costs by up to 81%.

Controlled Traffic Farming

Controlled traffic farming (CTF) is a specialised farm management system that is progressively gaining acceptance among the European countries. CTF completely eliminate soil wheel compaction within the cropped area by using permanent parallel wheel tracks (tramlines) for the agricultural vehicle. Wheels are restricted to permanent traffic lanes where compaction becomes a traffic-abillity advantage. The yield potential in various crops using CTF has been documented, but also energy savings is an important aspect of CTF (Tullberg et al. 2007). A 4 year trial by (Dickson and Campbell 1990) showed that the draught force for primary cultivation averaged 17 % more for conventional-traffic system than for the corresponding zero-traffic system. Conventional and low ground pressure systems increased shallow tine cultivator draught by 60 % compared to CTF (Chamen et al. 1992). On arable land, the elimination of wheel damage on the cropped area leads to substantial cultivation energy savings ranging from 37 % to 70 %. Presented Results from using a gantry system in arable cropping, showing up to 70 % tillage energy savings and a reduction of plough resistance by up to 50 % when removing all traffic from the cultivated area. (Braunack and McGarry 2006) based on experimental field trials results on sugarcane cropping suggested that future-farming systems for the sugar industry would benefit from the adoption of controlled traffic in conjunction with minimum tillage planting to minimize soil degradation in the crop row and maintain productivity in a sustainable manner. A 74 % increase in draught force for primary and secondary cultivations in the conventional traffic system in comparison with the zero traffic system was reported by (Dickson and Ritchie 1996). Besides the benefits arising from the yield increase and energy reduction, (Reicosky et al. 1999) reported a reduction in loss of carbon dioxide and water in CTF trails.

The major impediment to an enchased adoption of CTF is the lack of compatibility in equipment track, tyre and working widths. The lack of agreed width standards for CTF is a significant difficulty for the farm machinery industry, facing a situation where there is still not a large market for CTF-compatible equipment (Tullberg 2010). Furthermore, the traffic restrictions of the CTF system generate the need to examine, in a more comprehensive way, how the implementation of CTF affects field efficiency in terms of field operations involving material flow, such as harvesting and fertilising, which are carried out by machines that carry an on-board tank (i.e., for grain or fertiliser), and optimisation approaches that potentially increase the field efficiency in material handling operations that follow the CTF system has to be developed (Bochtis et al. 2010b). Also, the effect of the direction on the field efficiency as well as in the overlapped areas should especially be considered in the case of CTF since the tramlines are permanent and no modifications are allowed (Bochtis, Sørensen, Busato, Hameed, Rodias, Green, and Papadakis 2010a).

OPERATIONS MANAGEMENT

An important factor with a negative environmental impact is the agro-chemical loses occurred due to the area overlaps during spraying operations. Overlaps during spraying are directly affected by the driving direction of the spraying machine, that is, especially in the case of the CTF, the direction in which the tramlines have been established. Moreover, the effect of the direction should especially be considered in the case of CTF since the tramlines are permanent and no modifications are allowed. (Bochtis et al. 2010a) presented a targeted approach for the estimation of the operational machinery costs on an annual basis in CTF system. The method is based on a number of sub-models that are used to evaluate the consequences in terms of machinery performance for different driving directions when establishing tramlines in a CTF system.

The aims on the reduction of the environmental impact from field machinery operations benefit from the introduction of navigation aids and precision agriculture technologies. In recent years, navigation aids such as auto-steering systems have been introduced to accurately follow the tracks and to increase system efficiency (Batte and Ehsani 2006). Especially, positioning control for spraying operations is benefitted in terms of automatic and continuous records of the areas of a field that have been covered and turning on or off sections of the sprayer to prevent double coverage of previously sprayed field areas. Research has shown the significant potential for reducing pesticide by implementing, for example, patch spraying based on the combination of machine vision and image analysis algorithms with precision spraying systems or by reducing input application overlaps by adopting precision guidance and precision spraying control systems.

Beyond the direction of the fieldwork tracks, another factor that has to be taken into consideration during their establishment is the prevention of biodiversity. There is consensus amongst EU-politicians in charge of CAP (Common Agricultural Policy), that farming can play a major role in preserving and/or developing biodiversity in the agricultural landscape by integrating local, regional or national biodiversity schemes into farm management. To this end, (de Bruin et al. 2009) proposed a method for optimising the spatial configuration of cropped swaths (or field-work tracks, using another terminology) while creating space for field margins, and to assess its feasibility with respect to input data requirements. The position and orientation of the swath pattern are optimized so as to minimise a total cost arisen by the loss of net income for uncropped area, cost of an additional swath which includes the cost of turning and subsidy received for field margins.

Recently, a series of methods for optimal field work planning and scheduling have been demonstrated. (Bochtis 2008) introduced a new type of algorithmically computed optimal fieldwork patterns (*B-patterns*) where the sequence of the field-work tracks does not follow a standard motif, as in the traditional area coverage planning for field operations. Typically, the choice of the field tracks traversal sequence is based on the experience of the machine operator and consequently, it is strongly constrained by the presence of a human operator and its ability to distinguish the next track to be followed at the end of the track currently being taken. In some cases, this is a relatively simple task, due to the nature of the operation (e.g., harvesting or not harvesting), or due to various methods that provide traces on the field surface (soil-engaging discs, foam markers etc.). As a result, the routes followed by agricultural machines tend to form repetitions of standard motifs which are convenient for the operators, but on the other hand it may lead to patterns which are far from optimal in terms of field efficiency, compaction, etc.

In contrast, *B-patterns* are the result of an algorithmic approach, according to which, field coverage is expressed as the traversal of a weighted graph, and the problem of finding optimal traversal sequences is transformed into finding the shortest tours in the graph. The resulted fieldwork pattern, in terms of the traversal sequence of the field tracks, are the optimal one for the specific combination of the mobile unit kinematics (i.e., minimum turning radius) and dimensions, the operating width, and the field shape. This approach provided the potential of the implementation of combinatorial optimisation as part of the optimal operational planning for a single or multiple machinery systems operating in one or multiple geographically dispersed fields. From the available technology point of view, the introduction of commercially available auto-steering or navigation-aid systems for agricultural machines has made it possible, in principle, to enter arbitrary field pattern sequences into programmable navigation computers and then for the machine to follow them precisely.

Experimental results from the implementation of the approach for conventional agricultural machines with auto-steering systems showed that by using B-patterns instead of traditional fieldwork patterns the total non-working distance can be reduced significantly by up to 50%. The same approach has been implemented for the mission planning of an autonomous tractor for area coverage operations such as grass mowing, seeding and spraying (Bochtis et al., 2009). Within the FP7 EU project FutureFarm, research into the potential savings from the implementation of these methodologies to the operational planning of field machinery showed that the savings in terms of operational time could be in the range of 8.4% to 17.0%, while the mean savings of fuel consumption, and consequently of CO_2 emissions was estimated in the order of 18% (Bochtis et al. 2010c).

It has to be noted that the combinatorial problem that results the optimal B-patterns belongs to the same family of the well known vehicle routing problem, a well-known problem from the operational research scientific domain. The transfer of methodologies from the established operational research area has proved beneficial to the planning of agricultural operations as part of future envisioned fleet management systems. By using deliberated abstractions and representations, the specific planning of agricultural operations can be cast as instances of well-known operational research problems (Bochtis and Sørensen 2009; 2010). In this way, the established well proved solution methodologies from these problems can be seen to enhance agricultural field operations planning.

FERTILIZERS AND PESTICIDES

Every year between 2 and 3 million tons of various pesticides are put up for sale in the world. As a large fraction of these pesticides is aimed towards crop production, pesticide application for agricultural purposes, that may be large in places, is a major source of organic pollutants in the atmosphere. The European Commission in order to minimise the detrimental environmental impact of pesticides seeks to ensure their correct use. As it was reported by the European Commission in a memorandum on sustainable use of pesticides (Commission of the European Communities 2002) "The potential exposure of bystanders and residents to pesticides via the air might constitute an exposure route, which needs further attention by research and possibly regulatory measures".

Experimental data, collected over many crops with various application techniques, have demonstrated that pesticide spraying releases chemical contaminants into the atmosphere. During application the loss to the air usually stands from a few percent to 20–30%, although it can reach 50% of the total amount applied (Van-den Berg et al., 1999). This estimation is in good agreement with reported measurements of deposition on leaves and on the ground, that turn out to be of the order of 80% at least in normal conditions (Cross et al., 2001). The amount of atmospheric loss is influenced by several factors like the physico-chemical properties of the compounds, the environmental conditions and the agricultural techniques (Bedos et al., 2002).

Agricultural buildings, especially those used for intensive production like greenhouses, received a large amount of pesticides. Greenhouses are therefore likely to be a strong source of environmental pollution. The use of pesticides in greenhouse operations in order to control pests and diseases increases the potential risk exposure of workers (Bolognesi, 2003) and the pollution of environment since the application of pesticides is usually followed by natural ventilation. Although inhalation exposure to pesticides is considered more critical in greenhouse than outdoors, investigations on airborne residues in the greenhouse air are limited. In order to evaluate the exposure to pesticides, accurate, reliable and sensitive analytical methods for monitoring organic traces constituents in the air are necessary. For air sampling and analysis, methods of preconcentration as absorption on solid sorbents are becoming more widely used because of their advantages in the selection of the appropriate sorbent for a given group of air pollutants (Rudolf et al. 1980).

In recent years several research projects have been carried out to quantify the levels of pesticides in a greenhouse environment and the impact of these levels both to human health and the environment. Nevertheless, the relation between climatic conditions and the behaviour of pesticides substances in greenhouse and ambient environments still remain an important issue which attracts scientific circumspection. In addition, understanding the process of dispersion of pesticides from greenhouses will be a useful tool for authorities to specify the frame of pesticide legislation, integrating all necessary precautions to protect workers, bystanders, surrounding communities and the environment (FAO, 2005).

Concerning specific pesticides, analysis of lindane and of some endosulfan isomers in the greenhouse air indicates that 24 h after their application, concentration levels of about 8.5% of the initial values still remain in the air (Vidal et al., 1997). It was also found that the dissipation rate and the concentration decline rate were influenced by parameters such as vapour pressure, temperature and relative humidity. An application of metamidophos in a greenhouse showed that the concentration decreased dramatically the first hour after application, but in the following hours the diminution was slower and even 52 h after application metamidophos was detected in the air (Egea Gonzalez et al., 1998). The levels of chlorpyrifos in air, leaves and soil from a greenhouse were determined by Guardino et al. (1998) in order to evaluated three analytical techniques. The results indicated that chlorpyrifos levels in the air depend on greenhouse ventilation.

Experimental approaches are often limited by high costs, the time involved, and analytical detection limits. An alternative approach to the classical laboratory analysis is pesticide fate and exposure modelling. A greenhouse tomato model developed by Anton et al.

(2004), describes human exposure pathways for pesticides applied in greenhouses in Spain. For all pesticides, exposure via tomato intake represented the most important exposure pathway for humans. Pesticide exposure of indoor workers, and specifically exposure of greenhouse workers, has been assessed in numerous studies, by means of static and personal air samplers, skin pads and hand wipes or washes (Aprea et al., 2002).

In addition, understanding the process of emission and dispersion of pesticides from greenhouses will be a useful tool for responsible authorities in order to specify the frame of pesticide legislation, integrating all necessary precautions to protect workers, bystanders, surrounding communities and the environment.

The phenomenon of ventilation, through which pesticide pollutants are emitted from an experimental greenhouse vents openings, was analyzed by simulating the emission and dispersion of fungicide Pyrimethanil (a common fungicide used in greenhouse crops) by Kittas et al. 2010.

In Figure 2, the concentration of pyrimethanil inside the experimental greenhouse and in the outside environment 1.5 m above the ground was calculated for a N-S wind direction a) 30 sec, b) 1 min, c) 2 min, and d) 4 min after vents opening. The concentration of the pesticide decreased first in the windward part of the greenhouse and afterwards in the rest of the greenhouse volume. This distribution is due to the air movement inside the greenhouse. When air flows in parallel to the greenhouse long axis, the air enters in the greenhouse through the windward opening and exits through it through the leeward opening. Similar airflow pattern was measured in a greenhouse with a continuous roof vent (Boulard et al., 1997), and is also observed both experimentally and numerically in the same to the current study experimental greenhouse when, instead of the pesticide, a well know (N2O) tracer gas was used (Kittas et al. 2006) The wind direction has a major role in the dispersion of pesticide in the ambient environment. As can be seen in Figure 1, a N-S direction transfers the pesticide outside of the greenhouse and disperse it in the nearby greenhouses and buildings. The distance in which the pesticide can be found in high concentrations depends on the initial concentration of the pesticide inside the greenhouse and on the wind velocity (Kittas et al., 2010).

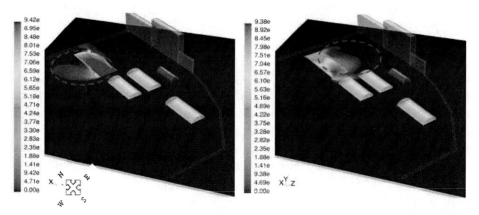

Figure 1. Simulated contours of pyrimethanil concentration inside the experimental greenhouse and at the ambient environment. Effect of wind direction on the dispersion of pesticide a) E-W wind direction; and b) N-S wind direction (Kittas et al. 2010).

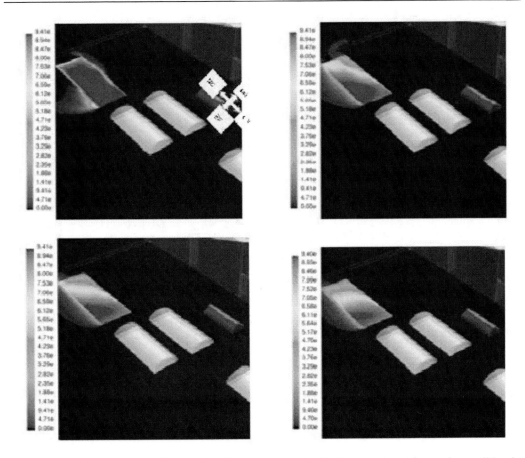

Figure 2. Simulated contours of pyrimethanil concentration inside the experimental greenhouse (30 s, 2 min, 3 min, and 4 min) after pesticide application for a S-N wind direction (Kittas et al. 2010). .

Tsiropoulos et al. (2008) experimentally monitored the decline of the residues of the fungicide Pyrimethanil in the indoor air of the greenhouse after its application with a low volume sprayer. The concentration after the low-volume spraying was ranged from 1242 to 1376 $\mu g/m^3$ and dropped rapidly to 116 -129 $\mu g/m^3$ within one hour and to 24-27 $\mu g/m^3$ within two hours after the application. The high concentration observed directly after application is attributed to the mode of application system, as after low-volume spraying the measured concentrations of pesticides were much higher than those after high-volume spraying. The day after the application (day 1) pyrimethanil concentration ranged from 5.1 to 18.2 $\mu g/m^3$ and the days after (days 2, 3 and 4) from 5.4 to 10 $\mu g/m^3$ for the second day and from 3.2 to 5.7 $\mu g/m^3$ for the third and forth days after the application (Figure 3). No constant dissipation was presented after the first hours after the application but a discontinuous course of concentration was observed. Similar behaviour was observed for malathion residues after summer application of the insecticide in a computer controlled ventilated glass greenhouse, where the concentration minima were correlate with the opening of the vents.

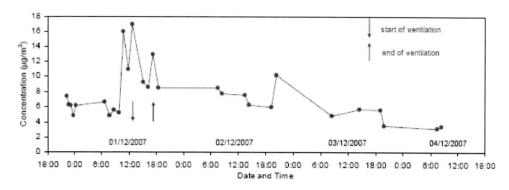

Figure 3. Evolution of pyrimethanil concentration ($\mu g/m^3$) in the air at the centre of the greenhouse after the application of 60 ml Scala (40% SC) with 1 L water. Application at 30/11/2007 (17.30 - 18.15). Measurements presented concern data taken 6 hours after the application until end of measurements. (Tsiropoulos et al., 2008)

Table 1. Health complaints of greenhouse workers

Complaint	Workers n
Fatigue	65
Headache	58
Changes in mood	50
Irritation in the eye	47
Dificulty breathing, pressure on the chest	43
Coughing, micous saliva	41
Skin itch, scars	39
Dizziness	29
Depression	28
Forgetfulness, memory disorders	28
Sleeplessness	27
Nausa, vomiting	22
Bleeding nose	12
Weight loss	8
Deformity of nails	4

The influence of pesticides on workers was analyzed by Ergonen et al. (2005). It has been shown that the amount of pesticide consumed per area, the period of use, the frequency of use and the total period of use play very important role in its exposure and harmful effects to health. Fatigue, headache and changes in mood were the common health complaints. Table 1 summarizes their main results in this field.

WATER

In many watersheds of the Mediterranean Countries, water resources are presently fully or overcommitted. Demand for water is likely to continue increasing due to population growth as well as increased demand from in-stream users. With the increasing water demand of other sectors and environmental constraints, water resources available for agriculture will decrease in the next decades. However maintaining or increasing the fraction of irrigated agriculture in the national food production is essential to reach or maintain food security and welfare in the developing countries of the World, and especially within the Mediterranean Basin. While the world as a whole may arguably have sufficient water to support its inhabitants, it is not equally the case for countries and regions everywhere. Lack of water resources is a common predicament in many Mediterranean countries, throughout North Africa, the Middle East and Southern Europe. Global climate change may exacerbate the situation. Water scarcity has become an increasing constraint to the economic development of such countries, particularly food production—the biggest water user. In this way, many Mediterranean countries have been exploiting their non-renewable fossil water to relieve the immediate pressure of water stress, depleting their resource base and undermining their long-term economic development and food security.

In all the Mediterranean countries there has been an intensification of water development and withdrawals through the building of dams/reservoirs and capturing or pumping groundwater (Casas, 1999). As a result, renewable water resources have been highly exploited. In Libya, the amount of water withdrawal is over eight times its renewable water resources, a gap filled largely by the pumping of non-renewable fossil groundwater (FAO, 2001). A similar situation is also evident in Egypt and Israel, though the overdraft there is not as serious as in Libya. In Tunisia and Morocco, the ratio of water withdrawal to renewable water resources is around 86% and 50%, respectively. The high withdrawal ratios leave them little potential for additional water supply and are clearly not sustainable. By the year 2025, it is estimated that 8.5 billion people will have to be fed and the protected agriculture will play an important role in meeting the world's food production requirements. So, it is clear that if we are to increase the supply of food during the next century, we must increase the output of the land.

With the intensification of water stress and the limited potential for additional water supply, in recent years great emphasis has been given to improving water use efficiency. In the agricultural sector, this has been expressed as "more crop and higher value per drop" (FAO, 2000). A classical pursuit of this goal has been to shift to higher-value cash crops, typically vegetables and fruits. It has been widely accepted that this shift is conducive to raising rural incomes and foreign exchange earnings (Barker and Koppen, 2001; Delgado, 1999; Wichelns, 2001). The annual rainfall in much of the Mediterranean countries ranges between 0 and 340 mm (FAO, 2001). With high potential evaporation, the rainfall is generally insufficient to meet the water demand of crops. Sharp variations between years and among different seasons within a year exacerbate the situation. Irrigation is of crucial importance for attaining a stable production and high yield. Agriculture is by far the largest water user, accounting for over 80% of the water withdrawals in Mediterranean countries' region (WRI, 2001). Poor maintenance/performance of the irrigation/drainage systems, inefficient distribution and on-farm application, low water prices (much below the real costs), private open access to aquifers, etc. are widespread problems. With the intensification of water stress, improving irrigation water use efficiency has been brought to the policy agenda in recent years.

In this critical context, there is an urgent need to foster the adoption and implementation of alternative irrigation systems and management practices that will allow increased crop water productivity

Recent advances in telecommunications, computer-based technologies and the increasing availability of a wide range of low-cost soil, plant and environmental sensors can allow the irrigation manager to access in real-time the water status within the Soil-Plant-Atmosphere Continuum (SPAC), leading to precision irrigation feedback and control. Precision irrigation aims to apply the minimum amount of water that matches the spatially and temporally distributed crop water demand by wetting the plant effective root-zone while ensuring that irrigation water is readily available for uptake and critical plant processes. In other words, precision irrigation could be achieved by fulfilling almost all the following tasks:

- Prevent irrigation-water from reaching non-cropped areas
- Use site-specific irrigation system where it is needed
- Wet only the plant-root zone with readily available water
- Estimate accurately the actual crop water needs
- Identify the plant growth-stages which are less-susceptible to water shortage.

On a plot scale, scientific approaches to precise irrigation management use either water requirements application maps or feedback data from individual sensors installed in the soil or on the plant. Water requirements maps are based on the soil texture and physical conditions, on soil elevation and topographical maps as well as electrical conductivity maps and the crop requirements. In the second case, soil water sensors help improving OFAE (open field application efficiency) by maintaining the soil water content at a readily available level and by restricting the wetting front within the plant root zone (Fares et al., 2006; Stirzaker, 2003; and Cary and Fisher, 1983). Plant sensors help to prevent severe water deficits and work by setting off alarms when a plant stress indicator is below a certain threshold (Ortuño et al., 2006; Intrigliolo and Castel, 2004; Goldhamer and Fereres, 2004; Fereres and Goldhamer, 2003; Goldhamer and Fereres, 2001). Sensors' feedback or application maps can also be introduced into mathematical models serving as decision support system or automatically triggering irrigation systems (Fernandez et al., 2008; Steppe et al., 2008; Luthra et al., 1996). On regional scale, remote sensing or satellite imagery are more appropriate for precise irrigation management as they integrate a greater number of parameters within the plant production ecosystem. Therefore, they provide the ability to assess plant water status at greater spatial resolution and they are useful techniques to identify spatially variable soils and irrigation system uniformity (Alexandridis et al., 2008; Calera et al., 2005; Fortes et al., 2005; Johnson et al., 2003; Bastiaanssen et al., 2000).

Conventionally, irrigation scheduling has aimed to meet full crop evapotranspiration (ET), as the relationship between ET and crop production of most of the major field crops is linear. Increasingly, the limited availability of irrigation water is forcing farmers in many areas of the Mediterranean to apply deficit irrigation (DI - applying less water than crop ET; Fereres and Soriano, 2007). This practice decreases irrigation demand, allowing water to be diverted to alternative uses. Much research has focused on two DI practices (although hybrids of these two techniques can also be exploited):

- Regulated deficit irrigation (RDI): varies the timing of crop water deficit
- Partial rootzone drying (PRD): varies the placement of irrigation within the rootzone

Regulated deficit irrigation (RDI) aims to exploit the differential sensitivity of yield-determining processes to water deficits, mainly in tree crops and vines. When crop phenology is sensitive to water deficit (eg. during fruit cell division), 100%ET is supplied but when crop phenology is less sensitive to water deficit (eg. during late fruit expansion) less irrigation is supplied. Such techniques have improved both agricultural water use efficiency and fruit quality.

Partial rootzone drying (PRD) aims to exploit plant root to shoot signalling to decrease crop ET while maintaining crop water status, by deliberately imposing soil moisture heterogeneity by independently watering different parts of the rootzone (Dry et al. 1996). For example, when the crop is irrigated by drip irrigation, only one side of the row receives water and the other is allowed to dry the soil. Theoretically, irrigated roots supply sufficient water to the shoots to prevent water deficits (eg. Stoll, Loveys and Dry 2000; Sobeih et al. 2004) while the remainder sense drying soil and produce chemical signals that are transmitted to the shoots to restrict water use. To maintain root signal output, the wet and dry parts of the rootzone are regularly switched (alternate PRD - PRD-A) instead of maintained throughout the growing season (fixed PRD - PRD-F), although the frequency of alternation has not been physiologically defined.

The success of any deficit irrigation technique can be judged by comparing its agronomic effects relative to conventional (optimal) irrigation. When irrigation is efficient, yield is linearly related to irrigation volumes. However, in certain crops decreased irrigation volume has improved crop quality with no significant yield penalty. Summarising over 20 literature reports has shown that in no cases did PRD significantly decrease yield compared to DI, and in about half the cases, PRD increased yield (Dodd in preparation). At least two mechanisms may account for the improved yield of PRD crops compared to conventional deficit irrigation: a decrease in evaporative losses from the soil (since a decreased area of soil is wetted during each irrigation event) and differences in plant physiological processes caused by differences in root-to-shoot signalling generated between the two irrigation techniques. To try to account for these signalling differences, a simple model of shoot signal intensity was developed, which quantified relationships between the fraction of sap flow from drying roots and soil water status of different parts of the root system during PRD (Dodd et al. 2008 a, b). This model revealed that a given signal could be higher or lower in PRD plants according to how dry part of the root system was, a result that has been experimentally verified in both controlled and greenhouse environments (Dodd et al. 2008 a, b).

Greenhouses, which give the possibility of programmed production of high quality products, constitute one from the most dynamic forms of agricultural production; they cover roughly 200 000 hectares around the Mediterranean See and constitute an important consumer of water (Sonneveld, 1995; FAO, 2006). Although protected cultivation should decrease crop water use by decreasing evaporative losses from the soil and plant evapotranspiration (as the aerial environment is humidified compared to the external environment), greenhouse growers routinely apply more irrigation water to the crops than the estimated water consumption (Bonachela et al. 2006). In general, Mediterranean growers are usually not trained in any subject concerning the greenhouse operation and management. They are usually managing their greenhouse according to their experience gained through time. There is an urgent need to inform and convince the growers, irrigation engineers and water authorities about the promising results of introducing the developed techniques and performing strategies in practice for greenhouse water management.

In (greenhouse) horticulture nutrients are usually supplied together with water. To prevent any shortage growers use excess amounts of nutrients and water. Even in the case a recirculation system is used, growers have to drain nutrients and water to the environment from time to time to prevent imbalances of the nutrient solution in the root zone. As in the present production systems nutrients and water are always supplied in excess, possibilities to control and to plan crop growth and product quality by a regulated water and nutrient supply are not used by growers. In greenhouse production systems in Mediterranean countries with severe water shortage and salinity problems the yearly losses are approximately 300-350 kg N, 125-300 kg P, 600 kg K and 3000-3500 L water per ha. As an example, for Mediterranean countries this means a total annual loss of 42000 tons of nitrogen and 600000 tons of water. In greenhouses, compared to open field production, above and below-ground conditions can be controlled to a large extent. This opens possibilities for a precise control of water and nutrient flows and crop production and quality. In order to ensure an optimal production of qualitative products, a high value of water use efficiency and a reduction of environmental pollution, high water losses have to be avoided.

Although most greenhouse crops grow in soil, soilless crops, especially closed hydroponic systems offer a great option for water saving in greenhouses (Kläring, 2001). Closed hydroponic systems are widely applied lowering the emission of nutrients to the environment and the water consumption. Since ground water contamination conflict with environmental considerations, closed growing systems, in the sense of systems avoiding run-off of drainage water to the environment, were introduced. Capture and recycling of the excess irrigation water that drains out of the root zone is possible in closed-cycle soilless growing systems and might considerably contribute to an improvement of the water use efficiency in greenhouse crops (Savvas, 2002). This technique might additionally restrict groundwater pollution by nitrates and phosphates. However, a closed hydroponic system requires precise control of fertigation to avoid ion accumulation or depletion in the root zone. For this reason, control strategies for water and nutrient supplies have to take into account the peculiarities of indoor climate, crop and growing system. Additional information about the root and shoot environment can be collected by modern, high-tech sensors, which enable accurate monitoring of several environmental parameters, thereby providing the basis for coupling the actual nutrient and water demands of the plants with the climatic conditions. As a consequence, imbalances in water and nutrient, supply, or salinization of the growth medium can be avoided. The latest is a major problem in substrate systems since currently there are no reliable means to monitor the nutrient and salt concentrations in the root environment.

Conclusion

Dedicated decision support systems and other types of assessment methods and tools for planning and controlling the environmental impacts from inputs necessary for plant production systems have been evaluated and described. Specifically, the state-of-the-art and future perspectives concerning the input factors like machinery, fertilizers, pesticides, and water have been considered. Decision support systems and technologies aimed at reducing CO_2 emissions steaming from the use of fossil fuels have been discussed and evaluated. This has involved describing and specifying planning and scheduling tools (route planning, field coverage planning, etc.), machinery sizing decision tools, automated navigation technologies (based on global and local systems) ICT technologies (e.g., telematics, web-based advisory systems), and fleet management tools.

In terms of input factors like fertilisers and pesticides, methods and technologies like the minimization of overlapped areas (web-based systems), variable rate application technologies (based on precision agriculture principles, e.g. map based applications), application planning tools were identified and evaluated.

Pesticides impact and related technologies especially for the case of greenhouse production were examined and presented. The impact of pesticides level both to human health and to the environment was analyzed. Previous and current research concerning pesticides residue evaluation in agricultural products and their emission and dispersion to the environment were presented and discussed including pesticides residues evaluation, impacts on human health and environment, emission and dispersion of pesticides to the environment, safe re-entry levels for workers, etc.

Issues related to sustainable use of water in irrigated agriculture were analysed and discussed. Improving water use efficiency at farm level have shown to imply a strategy involving a more efficient irrigation techniques improving water productivity that permit savings in water consumption and considering agriculture water demand, irrigation and environment, precise irrigation scheduling, etc.

REFERENCES

Alexandridis, T.K., Zalidis G.C., and Silleos, N.G. 2008. Mapping irrigated area in Mediterranean basins using low cost satellite Earth Observation. Computers and Electronics in Agriculture, 64, 93-103.

Anton, A., Castells, F., Montero, J.I. and Huijbregts, M. 2004. Comparison of toxicological impacts of integrated and chemical pest management in Mediterranean greenhouses. Chemosph. 54, 1225-1235.

Aprea, C., Centi, L., Lunghini, L., Banchi, B., Forti, MA. and Sciarra, G. 2002. Evaluation of respiratory and cetaneous doses of chlorothalonil during re-entry in greenhouses. J. Chromatogr B. Analyt. Technol. Biomed. Life Sci. 778, (1–2) 131–45.

Auernhammer, H., 2001. Precision farming – the environmental challenge. Computers and Electronics in Agriculture 30, 31-43.

Ball, B.C. 1989. Reduced tillage in Great Britain: practical and research experience. In: Proceedings of the Workshop: Agriculture: Energy Savings by Reduced Soil Tillage, Go¨ttingen, Germany, 10–11 June 1987 (Baumer K; Ehlers W, eds), pp 29–40. Commission of the European Communities, Luxemburg, EUR-Report. No. 11258.

Barker, R., and Koppen, B. 2001. Water scarcity and poverty. IWMI Water Brief 3, Online publication of the International Water Management Institute, Colombo. www.cgiar.org/iwmi.

Bastiaanssen, W.G.M., Noordman, E.J.M., Pelgrum, H., Davids, G, Torrezno, B.P., Allen, R.G. 2005. SEBAL Model with remotely sensed data to Improve water-resources management under actual field conditions. Journal of Irrigation and Drainage Engineering, 131, 85-93.

Batte, M.T. and Ehsani, M.R. 2006. The economics of precision guidance with auto-boom control for farmer-owned agricultural sprayers. Computers and Electronics in Agriculture, 53, (1) 28-44..

Bedos, C., Cellier, P., Calvet, R., and Barriuso, E. 2002. Occurrence of pesticides in the atmosphere in France. Agronomie 22, 35–49.

Bochtis, D. D., Vougioukas, S., Sørensen, C. G., Hameed, I. A., and Olesen, J. 2010c, Fleet management: Assessment of the potential savings. WP3 Analysis Influence of robotics and biofuels on economic and energetic efficiecies of farm production. EU project Futurefarm.

Bochtis, D., Vougioukas, S., and Griepentrog, H. 2009. A mission planner for an autonomous tractor. Transactions of the ASABE, 52, (5) 1429-1440.

Bochtis, D.D., and Sørensen, C.G. 2009. The vehicle routing problem in field logistics part I. Biosystems Engineering, 104, (4) 447-457.

Bochtis, D.D., and Sørensen, C.G. 2010. The vehicle routing problem in field logistics: Part II. Biosystems Engineering, 105, (2) 180-188.

Bochtis, D.D. 2008. Planning and control of a fleet of agricultural machines for optimal management of field operations. Ph.D. Thesis. AUTh, Faculty of Agriculturale, Dept. of Agricultural Engineering.

Bochtis, D.D., Sørensen, C.G., Busato, P., Hameed, I.A., Rodias, E., Green, O., and Papadakis, G. 2010a. Tramline establishment in controlled traffic farming based on operational machinery cost. Biosystems Engineering, 107, (3) 221-231.

Bochtis, D.D., Sørensen, C.G., Green, O., Moshou, D., and Olesen, J. 2010b. Effect of controlled traffic on field efficiency. Biosystems Engineering, 106, (1) 14-25.

Bolognesi, C. 2003. Genotoxicity of pesticides: a review of human biomonitoring studies. Mutation Res. – Reviews in Mutation Res. 543, 251–272.

Bonachela, S., González, A.M., Fernández, M.D. 2006. Irrigation scheduling of plastic greenhouse vegetable crops based on historical weather data. Irrigation Science, 25 (1), 53-62.

Boulard, T., Papadakis, G., Kittas, C. and Mermier, M. 1997. Air flow and associated sensible heat exchanges in a naturally ventilated greenhouses. Agr. For. Met. 88, 111-119.

Braunack, M.V. and McGarry, D. 2006. Traffic control and tillage strategies for harvesting and planting of sugarcane (Saccharum officinarum) in Australia. Soil and Tillage Research, 89, (1) 86-102.

Calera, A., Jochum, A.M., Cuesta, A., Montoro, A., and López Fuster, P. 2005. Irrigation management from space: Towards user-friendly products. Irrigation and Drainage systems 19, 337-353.

Cary, J.W. and Fisher, H.D. 1983. Irrigation decision simplified with electronics and soil water sensors (1983). Soil Science of America Journal, 47(6), 1219-1223.

Casas, J. 1999. Economy and agriculture of the WANA region: some basic data. http://www.icarda.cgiar.org/NARS/TOC.html. Aleppo.

Chamen, W.C.T., Vermeulen, G.D., Campbell, D.J., and Sommer, C. 1992. Reduction of traffic-induced soil compaction: a synthesis. Soil and Tillage Research, 24, (4) 303-318.

Cross, J., Walklate, P., Murray, R.and Richardson, G. 2001. Spray deposits and losses in different sized apple trees from an axial fan orchard sprayer: 1. Effects of spray liquid flow rate. Crop Protection 20, 13–30.

CTIC, 2002. Economic Benefits with Environmental Protection: No-till and Conservation Buffers in the Midwest. Conservation Technology Information Center, West Lafayette, Indiana, USA.

Davies, D.B. 1989. Objectives reduced tillage and perspectives on application. In: Proceedings of the Workshop: Agriculture: Energy Savings by Reduced Soil Tillage, Gö̈ttingen, Germany, 10–11 June 1987 (Bumer K; Ehlers W, eds), pp 1–6. EUR-Report. No. 11258.

de Bruin, S., Lerink, P., Klompe, A., van der Wal, T., and Heijting, S. 2009. Spatial optimisation of cropped swaths and field margins using GIS. Computers and Electronics in Agriculture, 68, (2) 185-190.

Delgado, C. 1999. Promoting growth and diversification through markets for high-value crop and animal products. On-line publication of IFPRI, www.ifpri.org. Washington DC.

Dickson, J.W. and Campbell, D.J. 1990. Soil and crop responses to zero- and conventional-traffic systems for winter barley in Scotland, 1982-1986. Soil and Tillage Research, 18, (1) 1-26.

Dickson, J.W. and Ritchie, R.M. 1996. Zero and reduced ground pressure traffic systems in an arable rotation 2. Soil and crop responses. Soil and Tillage Research, 38, (1-2) 89-113.

Dodd, I.C., Egea, G., and Davies, W.J. 2008a. ABA signalling when soil moisture is heterogeneous: decreased photoperiod sap flow from drying roots limits ABA export to the shoots. Plant, Cell and Environment, 31, 1263-1274.

Dodd, I.C., Egea, G., Davies, W.J. 2008b. Accounting for sap flow from different parts of the root system improves the prediction of xylem ABA concentration in plants grown with heterogeneous soil moisture. Journal of Experimental Botany 59, 4083-4093.

Dry, P.R., Loveys, B.R., Botting, D., and During, H. 1996. Effects of partial rootzone drying on grapevine vigour, yield, composition of fruit and use of water. In 'Proceedings of the 9th Australian Wine Industry Technical Conference'. (Eds CS Stockley, AN Sas, RS Johnstone, TH Lee) pp. 126-131. (Winetitles: Adelaide).

Dyer, J.R. and Desjardins R.L. 2003. The Impact of Farm Machinery Management on the Greenhouse Gas Emissions from Canadian Agriculture. Journal of Sustainable Agriculture, 22, (3) 59-74.

Egea Gonzalez, F.J., Martinez, Vidl, J.L., Castro, Cano, M.L. and Galera Martinez, M. 1998. Levels of methamidothos in air and vegetables after greenhouse applications by gas chromatography. Journal of Chromatography A, 829, 251-258.

Energy Action Plan, 1990. The Danish Energy Agency, Copenhagen.

Ergonen A.T., Salacin S. and Hakan Ozdemir M. 2005. Pesticide use among greenhouse workers in Turkey. Journal of Clinical Forensic Medicine 12, 205-208.

FAO, 2005. International Code of Conduct on the Distribution and Use of Pesticides. Publishing Management Service, FAO, Rome.

FAO, 2000. Crops and drops: making the best use of water for agriculture. On-line publication, www.fao.org. FAO, Rome.

FAO, 2001. FAOSTAT (Statistics Database). On-line information service, www.fao.org. FAO, Rome.

FAO, 2006. FAO statistical databases. Agriculture Data Collection. http://faostat.fao.org./ FAO, Rome Italy.

Fares, A., Hamdhani, H., Polyakou, V., Dogan, A., and Valenzuela, H. 2006. Real-time soil water monitoring for optimum water management. Journal of the American Water Resources Association, 42, 1527-1535.

Fereres, E., and Goldhamer, D.A. 2003. Suitability of stem diameter variations and water potential as indicators for irrigation scheduling of almond trees, J. Hortic. Sci. Biotechnol. 78, 139–144.

Fereres, E., and Soriano, M.A. 2007. Deficit irrigation for reducing agricultural water use. Journal of Experimental Botany 58, 147–159.

Fortes, P.S., Pereira, L.S., and Platonov, A.E., 2005. GISAREG—A GIS based irrigation scheduling simulation model to support improved water use. Agricultural Water Management, 77, 159-179.

Goldhamer, D.A., and Fereres, E. 2001. Irrigation scheduling protocols using continuously recorded trunk diameter measurements. Irrigation Science, 20, 115–125.

Goldhamer, D.A., and Fereres, E. 2004. Irrigation scheduling of almond trees with trunk diameter sensors. Irrigation Science, 23, 11–19.

Guardino, X., Obiols, J., Rosell, M.G., Farran, A. and Serra, C. 1998. Determination of chlorpyrifos in air, leaves and soil from a greenhouse by gas-chromatography with nitrogen–phosphorus detection, high-performance liquid chromatography and capillary electrophoresis. Journal of Chromatography A, 823, 91-96.

Harper, J.K. 1996. Economics of Conservation Tillage. Conservation Tillage Series No. 6, College of Agricultural Sciences, PennState University, USA.

Intrigliolo, D.S., and Castel, J.R. 2004 Continuous measurement of plant and soil water status for irrigation scheduling in plum. Irrigation Science, 23, 93–102.

Iversen, K.K., and Møller, A.S. 1987. Reduceret jordbehandling: driftsøkonomi og energieffektivitet. [Reduced soil tillage: operational economics and energy effectiveness.] Report No 29, Danish Research Institute of Food Economics, Frederiksberg C, Denmark.

Johnson, L., Roczen, D., Youkhana, S., Nemani, R., and Bosch, D., 2003. Mapping vineyard leaf area with multispectral satellite imagery. Computers and Electronics in Agriculture, 38(1), 37-48.

Kittas, C., Bartzanas, T., Sapounas, A., Katsoulas, N. and Tsiropoulos N. 2006. Numerical modelling of pesticides emission from a naturally ventilated greenhouse. Acta Hort. 719, 565-574.

Kittas, C., Katsoulas, N., Bartzanas, T., Tsiropoulos, N. and Boulard, T. 2010. Measurements and modelling of pesticides residues in the greenhouse air after low volume spray application. International Conference of Agricultural Engineering Towards Environmental Technologies – AgEng 2010, Clermont Ferrand 6-8 September France

Kläring, H.P. 2001. Strategies to control water and nutrient supplies to greenhouse crops: a review. Agronomie, 21, 311-321.

Koeller K (1989). Machinery requirements and possible energy savings by reduced tillage. In: Proceedings of the Workshop: Agriculture: Energy Savings by Reduced Soil Tillage, Go¨ttingen, Germany, 10–11 June 1987 (Baumer K; Ehlers W, eds), pp 7–16. Commission of the European Communities, Luxemburg EUR-Report. No. 11258.

Kordas, L.1999. Energoch"onnos´c` i efektywnos´c` ro´ z˙nych systemo´w uprawy roli w zmianowaniu. [Energy consumption and effectiveness of soil cultivation systems in crop rotation.] Folia Universitatis Agriculturae Stetinensis, Agricultura, Vol. 74, pp 47–52.

Lindwall, W., McConkey, B., Campbell, C., and Lafond, G. 1998. 20 Years of Conservation Tillage: "What Have We Learned?" Semiarid Prairie Agricultural Research Centre. Saskatchewan, Canada.

Luthra, S.K., Kaledhonkar, M.J., Singh, O.P., and Tyagi, N.K. 1996. Design and development of an auto irrigation system. Agricultural Water Management, 33, 169–181.

Nielsen, V., Luoma, T. 2000. Energy consumption: overview of data foundation and extracts of results. In: Agricultural Data for Life Cycle Assessments, Vol. 1: Second European Invitational Expert Seminar on Life Cycle Assessments of Food Products, 25–26 January 1999 (Weideman B P; Meeusen M J G, eds). Agricultural Economics Research Institute (LEI), The Hague, The Netherlands.

Ohlmer, B., Olson, K. and Brehmer, B., 1998. Understanding farmers' decision making processes and improving managerial assistance. Agricultural Economics, 18, 273-290.

Ortuño, M.F., Garcia-Orellana, Y., Conejero, W., Ruiz-Sanchez, M.C., Alarcon, J.J., and Torrecillas, A. 2006. Stem and leaf water potentials, gas exchange, sap flow and trunk diameter fluctuations for detecting water stress in lemon trees. Trees, 20, 1–8.

Philips R E; Blevins R L; Thomas GW; FogeWW; Phillips S H (1980). No-tillage agriculture. Science, 208, 1108–1113.

Reicosky, D.C., Reeves, D.W., Prior, S.A., Runion, G.B., Rogers, H.H., and Raper, R.L. 1999. Effects of residue management and controlled traffic on carbon dioxide and water loss. Soil and Tillage Research, 52, (3-4) 153-165.

Sandretto, C. 2001. Conservation tillage firmly planted in US agriculture. Agricultural Outlook, March 2001, 5–6.

Savvas, D., 2002. Nutrient solution recycling. In: Savvas, D., and Passam H.C. (Eds). Hydroponic Production of Vegetables and Ornamentals. Embryo Publications, Athens, Greece: 299-343.

Sobeih, W.Y., Dodd, I.C., Bacon, M.A., Grierson, D., Davies, W.J. 2004. Long-distance signals regulating stomatal conductance and leaf growth in tomato (Lycopersicon esculentum) plants subjected to partial rootzone drying. Journal of Experimental Botany, 55, 2353-2363.

Sonneveld, C., 1995. Fertigation in the greenhouse industry. In: Proced. Dahlia Greidinger Internat. Symposium on Fertigation. Technion – Israel Institute of Technology, Haifa, Israel, 121-140.

Steppe, K., De Pauw, D., and Lemeur, R. 2008. A step towards new irrigation scheduling strategies using plant-based measurements and mathematical modeling. Irrigation Science, 26, 505–517.

Stirzaker, R.J., 2003. When to turn the water off: scheduling micro-irrigation with a wetting front detector. Irrigation Science, 22, 177–185.

Stoll, M., Loveys, P., and Dry, P. 2000. Hormonal changes induced by partial rootzone drying of irrigated grapevine. Journal of Experimental Botany, 51, 1627-1634.

Sørensen, C.G., Fountas, S., Nash, E., Pesonen, L., Bochtis, D., Pedersen, S.M. Basso B. Blackmore S.B. 2010. Conceptual model of a future farm management information system. Computers and Electronics in Agriculture. Volume 72, Issue 1, June 2010, Pages 37-47

Sørensen, C.G.; Nielsen, V. 2005. Operational Analyses and Model Comparison of Machinery Systems for Reduced Tillage. Biosystems Engineering, 92(2): 143-155

Tsiropoulos, N., Papadi-Psyllou, A., Katsoulas, N., Kravvariti, K., Bartzanas, T., Boulard, T. and Kittas, C. 2008. Measurements of pyrimethalin residues in the greenhouse air after low volume spray application. Proceedings of the 5th European Conference on Pesticides and related micropollutants in the environment, 22-25 October 2008, Marseille, France, Pesticides 2008 pp: 73-77.

Tullberg, J. 2010. Tillage, traffic and sustainability--A challenge for ISTRO. Soil and Tillage Research, 111, (1) 26-32.

Tullberg, J.N., Yule, D.F., and McGarry, D. 2007. Controlled traffic farming--From research to adoption in Australia. Soil and Tillage Research, 97, (2) 272-281.

Van-den Berg, F., Kubiak, R. and Benjey, W. 1999. Emission of pesticides into the air. Water, Air, and Soil Pollution 115, 195–218.

Vidal Martinez, J.L., Egea Gonzalez, F.J., Glass, C.R., Galera Martinez, M. and Cano Castro, M.L. 1997. Analysis of lindane, α- and β-endosulfan and sulfate in greenhouse air by gas chromatography. J. of Chromatography A, 765, 99-108.

Weersink A; Walker M; Swanton C; Shaw J (1992). Costs of conventional and conservation tillage systems. Journal of Soil and Water Conservation, 47(4), 328–334.

Wichelns, D., 2001. The role of "virtual water" in efforts to achieve food security and other national goals, with an example from Egypt. Agricultural Water Management, 49, 131–151.

Wiedemann, H.T, Clark, L.E., Bandy, S.M., and Stout, B.A. 1986. Implement power requirement for reduced tillage. ASAE Paper No. 86-1588, St. Joseph, MI.

WRI, 2001. World Resources Institute (WRI). Environmental data tables. On-line Database of the World Resources Institute, www.wri.org. Washington DC.

In: Environmental Management
Editor: Henry C. Dupont

ISBN: 978-1-61324-733-4
© 2012 Nova Science Publishers, Inc.

Chapter 4

NATURAL ADSORBENTS FROM TANNIN EXTRACTS: NOVEL AND SUSTAINABLE RESOURCES FOR WATER TREATMENT

J. Sánchez-Martín[*] *and J. Beltrán-Heredia*[1]
Department of Chemical Engineering and Physical Chemistry
University of Extremadura, Badajoz, Spain

ABSTRACT

The adsorption of contaminants is one of the most popular processes in water treatment. The feasibility of a long variety of materials, such as clays or active carbons and multiple waste products such as biomass, tires or other kinds of industrial residues, makes the production of new adsorbents a very interesting researching subject.

This chapter is focused on a new type of natural adsorbent which is based on tannin extracts. Gelation of tannins is a recent chemical procedure that is often performed in order to obtain high quality adhesives. However, not many works have been carried out studying the ability of these gels for removing cationic pollutants from aqueous matrix. This is the scope of the current contribution.

Wastewater techniques should be improved in order to guarantee their applicability in a wide scenario: from developed countries, where no economical problems are detected; to less developed zones, where technology is usually a natural barrier. Global sustainability is evidently depending on the correct care of natural resources all over the world, not only where socioeconomic conditions allow to keep the environment clean.

Tannins from different trees, such as *Quebracho, Acacia, Pinus* or *Cypress*, have been gelified according to different gelation processes. Parameters such as aldehyde or NaOH concentration were varied and optimal conditions for gelification were applied in order to obtain efficient adsorbents for the removal of dyes, surfactants and heavy metals. Reliability of the results was technically confirmed by statistical methodologies.

---

[*] Corresponding author: Telephone number: +34 924289300 ext. 89033. E-mail addresses: jsanmar@unex.es (J. Sánchez-Martín) and jbelther@unex.es (J. Beltrán-Heredia)

1. INTRODUCTION

The increasing pollution level urges scientific community to research with more and more dedication in environmental remediation. In addition, natural resources are becoming relevant in this task because they present many advantages that fit very well to the sustainability scope [1]. Profiting the numerous chemical compounds that come from vegetal world may be an interesting way for the synthesis of useful products. In this scenario, tannins have been already used as natural adsorbents, after undergoing a gelation process [2,3,4].

The remediation of several pollution problems is a target of many researchers nowadays. Technical ways of solving environmental concerns and menaces such as the dumping of surfactants, dyes, pharmaceuticals and other hazards are available long time ago, but making them cheaper and sustainable is still a challenge. Natural raw materials are a possible source of low-cost adsorbents that could provide a successful solution [5,6].

Tannin structure involves multiple aromatic rings that provides a useful matrix in which active centers can be introduced by means of the adequate polymerization process [7]. Two classes of chemical compounds of mainly phenolic nature are included as vegetable tannins: condensed and hydrolysable tannins [8]. Inside the first group, one can present tannins from *Acacia mearnsii de Wild*, *Schinopsis balansae* or *Pinus pinaster*. Regarding the hydrolysable tannins, this group includes barks from *Castanea sativa* or *Caesalpinia spinosa*.

Tannin gelation is a chemical procedure that immobilizes tannins in an insoluble matrix [9] so that their properties of interest, e.g., metal chelation, are then available in an efficient adsorbent agent. In addition, the material resulting from their gelation (sometimes called *tannin rigid resin*) presents interesting properties in terms of resistance, non-flammability, and mechanical undeformability [10,11].

Gelation of tannins has been widely described in the scientific literature and in patents. The experimental conditions for gelation may involve the use of formaldehyde or other aldehyde in a basic or acid medium. One may find examples of basic gelation in the scientific literature [2,3,12], and in patents such as US patent 5,158,711 [13], and acid gelation is described by other workers [14,15].

The chemical basis of the tannin gelification are widely reported [7]. Formaldehyde and other aldehydes react with tannins to induce polymerization through methylene bridge linkages at reactive positions on the tannin flavonoid molecules. The reactive positions of the rings depend on the type of tannin, but mainly involves the upper terminal flavonoid units. For example, the A-rings (figure 1) of *Acacia mearnsii* and *Quebracho* tannins show reactivity towards formaldehyde comparable to that of resorcinol.

However, aspects such as size and shape make the tannin molecules lose mobility or flexibility at relatively low level of condensation, so that the available reactive sites are too far apart for further methylene bridge formation. The result may be incomplete polymerization and therefore poor material properties. It is needed to search for constitutional differences between several tannin extracts in order to predict the gelation product. E.g., among condensed tannins from mimosa bark (*Acacia mearnsii*) the main polyphenolic pattern is represented by flavonoid analogues based on resorcinol A-rings and pyrogallol B-rings. This is similar in the case of *Quebracho* bark extract, but no phloroglucinol A-ring pattern exists in the first or in the second type. The A-rings of pine tannin posses only the phloroglucinol type of structure, much more reactive towards formaldehyde than resorcinol-type counterpart. Elemental probable reaction mechanism in two steps between generic condensed tannin and formaldehyde is showed in figure 2.

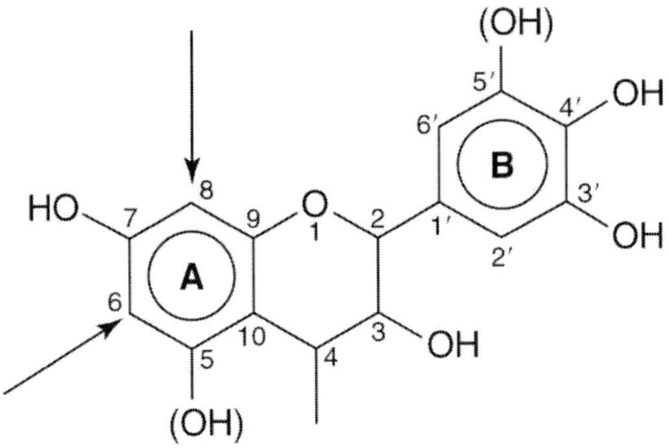

Figure 1. Tannin flavonoid basic structure.

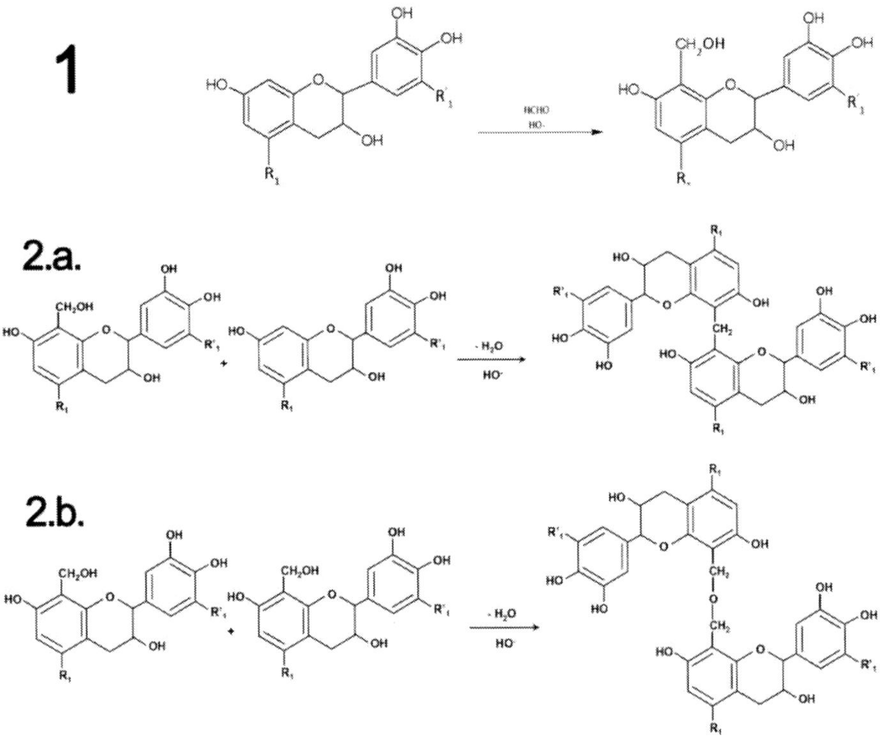

Figure 2. Probable generic gelation mechanism between condensed tannin and formaldehyde. 1) First step: hydroxy-methylation; 2a) Second step: Methylene bridge immovilization; 2b) Methylene ether bridge immovilization.

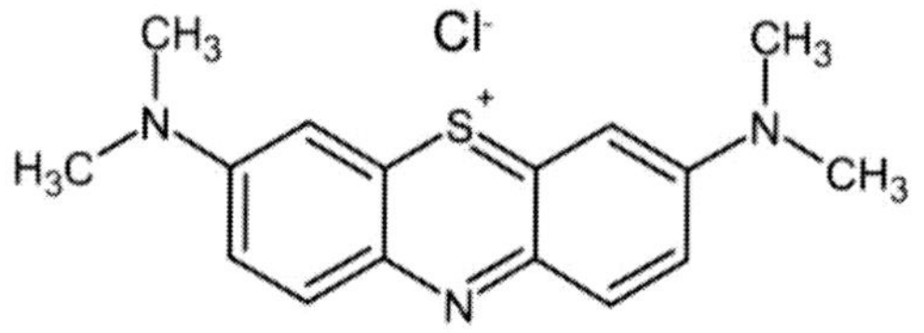

Figure 3. Chemical structure of Methylene Blue.

Therefore, it is very difficult to state the exact and real gelation mechanism for each type of tannin extract. Presumably, it depends on the real structure and composition of each one, which has to do not only with the tannin source (the vegetal bark), but also with the extraction process. In addition, researchers are not interested in working with pure tannin extracts but natural wood materials, the chemical characterization of gelified tannins is rather complex. However, researchers that have worked with these types of adsorbents performed several characterization procedures. This is the case of our own work [16], where a FTIR analysis was performed, or Yurtsever and Sengil [17], where SEM was presented.

This chapter is focused on so-called tanningels as water treatment agents. Polymerization (hence gelation) has been carried out from different tannin feedstock, including hydrolysable and condensed one, and the resulting products have been tested on dye, surfactant and heavy metal removal. *Tanningels* were found to be a very interesting new adsorbents since a high efficiency was obtained.

Three kinds of pollutants are presented in the current chapter: dye, surfactant and heavy metals.

- Industrial pollution involving dyes leads to colored water that can destroy environmental equilibrium. Many dyes are toxic and even carcinogenic. Although some of them are used in pharmaceutical production, the large exposure to them can cause several harmful effects. Methylene blue (structure showed in figure 3), which is the one selected in the current investigation, can cause increased heart rate, vomiting, shock, cyanosis, jaundice and many other dangerous injuries [18].
- Surfactants dumping into the environment represent a harmful and noxious practice. They may be useful and needed compounds, but they are also considered dangerous and undesirable substances because of their impact on water animal and vegetal life. Surfactants affect the environment in many aspects: groundwater and lakes pollution, pharmaceutical products binding (so pollution activity of these kinds of chemical compounds is considerably increased [19]) and animal and human toxicity [20]. Cationic ones enter into aquatic ecosystems together with polluted waters because they are widely used in many industries, including petroleum, oil refining, petrochemical and gas industry. They are also used in pest control in aquaculture for combating pathogenic organisms for fish [21]. Especially cationic surfactants presents higher biopersistance [22] and particular dangerous concerns such as negative impact on many species of bacteria and fungi [23]. Specifically, the risk of bioaccumulation of *Cetyltrimethyl ammonium bromide* (CTAB) has been fully characterized [24] and many studies have been found regarding the need of cationic surfactants degradation since long time ago [25] until present day [26]. CTAB

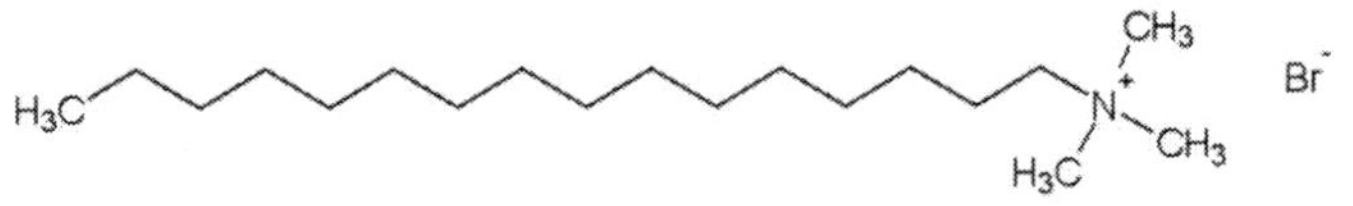

Figure 4. Chemical structure of Cetyltrimethylammonium bromide.

adsorption has been also studied previously [27]. Taking these risks into account, the investigation we have developed has been focused on this surfactant, so its chemical structure is shown in figure 4.

- Finally, the pollution with heavy metals is well known. Effluents from industrial facilities, mainly linked to minery, are highly charged with Cu^{2+}, Hg^{2+}, Pb^{2+}, Zn^{2+}, Ni^{2+}, or Cd^{2+}. These wastewaters represent cumulative pollution in the food chain [28], so the elimination of such harmful compounds has been searched by scientist from long time ago and it is achieved by different ways: chemical precipitation [29] coagulation and flocculation [30] or nanofiltration [31], among others.

MATERIALS AND METHODS

2.1. Tannin Extracts

Commercially available tannin extracts, such as *Acacia mearnsii* de Wild (Weibull black), *Schinopsis balansae* (Quebracho colorado), *Castanea sativa* (Chestnut) and *Caesalpinia spinosa* (Tara) were kindly supplied by TANAC (Brazil). They were products involved in the leather treatment and they are presented as powder.

Lab-extracted tannins, such as *Pinus pinaster* (Pine) and *Cupressus sempervivens* (Cypress) are extracted according to the following procedure [32]: 100 g of bark were milled in a cutting mill (RETSCH, SM 2000 model) and they were put in 600 mL of tap water. Then 5 g of NaOH (PANREAC) were added and the mixture was stirred in magnetic stirrer at 90° C for 1 h. Suspended solids were separated by filtration and liquid fraction was dried in oven (65° C) overnight. The resultant solid was considered the tannin extract.

2.2. Pollutants

Different pollutants were used in the current investigation. Methylene blue was used as dye model compound, although others were included in preliminary screening. It was provided by SIGMA-ALDRICH. Regarding heavy metal, Zn^{2+} was selected as model compound and it was supplied by PANREAC. Finally, Cetyltrimethylammonium bromide (CTAB) was the model compound regarding cationic surfactants. It was supplied by SIGMA-ALDRICH. Table 1 presents a summary of these pollutants.

Table 1. Chemical properties of pollutants

Pollutant	Chemical formula	Group	Supplier
Methylene Blue	$C_{16}H_{18}N_3SCl$	Cationic dye	SIGMA-ALDRICH
Zn^{2+}	$Zn(NO_3)_2$	Heavy metal	PANREAC
CTAB	$C_{19}H_{42}BrN$	Cationic surfactant	SIGMA-ALDRICH

2.3. Gelation Process

Tanningels were prepared according to the basis of Nakano [2]. 5 g of tannin extract were dissolved in 32 mL of NaOH (PANREAC) 0.125 mol L^{-1} and 30 mL of distilled water at 80° C. When mixture was homogeneous, 2 ml of formaldehyde (commercial purity grade) was added and reaction was kept at the same temperature for 8 hours until polymerization was considered completed. Then, the apparent gummy product was lead to complete evaporation of water remain and dried in oven (65° C) overnight.

After gelation, tannin rigid foams were crushed and sieved to produce 38-53 μm diameter particles. They were washed successively with distilled water and HNO$_3$ 0.01 mol L^{-1} (PANREAC) to remove unreacted sodium hydroxide. Finally, the adsorbent was dried again in oven. Differences are found between this preparation way and the description made by Yurtsever and Sengil [17], mainly concerning the amount of formaldehyde.

2.4. Physical Characterization of Tanningels

In order to characterize the main adsorbent aspect of the tannin gel scanning electron microscopy were carried out on tannin derived rigid foams. The BET surface area was determined with an AUTOSORB AS-1. SEM images were taken using Field Emission Gun (mod. HITACHI S-4800) microscopy. Samples were pre-treated with chrome sputtering.

FTIR spectra were recorded on a Thermo-Nicolet FTIR 300 spectrophotometer. A sample of tannin extract and the corresponding *tanningel* were dried at 60°C and stored under vacuum. Potassium bromide disk was prepared by mixing 1 mg of tannin with 250 mg of KBr (MERCK, spectrometry grade) at 10000 kg cm^{-2} presure for 5 min under vacuum. The spectra were recorded from 4000 to 400 cm^{-1}.

2.5. General Adsorption Tests

A stock solution of each pollutant was prepared. Different volumes of this initial solution were put into a 100 mL-flask, and a fixed quantity of *tanningel* was added (20 mg). Magnetic stirring was applied for 15 days on a magnetic multistirrer (SELECTA), until equilibrium was achieved. The kinetics was studied by collecting samples at regular intervals. The samples were centrifuged and the pollutant remnant was adequately analyzed according to APHA methods [33].

2.6. Mathematical and Statistical Procedure

Several statistical tests were applied in order to give consistency to the whole study. Categorical optimization of tannin gelation was carried out by using SPSS 14.0 for Windows [34]. A factorial Central Composite orthogonal and rotatable Design (CCD) was used for the optimization of the quantitative variables regarding Quebracho gelation. Each experiment of the planned series was made three times, so the average value was kept for the CCD analysis under Response Surface Methodology. Design of experiments section was statistically analyzed by using StatGraphics Plus for Windows 5.1 [35]. Non-linear multiparametric data adjustment was carried out by using SPSS 14.0 for Windows.

3. RESULTS

3.1. Classical Tanningels

Chemical polymerization of tannin extracts by means of formaldehyde is a well known process for producing *tanningels*. Our previous work on adsorbents derived from *Schinopsis balansae* (red quebracho) presented some interesting data about the removal of methylene blue from aqueous solutions. So obtained adsorbent is called *QTG*. As a first approach, the tannin gel were synthesized according strictly to Nakano et al. [2] even with the formaldehyde proportions (2 mL each 5 g of tannin extract). The first physical characterization of this preliminary product was made on the basis of FTIR. The spectra of both samples are shown in figure 5.

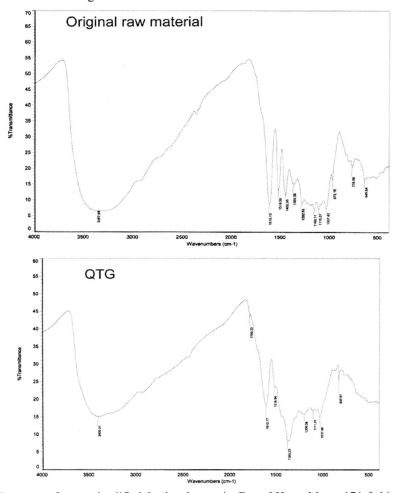

Figure 5. FTIR spectra of raw and gelified Quebracho tannin. From J Hazar Mater, 174, 9-16, 2010.

Wide bands in the range of 3600-3100 cm^{-1} correspond to -OH bridging groups in all systems and are attributed to O-H stretching (phenolic or alcoholic group) and to water molecules hydrogen bonded with -OH groups. The small peaks in the region of 2950-2850 cm^{-1} are associated with the methylene (-CH$_2$-) bridges. Also, stretching vibrations of C-H groups in the aromatic rings give absorption bands in this region. The absorption bands between 1620 and 1450 cm^{-1} are characteristic of the elongation of the aromatic -C=C- bonds. The deformation vibration of the C-C bonds in the phenolic group absorbs in the region of 1500-1400 cm^{-1}. The peak at 1390-1370 cm^{-1} is associated with the O-H deformation vibration of phenolic or alcoholic group. The peaks in the region 1280-1210 cm^{-1} are associated with the -CO stretchings of the aromatic ring and the methylene ether bridges formed by reaction with formaldehyde. The peaks at 1160-975 cm^{-1} are due to asymmetrical C-O-C stretching and C-H deformation. The deformation vibrations of the C-H bond in the aromatic rings give absorption bands in the range of 835 to 650 cm^{-1}.

3.1.1. Kinetics of Dye Removal

For confirming the validity of equilibrium periods (up to fifteen days), a kinetic study was carried out with Methylene blue (MB). A series of trials was performed with a fixed initial dye concentration (IDC) (100 mg L^{-1}) and with different proportions of QTG and MB (mmol per g of adsorbent). Figure 6 reports the decreasing concentration of dye in six experiments with 1.5, 1.8, 2.2, 2.5, 2.8 and 3.7 mmol g^{-1}. A rather rapid dye removal is achieved in the first 150 hours, although complete equilibrium dye concentration is reached at 350 hours.

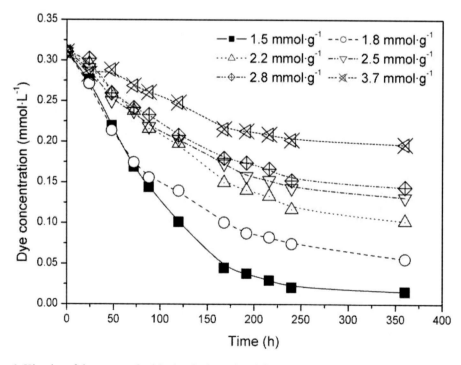

Figure 6. Kinetics of dye removal with classical tanningel from quebracho. From J Hazar Mater, 174, 9-16, 2010.

The Lagergren equation (1) is one of the most widely used adsorption rate equations for the adsorption of solute from a liquid solution. The modified first order kinetic model of Lagergren [36] may be represented by the following equation:

$$q = q_e - q_e e^{-k_1 t} \quad (1)$$

where q is the adsorption capacity, defined as mmol of dye per g of *tanningel*; k_l is the first order Lagergren constant (h^{-1}), q_e is the equilibrium q capacity (mmol g^{-1}) and t is the contact time (h).

Apart from this model, there is another one, which is attributed to Ho and McKay [37], where q follows a non-linear relationship according to equation (2).

$$q = \frac{t}{\left(\frac{1}{h}\right) + \left(\frac{t}{q_e}\right)} \quad (2)$$

where h is defined by equation (3):

$$h = k_h q_e^2 \quad (3)$$

and k_h is the second order constant (g mmol^{-1} h^{-1}).

Finally, Elovich model is the third one included in the current investigation [38] and it involves the following equation (4)

$$q = \frac{1}{\beta_E} \ln(\alpha_E \beta_E) + \frac{1}{\beta_E} \ln t \quad (4)$$

where α_e is the initial adsorption rate (mmol g^{-1} h^{-1}) and β_E is the desorption constant (g mmol^{-1}).

The application of these three models have been carried out by a non-linear adjustment and results are presented in figure 7.

As can be appreciated, the three of them fit reasonably well to the experimental situations, so regression coefficient r^2 may be considered in order to discriminate the goodness of each data fit. All of them presented r^2 levels above 0.95. According to this, the three models explain rather well the adsorption process. With little differences, Lagergren model gives a 0.98 regression coefficient. Due to its simplicity and to the goodness of the linear correlation, not only the non-linear regression, this hypothesis may be assumed as the best theoretical model in this adsorption case. Similar phenomena have been reported in Methylene blue adsorption on other natural products [39].

Bearing in mind these kinetic data, further experiments were carried out for the duration of 15 days in total to guarantee the chemical equilibrium was achieved.

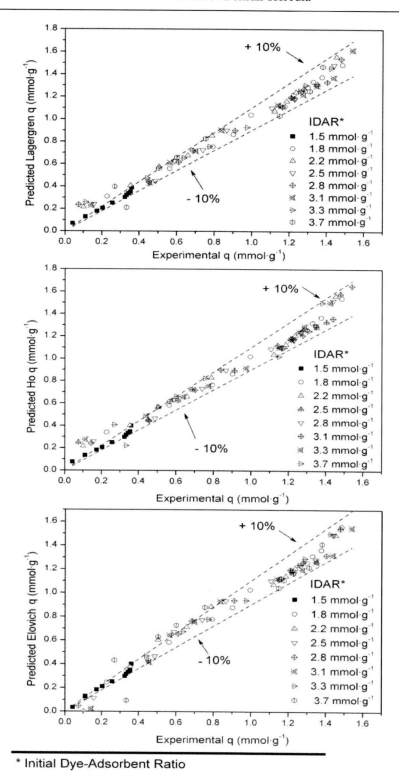

Figure 7. Kinetic of dye removal with classical *tanningel* from *quebracho* data adjustment. From J Hazar Mater, 174, 9-16, 2010.

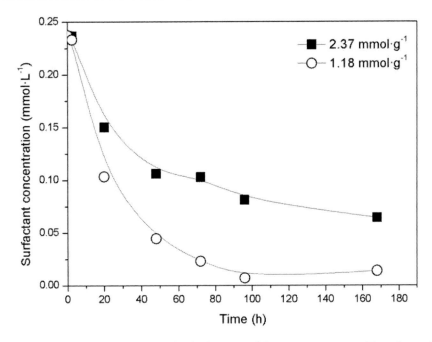

Figure 8. Kinetic of CTAB removal onto classical *tanningel* from *Pinus pinaster*. Experimental conditions: 20°C and pH 7.

3.1.2. Kinetics of CTAB Removal

Cationic surfactant CTAB was also a target pollutant. Kinetics of this removal process were obtained from a classical *tanningel* from *Pinus pinaster* (PTG). In this case, a series of trials was performed with a fixed initial surfactant concentration (ISC) (0.14 mmol L^{-1}) and two proportions of PTG and CTAB (mmol per g of adsorbent). Figure 8 shows the decreasing concentration of surfactant in the two experiments with 1.18 and 2.37 mmol g^{-1}. Fairly rapid surfactant removal is achieved in the first 48 hours, although the complete equilibrium surfactant concentration is reached only at 168 hours.

For the kinetics of the adsorption process, the previous three theoretical models were considered. The results of the nonlinear fits with these models are presented in table 2.

Table 2. Theoretical kinetics model parameters. Units in text

		1.18	2.37	Average[b]
Lagergren ISAR[a]				
k_l		3.11 10^{-3}	4.20 10^{-3}	3.60 10^{-3}
q_e		1.66	1.14	
r^2		0.98	0.99	0.98
Linear regression r^2		0.96	0.89	0.92
Ho ISAR[a]		1.18	2.37	
h		6.92 10^{-3}	6.51 10^{-3}	6.71 10^{-3}
q_e		2.01	1.34	

Table 2. (Continued)

k_h	$1.69\ 10^{-2}$	$3.60\ 10^{-2}$	
r^2	0.98	0.98	0.98
Linear regression r^2	0.99	0.99	0.99
Elovich			
ISAR[a]	1.18	2.37	
$\square_E$	$1.72\ 10^{-1}$	$1.42\ 10^{-1}$	
$\square_E$	2.42	3.53	2.97
r^2	0.99	0.97	0.98
Linear regression r^2	0.99	0.96	0.98

Notes:
[a] Initial Surfactant-Adsorbent Ratio, mmol g^{-1}.
[b] When appropriate.

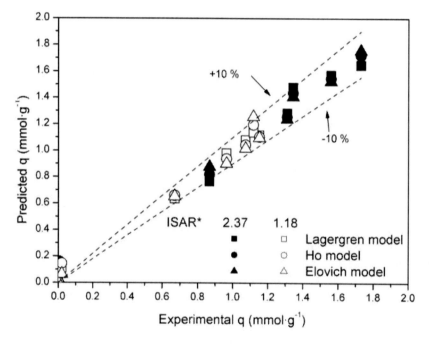

Figure 9. Nonlinear fit to the kinetic data (predicted versus experimental q). * Initial surfactant-adsorbent ratio.

The linear correlations between the predicted and experimental q values are shown in figure 9. As can be seen, by far most of the experimental points lie within ±10% of the predicted values.

The goodness of the correlations in all three cases can be appreciated mathematically by the high values of the r^2 coefficients given in table 2, and visually in the linearized plots shown in figure 10. According to the r^2 values, all of which are over 0.97, the three models explain the adsorption process fairly well. The differences between the three are very slight, with the simplest, the Lagergren model, giving a regression coefficient of 0.98. This was thus taken to be the best theoretical model for this case of adsorption.

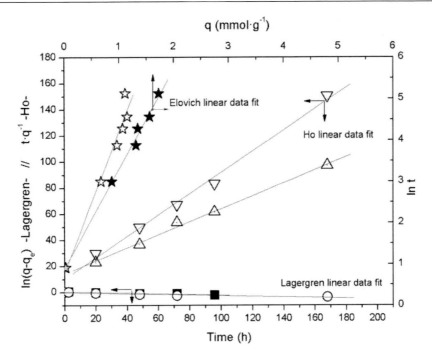

Figure 10. Linear fit to the kinetic data.

3.1.3. Equilibrium Studies

The Langmuir adsorption model assumes that the molecules striking the surface have a given probability of being adsorbed, and molecules already adsorbed similarly have a given probability of desorbing. At equilibrium, equal numbers of molecules adsorb and desorb at any time. The probabilities are related to the strength of the interaction between the adsorbent surface and the adsorbate [40]. That is the physical meaning of the equation (5):

$$q = k_{l1}\frac{c_l}{1+k_{l2}c_l} \qquad (5)$$

where k_{l1} is the first Langmuir adsorption constant (L [g of adsorbent]$^{-1}$), and k_{l2} is the second Langmuir adsorption constant (L [mmol of pollutant]$^{-1}$).

The linear form of Langmuir model can be expressed by equation (6):

$$\frac{c_l}{q} = \frac{1}{k_{l1}} + \frac{k_{l2}}{k_{l1}}c_l \qquad (6)$$

In addition, temperature can be included in a generalized expression of Langmuir hypothesis by considering Langmuir constants according to Arrhenius correlation. That assumes that k_{l1} and k_{l2} may have the following form (7):

$$k = k_0 exp\left(-\frac{E}{RT}\right) \quad (7)$$

where k_0 is the basic constant, whose units are equal to k,
E is the activation energy (J mol^{-1}),
R is the universal constant for perfect gases (8.314 J mol^{-1} K^{-1}),
and T is the temperature of the adsorption process (K).

The inclusion of definition (7) into Langmuir expression (6) leads to the expression (8):

$$q = \frac{k_{01} exp\left(-\frac{E_1}{RT}\right)C_l}{1 + k_{02} exp\left(-\frac{E_2}{RT}\right)C_l} \quad (8)$$

The adequacy of the multiparametric adjustment is given by a specific r^2 for each case.

The adsorption of MB onto classical QTG is presented on figure 11. The specific r^2 is 0.86 in this case. It stands near to the average value of r^2 in the case of non-linear individual adjustments for each temperature (0.84). Both non-linear procedures give a more accurate idea of the goodness of the model, while linear regression checks the adequacy of Langmuir's hypothesis in this adsorption process. In addition, multiparametric adjustment gives us two more data: activation energies. According to the mathematical results, the first of these parameters (E_1), which corresponds to adsorption energy (while E_2 corresponds to desorption energy) is equal to 9973.4 J mol^{-1}. E_2 is equal to zero, so desorption process is not temperature-depending. In every case, the values of each k belongs to similar magnitude order, as can be appreciated according to the adjustment of experimental to predicted data.

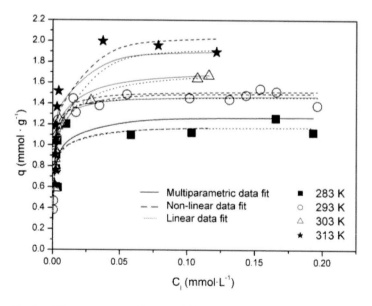

Figure 11. Data fit of multiparameter, nonlinear and linear adjustments of Langmuir equation. From J Hazar Mater, 174, 9-16, 2010.

Table 3. Fitting parameters for equilibrium adsorption process

Model	Parameters	r^2
Nonlinear Langmuir		
10 °C	k_{11}=0.69; k_{12}=19.4	0.52
20 °C	k_{11}=0.87; k_{12}=34.4	0.52
30 °C	k_{11}=0.77; k_{12}=105.2	0.77
40 °C	k_{11}=0.81; k_{12}=94.4	0.77
Linear Langmuir		
10 °C	k_{11}=0.67; k_{12}=21.8	0.84
20 °C	k_{11}=0.80; k_{12}=27.4	0.95
30 °C	k_{11}=0.82; k_{12}=48.2	0.97
40 °C	k_{11}=0.80; k_{12}=95.3	0.99
Multiparameter Langmuir		
	k_{01}=1.59; k_{02}=1.27	0.77
	E_1=0.18 10^4, E_2=5.37 10^4	

Notes: Units in text.

Regarding CTAB adsorption onto *Pinus pinaster* classical *tanningel*, the same three fitting procedures were performed: a linear and a nonlinear fit for each temperature, and a multiparameter fit for the entire temperature range. Table 3 lists the values of the parameters and the correlation coefficients r^2.

The nonlinear Langmuir fits for the specific temperatures gave lower correlation coefficients (mean of 0.65) because the error bars are wider. However, previous studies have shown this method to be more accurate than linear fits [41,42].

Finally, the goodness of the multiparameter fit is reflected in the reasonably high value of the r^2 coefficient (0.77). In addition, this model yields two further parameters: the activation energies. According to the mathematical results, one of these parameters (E_{02}), which corresponds to the equilibrium adsorption energy, is equal to 5.37 10^4 J mol^{-1}, while the other one is equal to 0.18 10^4 J mol^{-1}. Since the difference between them is positive, the adsorption process reaches equilibrium. The fact that its value is very low has to do with the enhancement of the adsorption process due to the thermal effect on the pores, i.e., to thermal activation. In each case of the linear and nonlinear fits, the values of the parameters are similar in order of magnitude (figure 12).

The energies involved in the adsorption process are high enough for it to be considered chemisorption (the links between adsorbate and adsorbent are stable and irreversible). The SEM images also led us to think that this chemisorption must be governed by the external diffusion stage, since they showed the material to have little porosity so that the adsorption process must take place on the surface at regularly distributed active centres.

Pinus tannin gel presents a reasonably high level of maximum *q* as shown by the nonlinear Langmuir model. Of the relatively few materials referred to in the scientific literature for CTAB removal, none belongs to the category of the so-called *waste* or *low-cost* materials. One observes that most of the adsorbents are synthetic in origin, such as perlite or modified silica gel. *Pinus tanningel* presents an intermediate situation in terms of CTAB removal, so its capacity is comparable to that presented by other materials.

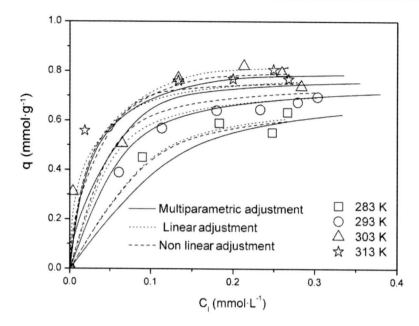

Figure 12. Multiparameter, nonlinear, and linear fits of the Langmuir equation to the CTAB adsorption onto *Pinus pinaster* classical tanningel data.

3.2. Optimization Studies on Tannin Gels

Two main optimization studies can be carried out on *tanningel* production: one regarding the optimal synthesis conditions from a unique tannin source and another one regarding different tannin sources and different aldehydes.

3.2.1. Optimal Synthesis Conditions on Quebracho-Formaldehyde Tanningel

As a first approach, we have developed a statistically significant study on the production of these kinds of adsorbents from *Schinopsis balansae* tannin extract. Design of experiments was carried out in order to evaluate the influence of parameters such as the amount of formaldehyde or the reaction temperature. Adsorbent products obtained by this procedure were tested on surfactant removal and dye elimination.

A factorial Central Composite (CCD) orthogonal and rotatable design was used for the optimization of the quantitative variables such as temperature and formaldehyde-tannin ratio. Each experiment of the planned series was made twice, so the mean value was kept for the CCD analysis under Response Surface Methodology. Reproducibility and consistency of these replicates are statistically confirmed through indistinguishability tests. Design of experiments section was statistically analyzed by using StatGraphics Plus for Windows 5.1 [35].

The basis of Design of Experiments (DOE) are well known. While the traditional experimental method, one factor at a time approach, can hardly be used to stablish relationships among all the experimental input factors and the output responses, DOE gives us an alternative and specific point of view on the phenomenon. Even through the traditional approach can be useful in finding predominant factors in this situation, it is difficult to observe an optimum value of the working parameters as no interaction among them is considered. To solve this problem and to obtain a probable optimum, DOE offers a better alternative to study the effect of variables and their response with minimum number of experiments [43].

The design of experiments is a common methodology in order to improve industrial and economical production processes [44]. By means of this mathematical procedure a lower number of experiments is needed and the results that are obtained from the investigation have the consistency of a statistical process. In fact, it has been used thoroughly in many related research, such as those reported by Bhatia et al. [45] or Sabio et al. [46].

Using design of experiments based on response surface methodology (RSM), the aggregate mix proportions can be arrived with minimum number of experiments without the need for studying all possible combination experiments. *StatGraphichs* software provides a useful and powerful mathematical and statistical tool in order to develop the experimental planning (in a random order for avoiding hidden effects) and to analyze the results, searching for conclusions.

In order to determine if there exist a relationship between the factors and the response variables investigated, the data collected must be analyzed in a statistically manner using regression. In developing the regression equation, the test factors were coded according to equation (9)

$$\chi_i = \frac{X_i - X_i^x}{\Delta X_i} \tag{9}$$

where χ_i is the coded value of the ith independent variable, X_i the natural value of the ith independent variable, X_i^x the natural value of the ith independent variable at the center point and ΔX_i is the value of the step change.

Each response Y can be represented by a mathematical equation that correlates the response surface (equation (10)).

$$Y = b_0 + \sum_{i \neq j; i=1}^{k} b_{ij} \chi_i \chi_j + \sum_{j=1}^{k} b_{jj} \chi_j^2 \tag{10}$$

where Y is the predicted response, b_0 the offset term, b_j the linear effect, b_{ij} the first-order interaction effect, b_{jj} the squared effect and k is the number of independent variables.

The optimization process at this point of the investigation may lead to optimal gelation conditions inside a working range which is presented in table 4. Formaldehyde concentration and temperature are varied and signification of these parameters are presented below.

Table 4. Working range in the optimization of the gelation process of Quebracho tannin extract

Variable	Upper level	Lower level
Formaldehyde ratio (g of HCHO per g of tannin extract)	0.22	0.07
Temperature (°C)	90	60

ANOVA report of this experimental design shows that formaldehyde ratio is the significant variable in both case of CTAB and MB removal, with a p-value over 0.05. Temperature seems to be not so influent in any case. Non-linear polynomic regression is carried out by taking into account equation (10). In this sense, this regression conrresponds to the following expressions for dye removal:

$$q = 1.30 + 0.04\,T - 0.36\,F + 0.013\,T^2 + 0.20\,F^2 - 0.09\,TF \qquad (11)$$

and equation (12) for surfactant removal:

$$q = 1.01 - 0.05T - 0.20F + 0.030T^2 + 0.02F^2 + 0.02TF \qquad (12)$$

where T are the values of temperature and F are the values of formaldehyde-tannin extract ratio (coded levels). The adjusted correlation coefficient r^2 are above 0.80 in both cases, and an optimum is also presented.

The results of this Design of Experiment can be presented in a various and complementary graphical ways. The most important graphical representation in this procedure based on Response Surface Methodology is the surface graphic (figure 13). It plots equation (10) and allows evaluating from a qualitative point of view how the behavior of whole studied system is. As can be appreciated, the response is a quite convex surface inside the studied region, but the ruling variable is the formaldehyde ratio. No clear maximum is obtained inside the working range, but an asymptotic behaviour is presented as formaldehyde ratio tends to be lower. The optimal combination of temperature and formaldehyde is set at minimum formaldehyde amount and an intermediate temperature, that is 75° C and 0.07 g of formaldehyde per g of tannin extract. The actual retention of MB and CTAB with this optimum adsorbent confirms the theoretical approach since deviations of less than 5% were found between experimental and predicted values.

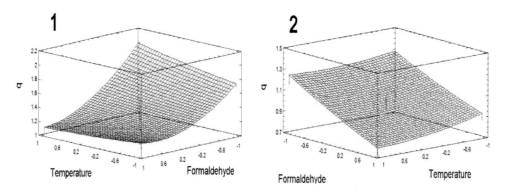

Figure 13. Response surface graphics for temperature and formaldehyde-tannin ratio. (1) Removal of Methylene Blue; (2) Removal of CTAB. Coded levels, q in mmol g^{-1}. From Ind Crops Prod 33, 409-417, 2011.

Table 5. Langmuir model non-linear fitting parameters

Compound	Parameter values	r^2
Methylene blue	k_{l1}=1.75; k_{l2}=2206	0.98
CTAB	k_{l1}=2.06; k_{l2}=54.9	0.93

Notes: Units in text.

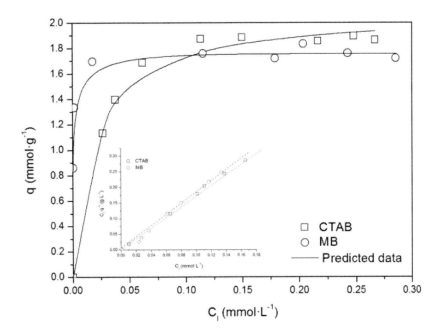

Figure 14. Adsorption isotherms at 20°C for (1) Methylene Blue and (2) CTAB. From Ind Crops Prod 33, 409-417, 2011.

Figure 14 depicts the adsorption isotherms for pollutant removal. Experimental points are placed onto the predicted path for Langmuir adsorption hypothesis, according to a very high regression level. Table 5 shows each regression parameter and r^2 for each case. The goodness of Langmuir model in the explanation of the phenomenon is confirmed by the adequacy of the linear approach, as can be observed from figure 14 as well. r^2 coefficient reached 0.98 in both cases (dye and surfactant elimination). Reliable parameters are obtained from non-linear adjustment.

As can be seen, the retention of CTAB and MB is very efficient since maximum q are significatively higher if compared to what other researchers referred in previous works. E.g, regarding MB retention, activated carbon maximum q was 1.32 mmol g^{-1} (*vetiver roots*) [47]), 1.36 mmol g^{-1} (coconut husk activated carbon [48]) or 1.51 mmol g^{-1} (non optimum *QTG* [16]). The removal of CTAB is clearly less efficient onto activated carbon (0.83 mmol g^{-1} [27]) or onto other materials, such as perlite (0.02 mmol g^{-1} [49]).

3.2.2. Other Parameters: Tannin Extract and Aldehyde

There are other parameters that should be studied in order to optimize the synthesis of these kinds of adsorbents. As a matter of fact, the tannin extract and the aldehyde involved in the gelation process can be easily taken into account in an optimization process.

Table 6. Gelation experiments

Tannin extract	Formaldehyde (mmol g^{-1})	Acetaldehyde (mmol g^{-1})	Does it gelify?	Symbol
Cypress	1	0	No	
Cypress	3.68	0	Yes	CFC
Quebracho	1	0	Yes	QFD
Quebracho	3.68	0	Yes	QFC
Quebracho	0	1.30	No	
Quebracho	0	4.85	No	
Pine	3.68	0	Yes	PFC
Pine	1	0	Yes	PFD
Pine	0	4.85	No	
Weibull Black	1	0	Yes	WFD
Weibull Black	3.68	0	Yes	WFC
Weibull Black	0	1.30	No	
Weibull Black	0	4.85	Yes	WAC

With this scope, up to four tannin extracts from different natural vegetal sources were gelified with formaldehyde and acetaldehyde. Tannins from *Acacia mearnsii* (Weibull black), *Schinopsis balansae* (Quebracho), *Cupressus sempervivens* (Cypress) and *Pinus pinaster* (Pine) underwent gelation with formaldehyde and acetaldehyde in two different doses each one (upper and lower ratio). Not every combination was feasible, as table 6 shows.

Two replicates of each *tanningel* combination were synthesized, and each one was tested twice with every target water (onto heavy metal, surfactant and dye removal).

The bases of the statistical method are found in the interaction of variables. Briefly, it can be summarized as follows:

- Up to thirteen different combinations regarding three variables (tannin extract, aldehyde type and aldehyde dosage) were attempted. The feasible combinations were just nine of them, the rest did not gelified (table 6).
- As long as it is not possible to test every combination since there are some of them that did not drive to a solid product, the whole system was considered as a categorical design, that is, no different variables are observed but nine different categories can be put into relationship, so an optimum point should be obtained.
- Inside this hypothesis, the two replicates of each product must be compared in order to establish if significative differences can be observed. If no, then one can work under the assumption that each replicate is indistinguishable from the other, so four tests of each *tanningel* on the adsorption of each model compound are taken into account.
- Categorical box-and-whisker plot of the whole system for each pollutant can be established, so an optimum appears for each situation.

Finally, we can develop an integrated statistical analysis by means of a new target variable, which includes the standardized responses of the previous ones. Hence, a general optimum can be presented.

Table 6 gives the main data of the experimental planification. Three qualitative variables were considered: tannin extract, type of aldehyde and concentration of this last one. Bearing in mind this general framework, not all the possible combinations drove to a satisfactory gelation product. It is clear that acetaldehyde presents a weak polymerization action if compared with formaldehyde: all acetaldehyde combination but WAC resulted unfeasible. That means it is needed a high tannin concentration from Weibull black (the most tannin-rich extract) and a high aldehyde concentration (4.85 mmol g^{-1}) for producing WAC.

Regarding the gelation process through formaldehyde polymerization, all the possible combinations are feasible and solid *tanningels* are produced. The immobilization of tannins with this aldehyde is previously referred, but no quantitative studies were carried out regarding the aldehyde-tannin ratio.

The particular analysis of each target variable can lead to the identification of several combinations that seem to work better regarding the removal of the studied contaminant agents. Graphically, this aspect can be shown in figure 15, where box-and-whisker plots are

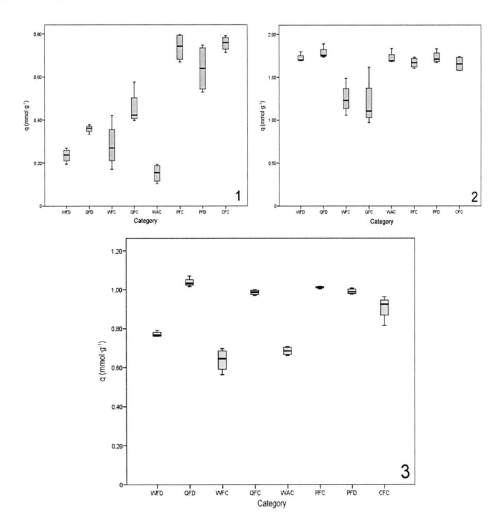

Figure 15. Box-and-whisker plot for the nine products in adsorption treatment of three model compounds. (1) Zn^{2+}; (2) Methylene blue; (3) CTAB. From Chem Engin J, 168, 1241-1247, 2011.

Figure 16. SEM images of 1) CFC, 2)QFC, 3)WFC and 4)PFC. The specific conditions of these photographs were Vacc.=1 kV; Mag.=x100 and working distance = 15 mm. From Chem Engin J, 168, 1241-1247, 2011.

presented for each system: metal, dye and surfactant removal. The depuration of metal-polluted solutions presented four subsests, the last one involves PFD, PFC and CFC categories with a mean q of ca. 0.7 mmol g^{-1} of Zn^{2+}. The rest of the groups presented an efficacy clearly lower in every case. The same behavior is appreciated in the removal of CTAB, with four subsets too, where the best one presented a mean capacity of ca. 1 mmol g^{-1}. This subset consisted of QFC, QFD, PFD and PFC. On the contrary, the elimination of MB presented just two subgroups, the best of them with a mean q of ca. 1.75 mmol g^{-1}.

According to these results, two main determinations were made on the representative samples of each different tannin sources: SEM and BET surface. As a matter of fact, just comparative studies can be made in order to explain the slow kinetics and the high adsorbent q capacity. Firstly, a graphical approach to the nature of the rigid foams was carried out with the SEM images that are shown in figure 16.

As can be appreciated, significant differences are presented in each case. While the three first *tanningels* are quite similar in their appearance (rigid and hard, non-porous aspect), PFC seems to be a different material, more brittle. This means the polymerization was carried out in a different way with *Pinus pinaster* and in the rest of the tannin resins, surely due to the polyphenolic nature of the tannin extracts.

BET surface was determined for the four types of adsorbents. The specific values of this parameter were 5.64 m^2 g^{-1} for WFC, 5.38 m^2 g^{-1} for PFC, 2.08 m^2 g^{-1} for CFC and 1.63 m^2 g^{-1} for QFC. These are very low surface values if compared with the normal BET surfaces that are obtained with other natural raw materials, e.g. such as *Moringa oleifera* or activated carbon from coconut husk. This is the reason of the low kinetics these adsorbent materials presented (up to 1 week to reach equilibrium). Further studies must be carried out in order to increase the BET surface by means of gasification or other procedures, as some relevant authors have recently proposed.

Table 7. Langmuir model non-linear fitting parameters

Compound	Parameter values	r^2
Methylene blue	k_{l1}=442.7 ; k_{l2}=309.6	0.70
CTAB	k_{l1}=90.91 ; k_{l2}=42.72	0.95
Zn2+	k_{l1}=51.07 ; k_{l2}=51.91	0.90

Notes: Units in text.

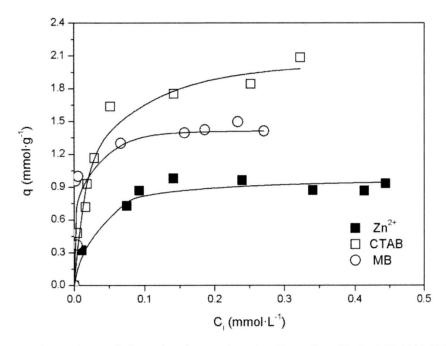

Figure 17. Isotherms for metal, dye and surfactant adsorption. From Chem Engin, 168, 1241-1247, 2011.

Figure 17 depicts the adsorption isotherms for pollutant removal with one of the best adsorbents: PFC. Experimental points are placed onto the predicted path for Langmuir adsorption hypothesis, according to a very high regression level. Table 7 shows each regression parameter, such as k_{l1}, k_{l2} and r^2 for each case.

As can be seen, MB adsorption is more efficient as maximum q is obtained. This is evident from the values of the Langmuir constants. On the contrary, Zn^{2+} presents lower adsorption level. These differences can be due to the molecular size (higher in the case of MB) and to the cationic character of the three model compounds, which probably affect to the sorption mechanism.

3.4. Preliminary Studies on Hydrolysable Tannins

Finally, a preliminary study on the gelation of hydrolysable tannins (*Castanea sativa* or Chestnut and *Caesalpinia spinosa* or Tara, mainly). This gelation process was carried out as expressed in section 3.1.

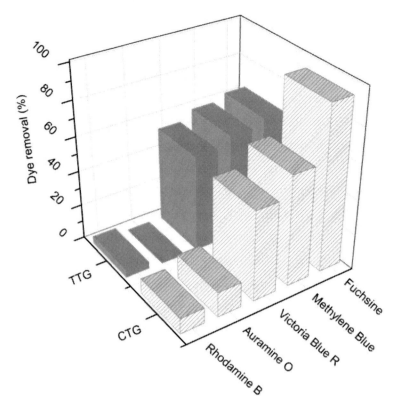

Figure 18. Preliminary screening for dye removal with hydrolysable tannin gel.

Hydrolysable tannins are reputed to be mixtures of simple polyphenols such as gallic and ellagic acids. They can form complex structures as well as condensed ones. Significative differences may be observed depending on the predominant gallic or ellagic building blocks.

As a first step, the ability of the two *tannin* gel (TTG and CTG) was tested with different cationic dyes. Trials were performed with a fixed initial dye concentration and with fixed mass of tannin gel. Figure 18 reports the percentage removal for each type of dye.

Similar rates of dye removal are obtained in both cases, so the adsorption process is quite analogous with each product. Chemical structures of dyes can support the fact that the affinity between tannin gel and Auramine O and Rhodamine B seems to be lower due to the lack of positive charges (Auramine O) and the large Rhodamine B molecule that surely difficults the adsorption due to steric reasons. The rest of dyes presents either positive charges (Methylene blue or Victoria blue R) or a very well-balanced small molecular weight (Fuchsine). This last case is removed by CTG in a significantly higher percentage.

Figure 19 shows the experimental q data and the predicted ones by the three Langmuir adjustments (linear, non linear and multiparameter). Both non linear and multiparameter procedures gives a more accurate idea of the goodness of the model, while linear regression checks the adequacy of Langmuir's hypothesis in this adsorption process. In addition, activation energies are also obtained. Table 8 presents these results.

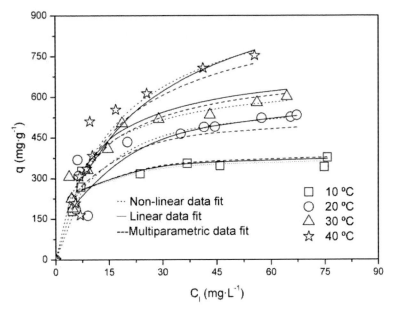

Figure 19. Langmuir non linear and multiparameter modelization of CTG-MB system.

Table 8. Fitting parameters for equilibrium adsorption process with tannin gels from Chestnut. Units referred to mg

Model	Parameters	r^2
Nonlinear Langmuir		
10 °C	k_{11}=108.01; k_{12}=0.284	0.78
20 °C	k_{11}=70.43; k_{12}=0.119	0.81
30 °C	k_{11}=83.74; k_{12}=0.127	0.92
40 °C	k_{11}=61.64; k_{12}=0.062	0.80
Linear Langmuir		
10 °C	k_{11}=105.48; k_{12}=0.27	0.99
20 °C	k_{11}=56.17; k_{12}=0.091	0.95
30 °C	k_{11}=78.74; k_{12}=0.11	0.99
40 °C	k_{11}=52.63; k_{12}=0.05	0.85
Multiparameter Langmuir		
	k_{01}=1.01 10^3; k_{02}=4.98 E_{01}=6.36 10^3, E_{02}=-5.35 10^3	0.80

CONCLUSION

Tanningels are presented in this chapter as a feasible option in the new generation of adsorbents. Their efficiency in widely checked in the removal of a large variety of pollutants, including heavy metals, cationic surfactants and dyes. It is easy to obtain such products from several tannin feedstocks, such as condensed (*Acacia mearnsii*, *Schinopsis balansa*e or *Pinus pinaster*) or hydrolysable ones (*Caesalpinia spinosa* or *Castanea sativa*). Optimization studies can be carried out either in the form of gelation (involving formaldehyde at a certain temperature level) or the tannin source itself with different aldehyde species (acetaldehyde or formaldehyde).

The results are promising attending to the general efficiency of these new adsorbents and to the competitive costs. They are definitively a new group of low cost agents in water treatment.

ACKNOWLEDGMENTS

Authors thank to Comisión Interministerial de Ciencia y Tecnología (CICYT) CTQ 2010-14823/PPQ project as well as to Junta de Extremadura under PRI-07A031 project.

REFERENCES

[1] Dorf, R.C. 2001. *Sustainable and Appropriate Technologies.* Technology, Humans and Society. San Diego, Academic Press.

[2] Nakano, Y., Takeshita, K. and Tsutsumi, T., 2001. *Adsorption mechanism of hexavalent chromium by redox within condensed-tannin gel.* Water Research, 35, 496-500.

[3] Tondi, G., Oo, C.W., Pizzi, A. and Thevenon, M.F., 2008. *Metal absorption of tannin-based rigid foams.* Industrial Crops and Products, 29, 336-340.

[4] Sánchez-Martín, J., Beltrán-Heredia, J. and Carmona-Murillo, C., 2011. *Adsorbents from Schinopsis balansae: Optimisation of significant variables.* Industrial Crops and Products, 33, 409-417.

[5] Mohan, D., Singh, K.P. and Singh, V.K., 2008. *Wastewater treatment using low cost activated carbons derived from agricultural byproducts - A case study.* Journal of Hazardous Materials, 152, 1045-1053.

[6] Demirbas, A., 2008. *Heavy metal adsorption onto agro-based waste materials: A review.* Journal of Hazardous Materials, 157, 220-229.

[7] Pizzi, A., 2008. *Tannins: major sources, properties and applications.* In: Galdini and Belgacem (eds), *Monomers, Polymers and Composites from Renewable Sources*, Elsevier, Amsterdam.

[8] Hagerman, A., 1995. *Tannin Analysis.* Miami University, Ohio.

[9] Pizzi, A., 1994. *Advanced Wood Adhesives Technology.* Marcel Dekker, New York.

[10] Tondi, G., Zhao, W., Pizzi, A., Du, G., Fierro, V. and Celzard, A., 2009. *Tannin-based rigid foams: A survey of chemical and physical properties.* Bioresource Technology, 100, 5162–5169.

[11] Zhao, W., Fierro, V., Pizzi, A., Du, G. and Celzard, A., 2010. *Effect of composition and processing parameters on the characteristics of tannin-based rigid foams. Part II: Physical properties*, Materials Chemistry and Physics, 123, 210-217.

[12] Kim, Y.-H. and Nakano, Y., 2005. *Adsorption mechanism of palladium by redox within condensed-tannin gel.* Water Research, 39, 1324-1330.

[13] Shirato, W. and Kamei, Y., 1992. *Insoluble tannin preparation process, waste treatment process and adsorption process using tannin.* US Patent 5,158,711.

[14] Vázquez, G., Antorrena, J., González, J and Doval, M.D., 1994. *Adsorption of heavy metal ions by chemically modified Pinus pinaster bark*. Bioresource Technology, 48, 251-255.
[15] Vázquez, G., González-Álvarez, J., Freire, S., López-Lorenzo, M.L. and Antorrena, G., 2002. *Removal of cadmium and mercury ions from aqueous solution by sorption on treated Pinus pinaster bark: kinetics and isotherms*. Bioresource Technology, 82, 247-251.
[16] Sánchez-Martín, J., González-Velasco, M., Beltrán-Heredia, J., Gragera-Carvajal, J. and Salguero-Fernández, J., 2010. *Novel tannin-based adsorbent in removing cationic dye (Methylene Blue) from aqueous solution. Kinetic and equilibrium studies*. Journal of Hazardous Materials, 174, 9-16.
[17] Yurtsever, M. and Sengil, I.A., 2009. *Biosorption of Pb(II) ions by modified quebracho tannin resin*. Journal of Hazardous Materials, 163, 58-64.
[18] Tamez, M.U., Islam, M.A., Mahmud, S. and Rukanuzzaman, M., 2008. *Adsorptive removal of Methylene blue by tea waste*. Journal of Hazardous Materials, 164, 53-60.
[19] Cserháti, T.; Forgács, E. and Oros, G., 2002. *Biological activity and environmental impact of anionic surfactants*. Environment International, 28, 337-348.
[20] Clara, M., Scharf, S., Scheffknecht, C. and Gans, O., 2007. *Occurrence of selected surfactants in untreated and treated sewage*. Water Research, 41, 4339-4348.
[21] Ostroumov, S.A., 2006. *Biological Effects of Surfactants*, CRC Press, Boca Raton.
[22] Beltrán-Heredia, J. and Sánchez-Martín, J., 2011. *Surfactants: an environmental menace. Removal form water by natural products*. In: Satinder Kaor Brar (ed.), Hazardous Materials: types, risks and control. Nova Science Publishers, New York.
[23] Blasco, J., Hampel, M. and Moreno-Garrido, I., 2003. *Analysis and Fate of Surfactants and the Aquatic Environment*. In: Knepper, T., Voogt, P. and Barcelo, D. (eds.), Toxicity of surfactants. Elsevier, Amsterdam.
[24] Basar, C.A., Karagunduz, A., Cakici, A. and Keskinler, B., 2004. *Removal of surfactants by powdered activated carbon and microfiltration*. Water Research, 38, 2117-2124.
[25] Sullivan, D.E., 1983. *Biodegradation of a cationic surfactant in activated sludge*, Water Research. 17, 1145-1151.
[26] Eng, Y-Y., Sharma, V.K. and Ray, A.K., 2006. *Ferrate (VI): Green chemistry oxidant for degradation of cationic surfactant*, Chemosphere. 63, 1785-1790.
[27] Yalcin, M. and Ahmet, G., 2004. *The adsorption kinetics of Cetyltrimethylammonium bromide (CTAB) onto powdered active carbon*. Adsorption, 10, 339-348
[28] Kurniawan, T.A., Chan, G.Y.S., Lo, W-H and Babel, S., 2006. *Physicochemical treatment techniques for wastewater laden with heavy metals*. Chemical Engineering Journal, 118, 83-98.
[29] Tünai, O. and Kabdasly, N.I., 1994. *Hydroxide precipitation of complexed metals*, Water Research. 28, 2117-2124.
[30] Charerntanyarak, L., 1999. *Heavy metals removal by chemical coagulation and precipitation*. Water Science and Technology, 39, 135-138.
[31] Yoon, J., Amy, G., Chung, J., Sohn, J. and Yoon, Y., 2009. *Removal of toxic ions (chromate, arsenate, and perchlorate) using reverse osmosis, nanofiltration, and ultrafiltration membranes*. Chemosphere, 77, 228-235.

[32] Vázquez, G., González-Álvarez, J., Freire, S., López-Suevos, F. and Antorrena, G., 2001. *Characteristics of Pinus pinaster bark extracts obtained under various extraction conditions*. European Journal of Wood and Wood Products, 59, 451-456.

[33] APHA, 1998. *Standard Methods for the Examination of Water and Wastewater*, 20th ed. American Public Health Association and American Water Works Association and Water Environment Association.

[34] SPSS Inc., 2005. *SPSS 14.0 Developer's guide*, Chicago, Illinois.

[35] StatPoint Technologies Inc., 2009. StatGraphics Centurion XVI User Manual, Richmond, Virginia.

[36] Lagergren, S., 1898. *Zur theorie der sogenannten adsorption geloster stoffe*. Kungliga Svenska Vetenskapsakademiens, Handlingar. Band, 24, 1-39.

[37] Ho, Y.S. and McKay, G., 1998. *The kinetics of sorption of basic dyes from aqueous solutions by sphagnum moss peat*. Canadian Journal of Chemical Engineering, 76, 822-827.

[38] Aroua, M.K., Leong, S.P.P., Teo, L.Y., Yin, C.Y. and Daud, W.M.A.W., 2008. *Real-time determination of adsorption of lead (II) onto palm shell-based activated carbon using ion selective electrode*. Bioresource Technology, 99, 5786-5792.

[39] Dogan, M., Abak, H. and Alkan, M., 2009. *Adsorption of methylene blue onto hazelnut shell: Kinetics, mechanisms and activation parameters*, Journal of Hazardous Materials, 164, 172-181.

[40] Langmuir, I., 1916. *The constitution and fundamental properties of solids and liquids. Part I. Solids*. Journal of American Chemical Society, 38, 2221-2295

[41] Kumar, K.V. and Sivanesan, S., 2006. *Pseudo second order kinetics and pseudo isotherms for malachite green onto activated carbon: Comparison of linear and non-linear regression methods*. Journal of Hazardous Materials, 136, 721-726.

[42] Kumar, K.V., Porkodi, K. and Rocha, F., 2008. *Isotherms and thermodynamics by linear and non-linear regression analysis for the sorption of methylene blue onto activated carbon: Comparison of various error functions*. Journal of Hazardous Materials, 151, 794-804.

[43] Montgomery, D.C., 2001. *Design and Analysis of Experiments, 5th ed*. John Wiley and Sons, New York.

[44] Milani, A.S., Wang, H., Frey, D.D. and Abeyaratne, R.C., 2009. *Evaluating three DOE methodologies: Optimization of a composite laminate under fabrication error*. Quality Engineering, 21, 96-110.

[45] Bhatia, S.; Othman, Z. And Ahmand, A.L., 2007. *Coagulation-Flocculation process for POME treatment using Moringa oleifera seed extract: Optimization studies*. Chemical Engineering Journal, 133, 205–212.

[46] Sabio, E., Zamora, F., Galán, J., González-García, C.M. and González, J.F., 2006. *Adsorption of p-nitrophenol on activated carbon fixed-bed*. Water Research, 40, 3053-3060.

[47] Altenor, S., Carene, B., Emmanuel, E., Lambert, J., Ehrhardt, J.J. and Gaspard, S., 2009. *Adsorption studies of methylene blue and phenol onto vetiver roots activated carbon prepared by chemical activation*. Journal of Hazardous Materials, 165, 1029-1039.

[48] Tan, I.A.W., Ahmad, A.L. and Hameed, B.H., 2008. *Adsorption of basic dye on high-surface-area activated carbon prepared from coconut husk: Equilibrium, kinetic and thermodynamic studies.* Journal of Hazardous Materials, 154, 337-346.

[49] Alkan, M., Karadas, M., Dogan, M. and Demirbas, Ö., 2005. *Adsorption of CTAB onto perlite samples from aqueous solutions.* Journal of Colloids and Interface Science, 291, 309-318.

In: Environmental Management
Editor: Henry C. Dupont

ISBN: 978-1-61324-733-4
© 2012 Nova Science Publishers, Inc.

Chapter 5

ISO 14001 RESEARCH: AN ACADEMIC APPROACH

G. Lannelongue[*], J. Gonzalez-Benito and O. Gonzalez-Benito

Universidad de Salamanca, Facultad de Economia y Empresa, Spain

ABSTRACT

An Environmental Management Systems (EMS) can be described by means of its primary measures; unfortunately, in contrast to the literature on quality management systems in which the most relevant aspects of the system are perfectly defined and compared, there is still lacking a consensus concerning the critical factors of an EMS. This has led to several different approaches in research on EMS, and has made it more difficult to compare scientific results.

In this chapter, we aim to deepen the theoretical bases that underpin the study of EMS. To this end, we first review the history of environmental management in firms and the appearance of formal EMS. Next, we examine nature of the standards and how they are employed within various organizations. Finally, we reconsider the literature on and the structure of the standard represented by ISO 14001:2004, and propose a model to study EMS that involves eighteen critical factors under four headings: Top Management Support, HR Management, Information System and Externals Factors. This model enables us to develop a tool that allows for a more systematic approach to empirical analyses of EMS, especially those based on ISO 14001.

1. INTRODUCTION

The relationship between business and the environment is a matter of prime importance for research in business economics. A widespread sensitivity to this area today has led to changes both in consumer demand and in the choices of other stakeholders, and above all among legislators. Taken together, this opens up a distinct horizon of competition, with new

---

[*] Corresponding Author: Gustavo Lannelongue, Organización de Empresas, Universidad de Salamanca, Facultad de Economía y Empresa, Campus Miguel de Unamuno, Edificio FES. 37007 Salamanca, Tel. +34 923294500 ext 3524. Fax +34 923294715. email: lannelongue@usal.es.

opportunities and risks. It is thus necessary to coordinate models for development with the protection of natural resources. Businesses have changed their strategies and their management in order to adapt to these new necessities. Despite the world crisis, 2009 saw more companies than ever gaining Environmental Management System (EMS) certification, this number reaching almost 225,000 companies. China, Japan and Spain headed the list of countries in terms of the number of companies achieving certification.

In this chapter we aim to introduce the concept of an EMS, analyse its origins, its main elements and the advantages of it. Firstly, we will look at the history of environmental management in companies and the appearance of the EMS concept. We will then examine the world of standards and their role within a company. We will move on to review the literature and the structure of the standard ISO 14001:2004, before finally proposing a model for analysing an EMS which includes eighteen critical factors, grouped into four categories, namely: *Management support*, *Personnel management*, *Information system* and *External factors*. This model allows us to construct a tool for the more systematic empirical analysis of an EMS, particularly one based on ISO 14001.

2. THE ENVIRONMENTAL MANAGEMENT SYSTEM

We have to go back to the 1980s to find the earliest mention of an environmental management system. In the United States, companies were facing legislation which was trying to keep up with reality, multiplying laws and standards and creating both confusion and compliance difficulties. In Europe, some companies were starting to see environmental issues as a business opportunity rather than a cost to be borne. This led to the need to create new management tools for implementing an environmental strategy, and at this stage the first environmental audits were performed and work was carried out on methods for calculating environmental risk.

In 1987, the United Nations Brundtland Report concluded that economic development had caused significant damage to the environment. The report also recommended adopting sustainable development measures as a way of jointly achieving economic and environmental objectives. In 1992, the United Nations Conference on Environment and Development ("The Earth Summit") approved what is known as the Rio Declaration, which included 27 sustainable development principles. On the basis of this declaration (and its influence on all other United Nations Conferences), various governments made efforts to support companies showing concern for the environment and treating it with more respect. One of the routes chosen was to establish guides to help with environmental management. In 1992, the British Standards Institute established the standard BS7750, the first ever environmental management standard. In 1993, the European Union drafted the voluntary standard "Eco-Management and Audit System" (EMAS). The International Standards Organisation (ISO) hurried to complete its own standard in 1995, the ISO 14001 standard.

The companies themselves created the Business Council for Sustainable Development (BCSD) in 1991, with 48 founding members. The BSCD is a platform for private companies to contribute to achieving a sustainable future, on the basis of economic efficiency aimed at adding value with lower resource use and pollution. Companies were also involved in 90 different environmental codes of conduct, such as that developed by the Coalition for

Environmentally Responsible Economies (CERES) or that by the Business Charter for Sustainable Development or the Global Environmental Management Initiative (GEMI). Despite the many studies linking environmental practices to financial results, there are still doubts about the direction of these relationships.

Awareness about the impact of economic activity on the environment has translated into a strengthening of the regulatory framework, the arrival of organised pressure groups, new market opportunities and into increasingly informed executives, who have traditionally been considered the promoters of change in terms of company strategy and management. This ecological transformation is taking place in companies to varying degrees, from the most reactive companies to the most proactive ones (Henriques and Sadorsky, 1999; Berry and Rondelli, 1998). Proactive environmental management is based on preventing (negative) environmental effects, while in contrast, reactive environmental management is based on repairing damage already inflicted (Aragón-Correa, 1998; Sharma and Vredenburg, 1998).

González-Benito and González-Benito (2006) identify different factors leading to a proactive approach, such as: (1) *Internal* components: Size, degree of internationalisation, position in the value chain, motivation of the management and strategic approach. (2) Pressure from the *stakeholders* (3) *External* components: Sector and location. Sharma and Aragón-Correa (2003) developed an instrument for measuring this based on the dynamic capabilities of the organisation. This approach argues that a company's results and competitive strategies depend largely on the specific nature of the organisation's resources and capacities (Wernerfelt, 1984; Barney, 1991; Rumelt, 1991). "Advanced or proactive environmental management has been defined as a capacity of the organisation because it allows heterogeneous resources (raw materials, technology, human resources, etc.) to be harmoniously coordinated inside and outside the company" (Aragón-Correa et al., 2005).

The way in which a company coordinates its resources is a fundamental part of its economic activity and one of the areas most intensively studied by academics. According to Cuervo (2008) "*the organisational structure must include a formal model for coordination and integration*". The use of a management system helps organisations to develop this *formal model for coordination* of its different elements so that it can achieve its desired objectives. Management systems are based on formalisation, which we define as the degree to which the activities, processes and procedures are standardised and put down on paper. A management system is based on a series of standards which regulate and programme these activities, processes and procedures (Mintzberg, 1984).

Environmental management can be defined as "all those technical and organisational activities carried out by companies to reduce the environmental effects generated by their operations" (Cramer, 1998) and its objective is to achieve environmentally sustainable development (Gupta, 1994). Therefore, it includes several parts of the company, will be more or less developed, and can be a subsidiary and costly occupation, based on corrective actions (Russo and Fouts, 1997; Aragón-Correa, 1998, Klassen and Angell, 1998), or alternatively integrated into the organisational structure through the implementation of an EMS (Gupta and Sharma, 1996, Azzone et al. 1997). The Regulation of the European Parliament and of the Council (EC No. 761/2001) defines an EMS as "the part of the overall management system that includes the organisational structure, planning activities, responsibilities, practices, procedures, processes and resources for developing, implementing, achieving, reviewing and maintaining the environmental policy".

3. CRITICAL FACTORS FOR AN EMS

An EMS can be described by means of its primary dimensions; unfortunately, however, there are few studies that identify them. In contrast to the literature on systems of quality management (QMS), in which the most relevant aspects of the system are perfectly defined and compared, there is still lacking a consensus concerning the critical factors of an EMS. We have tried to bring together various studies (see table 1) and to assemble the critical factors under four headings: Management support, HR management, Information system and External factors.

Table 1. Critical Factors (literature review

Hunt and Auster (1990)	(1) Top Level Support and Commitment, (2) Corporate Policies that Integrate Environmental Issues, (3) Effective Interfaces between Corporate and Business Unit Staff, (4) High Degree of Employee Awareness and Training, (5) Strong Auditing Program, (6) Strong Legal Base, (7) Established Ownership of Environmental Problems.
Newman and Breeden (1992)	(1) Clear vision, (2) Good corporate program, (3) Align organization's process, (4) Structure and resources to implement the vision, (5) Performance measurement, (6) Reward and recognition, (7) Training and Management development, (8) Communication and information Management, (9) Change Management, (10) Strategic program Management.
Welford (1994)	(1) Environmental policy, (2) Appropriate organizational structures, (3) Clear lines of authority, (4) Communications channels, (5) Activities should be indentifies and documented, (6) Environmental audits and review, (7) LPC.
Cairncross (1995)	(1) Corporate environmental policy, (2) Policy must have full support of the board of directors, (3) Top management must have the total involvement of employees to avoid (4) Information Management, (5) Audits.
Wilson (1997) (also in Chin et al. (1999); Hosseini (2007); Sambasivan and Fei (2008))	(1) Management attitude: Top management commitment and support, Appropriate environmental policy, Regular management reviews; (2) Organizational change: Structure and responsibility, Training and awareness, Communication, Documentation and control, Emergency preparedness; (3) External and social aspects: Environmental legislation, Market pressure, Employee relations; (4) Technical aspects: Environmental specialist assistance, Monitoring and measuring equipment, Production process enhancement.
Berry y Rondinelli (1998)	(1) Top Management Leadership, (2) Environmental Strategies and Policies (3) Goals, Targets, and Metrics (4) Participatory Decision-making and Implementation (5) Assessment and Communications.
Pun et al. (1998)	(1) Top management commitment, (2) Document control, (3) Training, (4) Formation, (5) Communication, (6) Community relationship.
Babakri et al. (2003)	(1) Identifying environmental aspects, (2) EMS documents, (3) Training, (4) EMS audits, (5) Operational control, (6) Environmental management program, (7) Objectives and targets, (8) Document control.

Zutshi and Sohal (2004)	(1) Management leadership and support: Top management commitment, cultural change and organizational vision, allocation of resources, appointment of a champion, importance of communication, avoidance of personality clashes. (2) Learning and training: Learning from other organization's experiences and benchmarking, reference to industry guidelines/standards, employee induction and training, general training and awareness for suppliers and other stakeholder.
	(3) Internal analysis: Conducting cost-benefits analysis, IER / gap analysis, identification of aspects and impacts and settings of objectives and targets, Necessity and usage of audits, Document control system, integration of existing management systems. (4) Sustainability: Life cycle analysis (LCA).
Wee and Quazi (2005)	(1) Top management commitment to environmental management, (2) Total involvement of employees, (3) Training (4) Green product/process design (5) Supplier management (6) Measurement (7) Information management.
Chavan (2005)	(1) Environmental policy, (2) Environmental impact identification, (3) Objectives and targets, (4) Consultation, (5) Operational and emergency procedures, (6) Environmental management plan, (7) Documentation, (8) Responsibilities and reporting structure, (9) Training, (10) Review audits and monitoring compliance, (11) Continual improvement.
Padma (2008)	(1) Top management commitment, (2) Environmental issues identification and legal compliance, (3) Environmental process management, (4) Emergency preparedness and response, (5) continuous improvement, (6) Measurement, monitoring and control, (7) Human resources management.

1) Management support (including top management commitment, strategic plan, environmental policy, objectives and goals, and leadership)

An EMS is a long-term project that involves resources and capabilities of different types and from multiple areas of an organisation. As a result, the senior management must recognise the importance of environmental management for the organisation and *lead* the tasks required for the system to work well (Hunt and Auster, 1990; Chin et al. 1999; Chavan, 2005; Wee and Quazi, 2005; Padma et al. 2008). This must start with the incorporation of environmental factors into *strategic planning*, considering the environmental strategy when formulating the corporate, competitive and operational strategy of the company (Wee and Quazi, 2005; Chin et al. 1999).

Top management has a fundamental role to play when it comes to implementing the *environmental policy*. This must include some realistic and achievable *objectives and targets* (Zutshi and Sohal, 2004), established with the help of the middle management, and it must also include the principles for environmental action, including compliance with legal requirements and a duty for continuous improvement.

Their commitment and support must not be limited to the implementation of the system, but must continue over time to ensure the best performance is obtained from the environmental system in terms of an improvement in the prevention of pollution, appropriate compliance with the legislation and the adequate supply of resources for the system (Chin et al. 1999; Padma et al., 2008).

In other words, the *commitment and support of the top management* must not be limited to implementing the system. The executives must also make an effort to encourage continuous improvements in environmental performance, and as a result it must review and evaluate the EMS at regular intervals (Chin et al. 1999; Wee and Quazi, 2005).

2) HR management (including the assignment of responsibilities, communication, motivation, training, responsible team)

The senior management must designate a team or *person as responsible* for the control and supervision of the EMS, delegating to this person the authority for environmental issues and supporting this person's decisions when these conflict with others in the company (Berry and Rondinelli (1998); Zutshi and Sohal, 2004). The control is based on an evaluation of the results obtained, comparing them with the expected results. For the EMS to function efficiently, the highest authority must provide this team with the long-term resources required to ensure that the system functions correctly at the moment and that it is adjusted for changes. This is particularly true for the *personnel in the company*. One of the central pillars of an EMS is that the environmental responsibility of an organisation is a commitment from each of its members. The employees must be authorised to solve environmental problems and must be actively involved in the process of determining environmental goals because they are the people who best know the procedures and tasks. The contribution of employees to improving environmental performance must be recognised (Wee and Quazi, 2005; Sambasivan and Fei, 2008). As a result, the leadership of the top management and its ability to *motivate* employees throughout the company is important (Zutshi and Sohal, 2004; Wee and Quazi, 2005). The employees must be familiar with the environmental policy, its targets and their particular responsibilities (Chavan, 2005). Resisting change is part of human nature (seeking the continuance of the status quo) and this can hinder the implementation of a management system. To reduce this resistance it is important to inform the employees about the basic operation of the system, and make them aware of the importance that putting it into practice has for them and for the organisation, telling them about the normal benefits and achievements obtained (NSF, 1996; Zutshi and Sohal, 2004; Chin et al., 1999). Equally, involving them in the decision making processes could improve the acceptance of this cultural change and in turn improve performance (Kinsella, 1994). The employees are the people who will implement the environmental programme, measure progress and meet the targets (NSF, 1996). As a result, the organisation must identify their *training* needs. This training must include the environmental policy and the requirements of the EMS, the most important objectives and targets, the specific environmental effects of their job, the benefits of improved performance and the consequences of non-compliance. Equally, the level of competence, training and experience required to ensure that they understand the importance of implementing an EMS must be determined (Ching et al, 1999). The responsibilities of each employee for environmental matters must be duly specified and correctly communicated. The support of the corporate management will only be effective if can be passed down to the business units through the appropriate channels (Hunt and Auster, 1990). As a result, communication (both internal and external) is a fundamental aspect of the system (Wilson, 1997; Sambasivan and Fei, 2008; Padma, 2008; among others).

3) Information systems (including documentation system, impact identification, emergency plan, continuous improvement, management review)

The company must consult the sector guides available to try to identify the best possible practices for its particular circumstances. In addition, the company must learn from other organisations (benchmarking) about how to manage its environmental problems, anticipating potential events (Zutshi and Sohal, 2004). An important part of an EMS consists of clearly *identifying and monitoring the environmental impacts* (actual and potential) generated by the organisation's operations (Zutshi and Sohal, 2004; Chavan, 2005; Samasivan and Fei, 2008 among others). Once identified, the impacts must be classified by evaluating their risks. This will help to establish the organisation's objectives and targets in environmental issues in a way which is more realistic, taking into account the organisation's resources. The objective in implementing this measurement system is to establish the organisation's environmental performance and to be able to check whether the company is making *continuous improvements* on environmental issues. This can be complemented with the use of other tools such as the analysis of the life cycle of the product to estimate its environmental effect or internal audits (Zutshi and Sohal, 2004; Wee and Quazi, 2005). Each element of the system must therefore be measured and controlled and there must be a record of this. This makes the *management of documents* an important part of the system, but this can act as a bureaucratic dead weight if it is not administered correctly (Wilson, 1997; Sambasivan and Fei, 2008; Padma, 2008; among others). It is also necessary to define the sequence of actions to be implemented in the event of possible environmental accidents through an *emergency plan* (Padma, 2008; Chavan, 2005 among others).

4) External factors (including stakeholders, legislation and audits)

The effective implementation of an EMS is inevitably influenced by a series of external and social aspects (Chin et al. 1999). The organization must identify the various *stakeholders* whose involvement in and contribution to the EMS are required. The organisation must open up communication channels to collect ideas and suggestions about changes and how they can be adopted to reduce possible resistance (Zutshi and Sohal, 2004). The company must take into account that the system will also affect its suppliers, contractors and customers, and therefore it must keep them informed about features of the system, including providing any training required on aspects about which they are unaware. This is particularly important in the relationships with small companies, which do not normally have resources allocated to this area (Zutshi and Sohal, 2004; Wee and Quazi, 2005). When choosing suppliers, environmental performance must be included as a selection criterion (Wee and Quazi, 2005). The company must be aware of the environmental *legislation* affecting it, and keep its knowledge up-to-date, so that it can comply with these obligations (Padma et al., 2008; Zutshi and Sohal, 2004; among others). This can be arranged through planning successive *internal audits* which will be the basis for management reviews and will guide the possible changes to the company's environmental policy, its objectives and other elements of the EMS (Chin et al. 1999). Table 2 includes a summary of these factors.

4. A WORLD OF STANDARDS

Management systems are based on the formalisation of activities and processes and this brings with it standardisation and recording. However, for a long time now the management systems themselves have been subject to standardisation.

In an economic environment characterised by globalisation and the integration of markets, the standardisation processes have become a coordination mechanism and an effective regulatory tool for business management (Brunsson and Jacobsson, 2000). These management system standards are also called metastandards (Yeung and Corbett, 2005) following Uzumeri (1997): "more than some detailed instruction manuals, they can create lists of design standards that guide the creation of different types of overall management systems. Given that the system theories use the term metasystems for lists of this type, one could refer to this type of standard as metastandards" (in Heras, 2006).

Table 2. Critical Factors

Headings	Critical Factors
1. Top Management Support	1.1. Top management commitment 1.2. Strategic plan 1.3. Environmental policy 1.4. Objectives and goals 1.5. Leadership
2. HR Management	2.1. Responsibilities allocation 2.2. Communication 2.3. Motivation 2.4. Formation 2.5. Team in charge
3. Information System	3.1. Documentation system 3.2. Impact identification 3.3. Emergency plan 3.4. Continual improvement 3.5. Management review
4. Externals Factors	4.1. Stakeholders 4.2. Legislation 4.3. Audits

Table 3. Types of standards and their economic effects

	Positive Effects	Negative Effects
Compatibility / Interoperability	Network externalities Avoidance lock-in Increased variety Efficiency in supply chains	Monopoly
Minimum Quality/ Safety	Avoidance adverse selection Reduction transaction costs Economy of scale	Raising rival's costs
Variety Reduction	Critical mass in starting industries	Reduction choice Market concentration
Information	Facilitate trade Reduction transaction costs	Raising rival's costs

Source: Blind (2004).

Generally, the use of standards should end the diseconomies found in a world in which each element is unique. However, their popularity and wide-scale use have led to them being seen as a third coordination route, alternative to the market and hierarchy (Brunsson and Jacobsson, 2000). This standardisation phenomenon affects many areas and is a new form of regulation (more or less formal), on an international scale, for which we voluntarily decide to follow some principles that create homogeneity between people and organisations in different parts of the world.

The diversity in the scope of the standards has led to the need to propose general classifications. Here we are referring to the work of Bling (2004), which was developed from earlier work by David (1987), who proposed a classification of standards based on their economic effects and the problems they resolve. Blind highlighted the positive and negative economic effects, classifying them into (1) Compatibility/interface standards; (2) Minimum quality/safety standards; (3) Variety reduction; and (4) Information standards (see Table 3).

Compatibility or interface standards ensure that goods and services can be exchanged between users. Trains, the telephone, software and mechanical pencils are examples of this type of standard. The effects obtained as a result of the number of users of the Standard, the *network effect* or network externalities, can be direct or proportional to the number of exchanges that take place among users, or indirect and induced by the total number of users. An added problem for these standards is the *cost of switching* from one standard to another. Before deciding on a standard, the users and producers are relatively free to choose between one system and another, but once they have made the investment required to implement that system or standard there will be a cost associated with changing it, and this will increase the longer the standard is in place. When the *network effect* and *switching costs* are taken together, the possibility of being stuck with a sub-optimal system or standard exists *(lock-in)*. From a static efficiency point of view, a public and open standard (non-proprietary) would be preferable. However, we must not forget that proprietary standard companies have greater incentives to develop their standards, which means that from a dynamic perspective proprietary standards may be preferable to open ones. In this case, the users of the standard will face concerns about possible monopolistic behaviour by the standard's owner.

Minimum quality standards create a lower quality boundary above which the members of the standard must operate. The markets characterised by asymmetric information, and particularly the markets in which product quality can only be assessed through product use (experience goods), are exposed to problems of adverse selection (Akerlof, 1970). Minimum quality or quality discrimination standards can be a useful tool for helping to solve these problems. This is particularly true for markets that are very sensitive to quality changes and are subject to relatively inelastic demand (Leland, 1979). On using these standards as announcements of belonging to a "quality club", there is also a reduction in the cost of searching for information to allow the consumers to discriminate between qualities, and this leads to a reduction in the *transaction costs* in the market. Safety standards, in the sense of a guaranteed minimum effect in the area of production/consumption of goods and services, in

Table 4. Taxonomy of Management Standards

Criteria	Typology
Geographic scope	Nationals (i.e. UNE in Spain or DIN in Germany). Internationals (i.e. ISO or EU).
Standard developer	Organizations with a long tradition to set standards (i.e. ISO, EU, UNE). Specific organization created *ad hoc* to set one standard (i.e. SA 8000 o *Investor in People*). Firms, alliances, industry or trade associations (i.e. automotive industry: EAQF or VDA standard).
Industry	Apply to all sectors (i.e. ISO 9001). Apply to a specific sector (i.e. ISO/TS 16948 automotive).
Criteria	Typology
Organizational scope	Apply to all the organization (i.e. ISO 14001). Apply to specifics processes (i.e. ISO 10002 - guidelines for complaints handling in organizations).
Certifiabillity	Can be certified (i.e. ISO 9001). Cannot be certified (i.e. ISO 10002).
Contents	Management systems standards (i.e. ISO 14001). Standards of procedures (i.e. ISO 14004). Performance standards (i.e. SA 8000). Standards of indicators (i.e. Investors in People).

Source: Adapted from Heras (2006).

turn reduce the negative externalities that arise as a result of controlling these effects. In addition, they can lead to pressure from the producers of higher quality/safety products to increase the level demanded by the standard and thereby exclude the lower quality/safety competitors (*raising rival's costs*). This would cause a "regulation" for the producers instead of one for the customers.

Certain standards reduce the variety of products, limiting the range of certain characteristics such as size or quality. The reduction in the number of variations in the product allows for greater economies of scale. It also allows a sector to focus and draw together around a reduced number of parameters and avoid the dispersion of supplier and customer demands from preventing a critical mass being reached which can consolidate a market. These standards are the most complex to analyse because they could strengthen or inhibit innovation. Information standards could be included in the three previous categories since they are standards which describe products, services or processes. In this sense they reduce the risks to the customers who do not need to test the attributes of a product, service or process. They also reduce the costs of seeking information and therefore reduce *transaction costs*.

Management standards (such as ISO 14001) are standards that ensure a minimum quality and provide information about different aspects of the organisation. In addition, they help the exchange of information between companies, reducing variety and increasing compatibility between management styles. Heras (2006) provides us with an analysis of management standards, categorising them according to geographical scope, the standards body, the activity sector, organisational scope, certifiability, and the content of the standard (see Table 4).

5. THE ISO 14000 SERIES

There are different standards on which to base a company's EMS. In Europe, the two most popular standards are the Eco-Management and Audit Scheme 1836/93 (EMAS) and the ISO 14001 standard. The EMAS scheme requires a greater commitment from the company than the ISO 14001 standard as it makes the publication of a report on the main aspects of its environmental impact obligatory and demands full compliance with the current environmental legislation (see Table 5). This last demand is one of the main reasons why ISO 14001 has been much more widely used than EMAS. The literature also indicates that the cost of certification and international recognition are factors which explain the difference in the number of certifications (at a European level): 89,237 (ISO 14001) compared to 7,404 (EMAS) (ISO Survey, 2009; emas-register, 2010). In Spain there are other, simpler standards, of an autonomous nature, launched to facilitate certification of small and medium-sized companies.

The ISO 14000 family consists of a series of standards, one of which is ISO 14001. This last standard is the reference standard which sets out the minimum requirements for implementing an EMS. These series can be called *augmentative standards* or enlargement standards, since they address the area of application in more depth (Karapetrovic and Casadesús, 2006). The ISO 14000 family of standards covers the different aspects of an EMS,

Table 5. ISO 14001 and emas differences

ISO 14001	EMAS
International scope.	European scope.
Only commitment to comply with applicable legal requirements. There is no compliance-audit.	Obligatory to demonstrate it. Required full legal compliance. There is a compliance-audit.
Does not go towards entities or sites.	The entity to be registered shall not exceed the boundaries of the Member State, and it is intended to go towards entities and sites.
Initial review is recommended, but not required.	Obligatory preliminary review, when is the first time that the organization sets its environmental status.
Included only system audit against the requirements of the standard.	Includes: system-audit, a performance-audit (= evaluation of environmental performance) and an environmental compliance-audit (= determination of legal compliance)
Only is required to respond to relevant communication from external interested parts.	Public Environmental Statement (validated for verifiers).

Source: Heras et al., 2008 and EMAS Factsheet.

such as: general requirements and principles, eco-design, ecological labeling, audits, integration of management systems, life cycle assessment, vocabulary, greenhouse gases and carbon footprint of products (see Table 6).

The ISO 14001 standard has been developed by the International Standards Organisation (ISO). This international organisation was founded in 1947 and its mission is to identify the standards required by governments, companies and society and develop them in collaboration with the various groups so that they can be applied in the greatest possible number of countries around the world. As a result, the organisation started to create standards aimed at supporting companies in the management of their environmental issues, by setting up the technical committee (TC 207) in 1993.

It was this body that laid the foundations for the first ISO 14001 standard in 1996. This standard was then updated[1] with the publication of ISO 14001:2004, which replaced the previous standard and remains in force today. The changes introduced by the new version did not involve additional requirements, but instead the text of the standard was altered to reflect the real features of the systems certified by accredited bodies. The ISO 14001 standard built on the success of the ISO 9000 family of standards, which in part explains why this environmental standard was quickly taken up. Corbett and Kirsch (2001) argue that the success of the ISO 14001 standard in a country is proportional to the level of companies

---

[1] Please see Annex A for changes from one Standard to another.

Table 6. ISO 14000 Family

Description	Standard
Environmental management systems – Requirements with guidance for use.	ISO 14001:2004
Environmental management systems – General guidelines on principles, systems and support techniques.	ISO 14004:2004
Environmental management systems – Guidelines for the phased implementation of an environmental management system, including the use of environmental performance evaluation.	ISO 14005
Environmental management systems – Guidelines on eco- design.	ISO/CD 14006
Environmental labels and declarations – General principles.	ISO 14020:2000
Environmental labels and declarations – Self-declared environmental claims (Type II environmental labeling).	ISO 14021:1999
Environmental labels and declarations – Type I environmental labeling – Principles and procedures.	ISO 14024:1999
Environmental labels and declarations – Type III environmental declarations – Principles and procedures.	ISO 14025:2006

Description	Standard
Environmental management – Quantitative environmental information – Guidelines and examples.	ISO/AWI 14033
Environmental management – Life cycle assessment – Principles and Framework.	ISO 14040:2006
Environmental management – Life cycle assessment – Requirements and guidelines.	ISO 14044:2006
Environmental management – Life cycle impact assessment – Examples of application of ISO 14042.	ISO/TR 14047:2003
Environmental management – Life cycle assessment – Data documentation format.	ISO/TS 14048:2002
Eco-efficiency assessment – Principles and requirements	ISO/WD 14045
Environmental management – Vocabulary	ISO 14050:2009
Greenhouse gases - Quantification of greenhouse gas emission.	ISO 14064:2006
Greenhouse gases – Requirements for greenhouse gas validation and verification bodies for use in accreditation or other forms of recognition.	ISO 14065:2007
Greenhouse gases – Competency requirements for greenhouse gas validators and verifiers document.	ISO/CD 14066
GHG – Quantification and reporting of GHG emissions.	ISO/AWI 14069
Carbon footprint of products – Quantification and Communication	ISO/WD 14067

Source: ISO (2010).

certified with ISO 9001 in that country, the propensity to export and concern about the environment. However, Neumayer and Perkins (2004) state that the number of certifications per head of population is correlated with direct foreign investment and the level of exports to Europe and Japan. Delmas (2002), in contrast, refers to legitimisation and minimisation of costs. Jiang and Bansal (2003) argue that companies certify their EMS systems if there is greater visibility of their operations and a greater lack of transparency about the environmental impact they have. All of these studies support the proposition by Bansal and Roth (2000), who conclude that the determinants of the environmental transformation in the companies do not necessarily have anything to do with social responsibility. According to the ISO Survey (2009), more than 220,000 companies in the word have obtained ISO 14001 certification, and the number is increasing by 20,000 each year[3].

The ISO 14001 standard is an internationally used standard, established by ISO, a private organisation, and it is aimed at all sectors of the economy, affects the entire organisation in which it is implemented, and can be certified. In addition, since it is a management standard, it facilitates exchanges due to providing and homogenising the environmental information for companies. On the one hand, the companies certified demonstrate that they have a guaranteed level of environmental commitment given that they have implemented a management system which helps to control and manage the organisational processes and activities with an environmental impact. This signal is a guarantee that reduces the information asymmetry and corrects, in part, the problem of adverse selection.

Let's suppose that we have two groups of companies operating in a market. The first is a group of companies strongly committed to the environment and the second group contains companies with a weak environmental commitment. In this market the customers cannot tell which group the companies belong to. However, their decision rule is that they prefer companies in the first group to those in the second group. In these circumstances the not very ecological companies try to pretend that they are ecological companies by sending false signals to the market. In this scenario, a guarantee of ecological commitment such as the ISO 14001 standard reduces the problem of adverse selection.

This problem also exists in reality, with companies trying to build an ecological reputation, and it is therefore easy to find companies that are not very environmentally committed but which have an ecological product or range. Without a way of testing their degree of ecological responsibility (such as certification by a third company) it is impossible to distinguish between an environmentally committed company and one that is not, as a result of which companies have no incentive to be environmentally committed if they cannot demonstrate or publicise this. In this sense, the standard is an announcement that the company belongs to an "Ecological Club", and this reduces the search for environmental information in the market and also provides certification. In addition, the certified companies use similar management tools and documentation, helping to identify environmental aspects in the coordination among companies. As a result, the guaranteed ecological management and homogeneity of the parameters for measuring environmental performance reduces transaction costs.

These effects multiply as the number of companies using the standard increases. In other words, there is a network effect. On the one hand, there is a direct effect in the companies that can benefit from this homogenisation of the environmental parameters. On the other hand, there is an indirect effect in terms of the publicity and importance of the standard, or in other words, the importance that belonging to the Club takes on. Könnölä and Unruh (2007) have studied the *lock-in* effect of the ISO 14001 standard, concluding that after an initial

---

[3] According to the forecasts for the different standards, we are in the phase with the highest growth rate, coming closer to the saturation point due to their intense dissemination (Marimón et al., 2006).

improvement in environmental performance, the system can constrain the focus of the organisation to the implementation of the current production system, instead of exploring discrete (radical) innovations which would involve new management systems with a better environmental performance. These conclusions are consistent with the general criticism of management standards, which argues that companies have a tendency to become more bureaucratic and excessively rigid in their processes and procedures (Seddon, 1997; Dick, 2000).

6. STRUCTURE OF THE ISO 14001 STANDARD

The ISO 14001 standard is structured into 4 sections: Scope and field of application, Normative references, Terms and definitions and Requirements of the environmental management system. This final section is the body of the standard and is in turn divided into 6 sections (see Table 7) which detail the model of continuous improvement (figure 1) using the methodology known as Plan-Do-Check-Act (PDCA)[4] on which it is based.

Table 7. Environmental management system requirements (STRUCTURE)

1. General requirements	1.1 EMS model 1.2 Top management commitment and leadership 1.3 EMS scope 1.4 Initial review
2. Environmental policy	
3. Planning	3.1 Environmental aspects 3.2 Legal and other requirements 3.3 Objectives, targets and programme(s)
4. Implementation andoperation	4.1 Resources, roles, responsibility and authority 4.2 Competence, training and awareness 4.3 Communication 4.4 Documentation 4.5 Control of documents 4.6 Operational control 4.7 Emergency preparedness and response
5. Checking	5.1 Monitoring and measurement 5.2 Evaluation of compliance 5.3 Nonconformity, corrective action and preventive action 5.4 Control of records 5.5 Internal audit
6. Management review	

Source: ISO 14001.

---

[4] PDCA (Plan, Do, Check, Act) cycle invented by Walter Shewhart (1931) and subsequently developed by W. Edwards Deming (1986), from who it takes the name Deming Cycle.

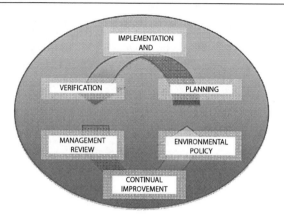

Source: ISO 14001.

Figure 1. ISO 14001 Estructure.

Table 8. Correspondence between ISO 14001:2004 and ISO 9001:2008

ISO 14001:2004			ISO 9001:2000
Environmental management system requirements (title only)	4	4	Quality Management System (title only)
General requirements	4.1	4.1	General requirements
Environmental policy	4.2	5.1 5.3 8.5.1	Management commitment Quality policy Continual improvement
Planning (title only)	4.3	5.4	Planning (title only)
Environmental aspects	4.3.1	5.2 7.2.1 7.2.2	Customer focus Determination of requirements related to the product Review of requirements related to the product
Legal and other requirements	4.3.2	5.2 7.2.1	Customer focus Determination of requirements related to the product
Objectives, targets and programme(s)	4.3.3	5.4.1 5.4.2 8.5.1	Quality objectives Quality management system planning Continual improvement
Implementation and operation (title only)	4.4	7	Product realization (title only)

ISO 14001:2004			ISO 9001:2000
Resources, roles, responsibility and authority	4.4.1	5.1 5.5.1 5.5.2 6.1 6.3	Management commitment Responsibility and authority Management representative Provision of resources Infrastructure
Competence, training and awareness	4.4.2	6.2.1 6.2.2	(Human resources) General Competence, training and awareness
Communication	4.4.3	5.5.3 7.2.3	Internal communication Customer communication
Documentation	4.4.4	4.2.1	(Documentation requirements) General
Control of documents	4.4.5	4.2.3	Control of documents
Operational control	4.4.6	7.1 7.2 7.2.1 7.2.2 7.3.1 7.3.2 7.3.3 7.3.4 7.3.5 7.3.6 7.3.7 7.4.1 7.4.2 7.4.3 7.5 7.5.1 7.5.2 7.5.5	Planning of product realization Customer-related processes (title only) Determination of requirements related to the product Review of requirements related to the product Design and development planning Design and development inputs Design and development outputs Design and development review Design and development verification Design and development validation Control of design and development changes Purchasing process Purchasing information Verification of purchased product Production and service provision (title only) Control of production and service provision Validation of processes for production and service provision Preservation of product
Emergency preparedness and response	4.4.7	8.3	Control of nonconforming product
Checking (title only)	4.5	8	Measurement, analysis and improvement (title only)

Table 8. (Continued)

ISO 14001:2004			ISO 9001:2000
Monitoring and measurement	4.5.1	7.6	Control of monitoring and measuring devices
		8.1	(Measurement, analysis and improvement)
		8.2.3	General
		8.2.4	Monitoring and measurement of processes
		8.4	Monitoring and measurement of product Analysis of data
Evaluation of compliance	4.5.2	8.2.3	Monitoring and measurement of processes
		8.2.4	Monitoring and measurement of product
Nonconformity, corrective action and preventive action	4.5.3	8.3	Control of nonconforming product
		8.4	Analysis of data
		8.5.2	Corrective action
		8.5.3	Preventive action
Control of records	4.5.4	4.2.4	Control of records
Internal audit	4.5.5	8.2.2	Internal audit
Management review	4.6	5.1	Management commitment
		5.6	Management review (title only)
		5.6.1	General
		5.6.2	Review input
		5.6.3	Review output
		8.5.1	Continual improvement

Source: ISO 14001.

The standard has its roots in other ISO standards, above all in ISO 9001[5] for quality management with which it shares the creation methodology, implementation structure and process, and verification by a third party (see Table 8).

The main objectives of ISO 14001 are to:

- provide assurance to management that it is in control of the organizational processes and activities having an impact on the environment
- assure employees that they are working for an environmentally responsible organization.
- provide assurance on environmental issues to external stakeholders – such as customers, the community and regulatory agencies
- comply with environmental regulations

---

[5] Although the standard ISO 9001 has a process-based focus, the ISO considers that this focus is compatible with the PDCA approach since PDCA can be applied to all processes.

- support the organization's claims and communication about its own environmental policies, plans and actions
- provides a framework for demonstrating conformity via suppliers' declarations of conformity, assessment of conformity by an external stakeholder - such as a business client - and for certification of conformity by an independent certification body.

It is important to note that the ISO 14001 standard does not require certification to achieve compliance, but it is generally agreed that most companies rigorously applying the standard have certified their operations. The standard is based on the premise that the organisation will periodically review and evaluate its EMS to identify improvement and implementation opportunities.

The following is a summarised list of the requirements for the standard:

General Requirements

This is a summary of the standard in which is says that "the organisation must establish, document, implement, maintain and continuously improve an environmental management system in accordance with the requirements of this international standard, and determine how it will comply with these requirements."

Environmental Policy

This is what "drives the implementation and improvement" of the EMS. It must demonstrate commitment from the senior management to comply with legal requirements, prevent contamination and continuously improve. "It is the starting point for setting the organisation's EMS objectives and targets". It must be communicated to and understood by the company's "interested parties", both internal and external. The scope of the environmental policy must be clearly defined, identifying "the impact of the products and services" within its scope of application.

Planning

A company must "identify the environmental aspects of its activities, products and services" and determine which are the most significant. This information must be documented and updated. Similarly, the organisation must *"identify the applicable legal requirements" and determine how these apply in terms of their environmental aspects. They must also establish "environmental objectives and targets.... throughout the organisation" as well as a programme to achieve these. "*

The objectives and targets must be measurable where possible" and they will be reviewed taking into account the criteria established by the standard.

Implementation and Operation

The management will ensure the "availability of resources for.....the environmental management system", be these human, financial or technological resources. "Roles, responsibilities and authorities need to be defined, documented and communicated to facilitate efficient environmental management". Senior management will appoint a representative as the highest authority on environmental issues, ensuring that the system is implemented in accordance with the standard and reporting the performance of the EMS to the senior management for review and improvement. The organisation needs to identify the training needs of personnel whose work may create a significant impact upon the environment and ensure that these people have the experience required to be competent at completing that task. The organisation must identify and cover the training needs associated with the system. They must also make all employees aware of the "importance of conforming to the environmental policy" and of "the significant environmental effects", of their role within the system and of "the potential consequences of not following the specific procedures". The internal and external communication processes must be specified. The organisation will decide whether or not to make public "information about its most significant environmental effects".

The system documentation must include: environmental policy, objectives and targets; scope of the system; description of the main elements and their interaction; and the records required by the standard. This documentation must be controlled, which means that documents must be approved, reviewed and updated. There must be specific control of the most significant environmental effects, and this brings with it the need to establish some "operational criteria for the procedures". The organisation must implement "procedures to identify potential accident and emergency situations" and indicate how it would deal with them.

Checking

The organisation must implement the procedures required to complete "regular monitoring and measurement" of the most significant environmental operations. In keeping with the commitment established, the company must periodically evaluate legal compliance. It must also implement the processes for dealing with nonconformity[6] and take the corrective[7] and preventative[8] actions required. The organisation must identify and maintain the records required to demonstrate conformity with the requirements of the EMS and of the standard and provide evidence of the results obtained. It must ensure that the internal audits of the system are carried out at the planned intervals to determine whether the system conforms to the planning and the standard and to provide information for management. The selection of the auditors must be carried out to ensure the objectivity and impartiality of the audit processes.

---

[6] Non-conformity: Breach of a requirement, ISO 14001 (2004).
[7] Corrective action: action to eliminate the cause of a detected nonconformity, ISO 14001 (2004).
[8] Preventative action: action to eliminate the cause of a potential nonconformity, ISO 14001 (2004).

Management Review

"The organisation's management needs to regularly review their EMS, at planned intervals, to ensure that it is operating effectively and provide the opportunity to address changes that may be required". The company must evaluate "the opportunities for improvement and make the required changes to the EMS, including the environmental policy, objectives and targets."

Table 9. Concordances

	Critical Factors	ISO 14001
1. Top Management Support	1.1. Top Management Support	1.2 Top management commitment and leadership
	1.2. Strategic plan	1.1 EMS model 3. Planning
	1.3. Environmental policy	1.3 EMS scope 1.4 Initial review 2. Environmental policy
	1.4. Objectives and goals	3.3 Objectives, targets and programme(s)
	1.5. Leadership	1.2 Top management commitment and leadership
2. HR Management	2.1. Responsibilities allocation	4.1 Resources, roles, responsibility and authority
	2.2. Communication	4.3. Communication
	2.3. Motivation	4.2 Competence, training and awareness
	2.4. Formation	4.2 Competence, training and awareness
	2.5. Team in charge	4.1 Resources, roles, responsibility and authority
3. Information System	3.1. Documentation system	4.4 Documentation 4.5 Control of documents 5.4 Control of records
	3.2. Impact identification	5.1 Monitoring and measurement 4.6 Operational control
	3.3. Emergency plan	4.7 Emergency preparedness and response
	3.4. Continual improvement	6.2 Continual improvement
	3.5. Management review	6.1 Management review
4. Externals Factors	4.1. Stakeholders	4.3.2 External communication
	4.2. Legislation	5.2 Evaluation of compliance
	4.3. Audits	5.5 Internal audits

7. CORRELATION BETWEEN ISO 14001:2004 AND THE CRITICAL FACTORS EXTRACTED FROM THE LITERATURE REVIEW

In section 3 of this chapter we analysed the critical factors of an EMS included in the literature. These were (1) Management support, (2) HR management, (3) Information systems and (4) External factors. The ISO 14001:2004 standard structures the requirements of the EMS as discussed in the previous section, namely: (1) General requirements; (2) Environmental policy; (3) Planning; (4) Implementation and operations; (5) Checking and (6) Management review. With both classifications, we can easily establish the points of agreement between the critical factors of an EMS and the elements that make up an EMS based on ISO 14001:2004. This is shown in Table 9.

As is to be expected, the correlation between the critical factors identified in the literature and the elements making up ISO 14001:2004 is almost total. We have only found two areas of disagreement. The first is that referred to as identifying and applying the best practices known in the sector, and the second relates to the corrective and preventative actions which accompany nonconformity with a requirement of the standard.

This second area is a general part of ISO standards and it is therefore normal for it not to be one of the critical factors included in the literature. The reason for leaving the identification and application of the best practices in the sector out of the ISO 14001:2004 standard may be because the standard only includes requirements that can be objectively audited.

This demonstrates that the most significant environmental management factors and the elements making up an EMS certified according to the standard ISO 14001:2004 are reasonably similar and we can confirm that if we measure the efforts made by a company in each of these factors we will also be measuring the effort that the company is making in the elements of its EMS based on ISO 14001:2004, because they are the same thing.

CONCLUSION

We can define environmental management as the collection of activities that companies perform to reduce the environmental impact of their actions. The importance of these aspects to the business world has been growing exponentially since the proclamation of the principles in the 1992 Rio Declaration. This management can be more or less systemised within the organisation. The most popular international standard for managing the environmental impact of a company is ISO 14001. In this chapter we have proposed a systematic way of studying the environmental management systems based on this standard. To do so, we have reviewed the academic literature and analysed the documentation for the standard, proposing 18 critical factors that can be grouped into four categories:

- Management support: commitment of the senior management, strategic planning, environmental policy, objectives and goals, and leadership.
- HR management: assignment of responsibilities, communication, motivation, training, responsible team.

- Information systems: system for documentation, identifying effects, emergency plan, ongoing improvements, management review.
- External factors: interest groups, legislation, audits.

We have reviewed the different types of standard on the basis of their effects and completed a more in-depth analysis of the ISO 14000 family. The main conclusion that we can draw about the reference standard, ISO 14001, is that this management standard improves and homogenises the exchange of environmental information by the companies. In addition, it helps to control and manage the organisational processes and activities that have an effect on the environment. Lastly, it serves as a signal, thereby reducing possible informational asymmetries.

REFERENCES

Aragon-Correa, J.A. (1998): "Strategic proactivity and firm approach to the natural environment", *Academy of Management Journal*, Vol. 41, No. 5.

Aragón-Correa, J.A. et al. (2005): "Un modelo explicativo de las estrategias medioambientales avanzadas para pequeñas y medianas empresas y su influencia en los resultados", *Cuadernos de Economía y Dirección de la Empresa*, No. 25, pp. 029-052.

Akerlof, G. A. (1970): "The Market for 'Lemons': Quality Uncertainty and the Market Mechanism", *Quarterly Journal of Economics*, Vol. 84, pp. 488-500.

Azzone, G., Bertelè, U. and Noci, G. (1997): "At last we are creating environmental strategies witch work", *Long Range Planning*, Vol. 30, pp. 562-571.

Babakri, K.A., Bennett, R.A. and Franchetti, M. (2002): "Critical factors for implementing ISO 14001 standard in United States industrial companies", *Journal of Cleaner Production*, Vol. 11, pp. 749-752.

Bansal, P. and Roth, K. (2000): "Why companies go green: a model of ecological responsiveness", *Academy of Management Journal*, Vol. 43, No 4, pp. 717-36.

Barney, J. B. (1991): "Firm resources and sustained competitive advantage". *Journal of Management*, vol. 17, nº 1, pp. 99-120.

Berry, M.A. and Rondinelli, D.A. (1998): "Proactive corporate environmental Management: a new industrial revolution", *The Academy of Management Executive*, Vol. 12, No. 2, pp. 38-50.

Blind, K. (2004): "*The Economics of Standard: Theory, Evidence, Policy*", Edgard Elgar Publishing, Glos, England.

Brunsson, N. and Jacboon, B. (2000): "The Contemporary Expansion of Standardization", en Brunsson et al. (Eds): *A World of Standards,* Oxford University Press, pp. 52-70.

Cairncross, F. (1995): "Costing the Earth: The Challenge for Governments, the Opportunities for Business", *Harvard Business School Press*, Boston, MA.

Chavan, M. (2005): "An appraisal of environment Management Systems A competitive advantage for small businesses", *Management of Environmental Quality: An International Journal*, Vol. 16, No. 5. pp. 444-463.

Chin K. S., Chiu S., Tummala V.M.R. (1999): "An evaluation of success factors using the AHP to implement ISO 14001-based EMS", *The International Journal of Quality and Reliability Management*, Vol. 16, No. 4, pp. 341.

Corbett, C. J. and Kirsch, D. A. (2001): "International diffusion of ISO 14000 certification", *Production and Operations Management*, Vol. 10, No. 3, pp. 327-342.

Cramer, J. (1998): "Environmental Management: From "fit" to "stretch", *Business Strategy and the Environment*, Vol. 7, No. 3, pp. 162-172.

Cuervo, A. (2008): "Introducción a la Administración de Empresas", *Thomson/Civitas*, Cizur Menor, Navarra.

David, P.A. (1987): "Some New Standards for the Economics of Standardization in the Information Age", in Dasgupta, P. and P. Stoneman (eds), Economic Policy and Tchnological Performance, Cambridge: Cambridge University Press.

Delmas, M. (2002): "The diffusion of environmental management standards in Europe and in the United States: An institutional perspective", *Policy Sciences*, Vol. 35, No.1, 91-119.

Deming, E. (1986): *"Out of the Crisis: Quality, Productivity and Competitive Position"*, Cambridge University Press.

Dick, G.P.M. (2000): "ISO 9000 certification benefits, reality or myth?", The TQM Magazine, Vol. 12, No. 6, pp. 365-371.

Emas-register (2010): http://www.emas-register.eu

González Benito, J. and González Benito, O. (2006), "A Review of Determinant Factors of Environmental Proactivity", *Business Strategy and the Environment*, Vol. 15, pp. 87-102.

Gupta, M.C. (1994): "Environmental Management and Its Impact on the Operations Function", *International Journal of Operations and Production Management*, Vol. 15, No. 8, pp. 34-51.

Gupta, M.C. and Sharma, K. (1996): "Environmental operations management: an opportunity for improvement", *Production and Inventory Management Journal*, Vol. 37, No. 3, pp. 40-46.

Heras, I. (2006): "Génesis y auge de los estándares de gestión: una propuesta para su análisis desde el ámbito académico" in "ISO 9000, ISO 14001, y otros estándares de gestión: pasado, presente y futuro. Reflexiones teóricas y conclusiones empíricas desde el ámbito académico." Thomson. Cizur Menor (Navarra).

Heras, I., Arana, G., Molina, J.F. (2008): "EMAS *versus* ISO 14001", Boletín Económico del ICE, n° 2936.

Henriques, I. and Sadorsky, P. (1999): "The relationship between environmental commitment and managerial perceptions of stakeholders importance", *Academic Management Journal*, Vol. 42, No. 1, pp. 87-99.

Hosseini, A. (2007): "Identification of green Management system's factors: A conceptualized model", *International Journal of Management Science and Engineering Management*, Vol. 2, No. 3, pp. 221-228.

Hunt, C. and Auster, E. (1990), "Proactive environmental management: avoiding the toxic trap", *Sloan Management Review*, Vol. 31, No. 2, pp. 7-18.

ISO (2009): "The ISO Survey is ISO 9000 and ISO 14000 Certificates", Ginebra, Suiza.

Jiang, J. and Bansal, P. (2003): "Seeing the Need for ISO 14001", *Journal of Management Studies*, Vol. 40, No. 4, pp.1047-1067.

Karapetrovic, S. and Casadesús, M. (2006): "A Future of ISO Standards in Quality Management: Augmenting ISO 9001", in Foley, K., Hensler, D., Jonker, J. (ed.): *Quality*

Management and Organizacional Excellence: Oxymorons, Empty Boxes, or Significant Contributions to Management Thought and Practice, Sydney, Australia.

Klassen, R. and Angell, L. (1998): "An international comparison of environmental management in operations: the impact of manufacturing flexibility in the U.S.", *Journal of Operations Management,* Vol. 16, No. 2/3, pp. 177-194.

Klassen, R. and McLaughlin, C. (1996): "The impact of environmental management on firm performance", *Management Science,* Vol. 42, No. 8, pp.1199-1214.

Könnölä, T. and Unruh G.C. (2007): "Really Changing the Course: the Limitations of Environmental Management Systems for Innovation", *Business Strategy and the Environment*, Vol. 16, pp. 525–537.

Leland, H.E. (1979): "Quacks, Lemons, and Licensing: A Theory of Minimum Quality Standards," *Journal of Political Economy, University of Chicago Press*, Vol. 87, No.6, pp. 1328-46.

Mintzberg, H. (1984): "La Estructuración de las Organizaciones", *Ariel*, Barcelona.

Newman, J. and Breeden, K. (1992), "Managing in the environmental era: lessons from environmental leaders", *The Columbia Journal of World Business*, Vol. 27, No. 3/4, pp. 210-21.

NSF International (1996): "Environmental Management System Demonstration Project. Final Report", *NSF International*. Ann Arbor, MI.

Padma, P., Ganesh, L.S. and Rajendran, C. (2008): "Astudy on the ISO 14000 certification and organizational performance of Indian manufacturing firms", *Benchmarking: An International Journal*, Vol. 15, No. 1, pp. 73-100.

Pun K.F., Fung, Y.K. and Wong, F.Y. (1998): "Identification of critical success factors for total quality environment Management", *Proceedings of the 3rd annual International conference on industrial engineering theories, applications and practice*, Hong Kong.

Rumelt, R.P. (1991): "How Much Industry Matter?", *Strategic Management Journal*, vol. 12 pp. 167 -185.

Russo, M. and Fouts, P. (1997): "A Resource-Based Perspective On Corporate Environmental Performance and Profitability", *Academy of Management Journal,* Vol. 40, No. 3, pp. 534-559.

Sambasivan, M. and Fei, Y. (2008): Evaluation of critical success factors of implmentation of ISO 14001 using analytic hierarchy process (AHP): a case of study from Malaysia", *Journal of Cleaner Production*, Vol. 16, pp. 1424-1433.

Sedon, J. (1997): "Ten arguments against ISO 9000", *Managing Service Quality*, Vol. 7, No. 4.

Sharma, S. and Vredenburg, H. (1998): "Proactive corporate environmental strategy and the development of competitively valuable organizational capabilities", *Strategic Management Journal*, Vol. 19, pp. 729-753.

Sharma, S. and Aragón-Correa, J.A. (2003): "A contingent resource-based view of proactive corporate environmental strategy", *Academy of Management Review*, Vol. 28, No. 1, pp. 71-88.

Shewhart, W. (1931): *Economic control of quality of manufactured product*, D. Van Nostrand Company, New York.

Uzumeri, M. (1997): "ISO 9000 and Other Metastandards: Principles for Management Practice?", *Academy of Management Executive*, Vol. 11, No. 1, pp. 21-36.

Wee, Y.S. and Quazi, H.A. (2005): "Development and validation of critical factors of environmental management", *Industrial Management and Data Systems*, Vol. 105, No.1, pp. 96-114.

Welford, R. (1994): "Cases in Environmental Management and Business Strategy", *Pitman Publishing*, Boston, MA.

Wernerfelt, B. (1984): "A resource-based view of the firm", *Strategic Management Journal*, vol. 5(7), pp. 171-180.

Wilson, R.C. (1997): "ISO 14000 Insight", *Pollution Engineering*, September, pp. 53-57.

Yeung, G. and Corbett, C. J. (2005): "Meta-Standards in Operations Management: Cross-disciplinary Perspectives", *Call for papers for special number of International Journal of Production Economics*.

Zutshi, A. and Sohal A.S. (2004): "Adoption and maintenance of environmental management systems: Critical success factors", *Management of Environmental Quality: An International Journal*, Vol. 15, No. 4, pp. 399.

In: Environmental Management
Editor: Henry C. Dupont

ISBN: 978-1-61324-733-4
© 2012 Nova Science Publishers, Inc.

Chapter 6

MEASURING SUSTAINABLE CULTURE AMONG CONSTRUCTION STAKEHOLDERS IN HONG KONG

Robin C. P. Yip[1], C. S. Poon[1] and James M. W. Wong[2]
[1]Department of Civil and Structural Engineering,
The Hong Kong Polytechnic University, Hung Hom, Kowloon, Hong Kong
[2]Department of Civil Engineering, The University of Hong Kong,
Pokfulam Road, Hong Kong

ABSTRACT

The construction industry is a leading contributor in improving the quality of the built environment, but concurrently it is a main producer of solid waste and greenhouse gas that damage the environment. Stakeholders of the construction industry thus have a decisive role to play in enhancement of sustainability and suppression of environmental damages. In the process of performing sustainable construction, stakeholders changed subconsciously their attitudes and behaviours towards a more sustainable culture. This paper aims to examine the extent of these attitudinal and behavioral changes by caterizing these changes in four sustainable cultural components. The attitudinal changes are classified into *awareness* and *concern*, while the behavioral changes are classified into *motivation* and *implementation*. The investigation was carried out by means of two surveys conducted in years 2004 and 2006 among different stakeholder groups of various disciplines including the Government, Developer, Consultant, Contractor and the frontline construction supervisors embracing site agents, site supervisors and foremen. The findings indicated that different stakeholder group carries different influential power to contribute sustainability. The consultant group and the frontline participants group demonstrated readiness in compliance by their willingness to adopt new practices favorable to sustainable construction. On the other hand, although embracing high influential power, the developer group had yet a relatively lower apprehension on sustainability, particularly in motivation and implementation aspects. Holding the highest influential power, the government group had a remarkable awareness and motivation on sustainable construction but inadequate in implementation when compared with other industry stakeholders. Although the contractor group exhibited an overall improvement in sustainable culture, but the improvement of various cultural components are relatively low. The results of investigation that reflect such a social phenomenon is an important reference for decision-makers in the government and in private sectors to formulate policies that couple with universal demands for sustainable development. The means of measurement so developed may also serve as a valuable reference for other industries.

Keywords: Culture, Cultural Component, Stakeholder group Sustainable Culture, Sustainable development, Sustainability.

INTRODUCTION

Sustainability per se is the outcome of human culture and embodies entirely new values and consciousness through sustainable development activities. The sustainable development concept established in the Brundtland Report (WCED, 1987) attempts to support continuous development of human society for the present and future generations. Brundtland's concept of sustainable development emphasizes balance of growth in social, economic and environmental aspects and this balanced growth is regarded as the core value that supports building a sustainable society (Figure 1).

To genuinely achieve sustainable development is to nourish a sustainable code of conduct favorable to sustainability requirements in a society, this code of conduct is recognized as the sustainable culture (Yip and Poon, 2009). Stated by the United Nations Educational, Scientific and Cultural Organization (UNESCO, 2002), culture is the *"set of distinctive*

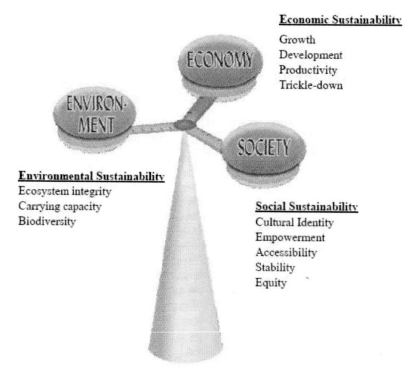

Source: Sustainable Development Unit, http://www.susdev.gov.hk, April 2002.

Figure 1. Three Attributes of Sustainable Development.

spiritual, material, intellectual and emotional features of society or a social group, and that it encompasses, in addition to art and literature, lifestyles, ways of living together, value systems, traditions and beliefs". Culture is the guiding principle for everyone to think and act within a framework that improves the well being of human life and the integrity of nature (King, 2004). Sustainable culture that built up among construction participants is therefore the driving force of a society towards sustainability.

The construction industry is one of the key economic sectors which brings both positive and negative impacts on the sustainability and the cultural settings to all its participants. This is no exception in Hong Kong. While the construction industry of Hong Kong has been a major contributor to economic growth and a key provider of employment that maintains social stability, the industry is concurrently a main consumer of the earth resources (energy, water, materials and land), a major producer of solid waste; and a polluter of the atmosphere with gaseous emission (Spence and Mulligan, 1995). Extensive extraction of natural resources for building construction jeopardizes the principle of sustainability and has received increasing objections from environmentalists (Tam, 2009). Thus stakeholders of the construction industry have a decisive role in pursuing sustainability and improve the built environment. It is therefore worth to investigate the sustainable culture built up among stakeholders in the construction industry.

Abundant studies have been conducted to enhance sustainability by exploring new materials, equipment and construction management systems aiming at lesser consumption of earth resources, higher production rate, better product quality and lower waste generation (e.g. Lozar, 1994; Wyatt *et al*, 2000; Poon *et al*, 2003; Yip and Wong, 2004). The results of these research activities, no matter tangible or intangible, are deliverables of new construction technologies and management systems that enhance sustainability in economic, social and environmental aspects. However, there are very limited research activities measuring the sustainable culture in the construction industry. Using numerical measurements to investigate the accomplishment of sustainable culture among the industry stakeholders is still under-explored.

A change of sustainable culture within the construction industry is a reflection of the attitudes and practices among various participants in project development, project design and construction operations. This study focuses on the investigation of the change of sustainable culture among construction participants in Hong Kong and identifies disparities in attitudes and behaviours of stakeholders in sustainability. The findings could serve as important references for decision-makers in the government and in private sectors to take remedial action to rectify the shortcomings.

The participants of the construction industry of Hong Kong can be categorized into five different discipline groups according to their work duties and functional output, viz. Government, Developer, Consultant, Contractor and Non-professional Frontline Participants. Specifically, the following key research questions are investigated:

1. How to measure the sustainable culture and its change within the construction industry?
2. What is the influential level of each discipline group stakeholder affecting the sustainable output of the built environment?
3. Do the industry stakeholders have a unified intensity on the sustainable development?

To study the above issues would effectively help develop the sustainable culture of the chosen field - the construction industry of Hong Kong. In this connection, guidance and action plans specific to an industry stakeholder could be set up to pursue a sustainable society. The means of measurement may also serve as a reference for continuous evaluation of the evolution and change of sustainable culture across time. The extent of change in sustainable culture (cultural shift) is represented by the essence and magnitude of the change of attitudes and behaviours within a designated time frame. Based on the defined cultural shift, two surveys have been conducted to identify discrepancies among key components on sustainable culture. The findings of these discrepancies are valuable references for remedial actions to rectifying the specific shortcomings in cultural development in sustainability. This paper is organized as five sections. Following this introduction, the next section reviews the development of sustainable culture reported from the literature. Details of the research methodology are then elaborated. This is followed by the empirical results and the implications of the findings. Concluding remarks and recommendations are drawn in the last section.

DEVELOPMENT OF SUSTAINABLE CULTURE

As a guideline, Kibert (1994) has introduced sustainable construction which conceptualized minimizing consumption of energy, water, materials and land throughout the life cycle performance of built facilities. The growth of recognition in the direction of sustainable construction is the increase of knowledge among construction participants that enriched their value and belief in sustainability. Thus the concept of sustainable construction and the increased knowledge of sustainability inevitably changed the attitude and behaviour among construction participants and became part of their inherent culture. Hungerford and Volk (1990) developed in their educational research to reveal changes in learner behaviour through environmental education, *"increasing knowledge could increase awareness or change attitudes, which in turn change behaviour"*.

Subsequent studies demonstrated that there is moderate to limited relationship between awareness, attitudes and participatory actions. Ajzen (1991) in his behavior theories, the theory of reasoned actions (TRA) and the theory of planned behavior (TPB) identified that from acquisition of knowledge to taking an action is not a linear approach but a combination of a number of factors including attitudes, subjective norms and perceived behaviour control (Figure 2). Using Ajzen's behavioral theories as a basis, Teo and Loosemore (2001) have studied behaviour of construction participants in handling construction waste.

The process of judgment and evaluation is a diagnostic process. Fazio and Roskos-Ewoldsen (1992) theorized that "attitudes themselves may be diagnostic of a variety of judgments and behaviour; they serve as heuristics, saving capacity especially in low-motivation or capacity-limited situations". Eagly and Chaiken (1993) concurred with their study and defined attitudes are evaluative responses to some object of judgment. Attitudes would be constructed from cognitions, affective responses and behaviour, all may be diagnostic of an evaluative judgment.

Lewin's model of social behaviour (1936) was further expanded by Bordens and Horowitz (2001) as shown in Figure 3. The process "evaluation of situation" is the process of diagnostic and judgment that influence the resultant social behaviour. The change of attitudes from acceptance to implementation consumed a period of time to accommodate the process of diagnostic. Therefore, it would be proper to place an incubation process in Ajzen's model prior to the process of behavioural intention (Figure 4) as a representation of culture movement model in the construction industry.

The culture movement model for sustainability thus formed is based on the processes of *awareness, concern, motivation* and *implementation* of sustainability. These key changes drive the development of attitudes and behaviours of construction participants towards sustainability and are defined as the cultural components for sustainable construction. The sustainable culture of construction starts with cognition of the concept and knowledge (*awareness*) of sustainability. Awareness is defined as the sense of detection about the needs to change an unsatisfied condition or an unease state of mind (Blank, 1996). The increase of awareness results in stronger diagnostic of evaluative judgment (*concern*) for any change that may cause negative sustainable effect (Fazio, 1990). Concern thereby arises as a result of awareness on scenarios which arouse desires for improvement (Eagly and Chaiken, 1993).

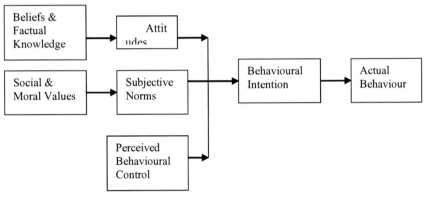

Source: Ajzen (1991).

Figure 2. The Theory of Planned Behaviour.

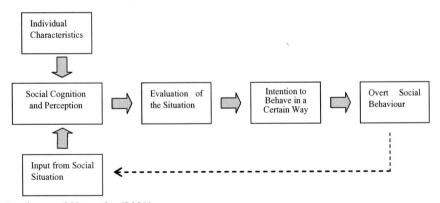

Source: Bordens and Horowitz (2001).

Figure 3. An Expanded Model of Social Behaviour.

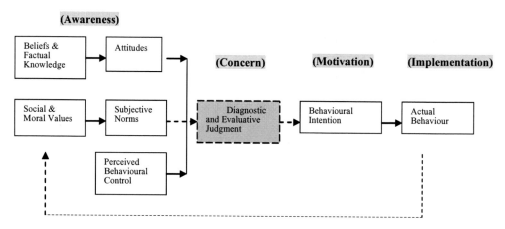

Figure 4. The Culture Movement Model of the Construction Industry.

These are cognitive attitudes that constitute beliefs of sustainability. The theories of TRA and TPB illustrated behavioral intention to respond (*motivation*) follows a subjective norm in sustainable consciousness and results in performance of corresponding behaviour (*implementation*).

Through implementation of sustainability, participants realize more merits of sustainable construction and further enhance the sustainable knowledge within themselves. As a result, they further reciprocate their awareness in sustainability by activating diagnosis and judgment. The entire process keeps on recurring and synchronizes with the change of the global atmosphere and social demand. Hence, a responsible sustainable-conscious participant in the construction industry is the one who possesses basic understanding of sustainable construction issues; has an awareness of and sensitivity to sustainability; concerns changes which could have negative effect; motivates autonomously to initiate improvement; and actively involves in a certain level of implementation toward sustainable construction. The earlier process of the sustainable culture is represented by the change of cognitive attitudes in *awareness* and *concern*, followed by behavioral actions in *motivation* and *implementation*. A synthesis of these four components is the combined effect of the movement of sustainable culture.

RESEARCH METHODOLOGY

Figure 5 summarizes the research procedures and methods to achieve the stipulated research objectives. Research methodology adopted in this study includes literature review, statistical analysis based on questionnaire survey as well as semi-structured interviews to collect necessary data.

A comprehensive literature review was conducted initially to uncover fundamental information related to sustainable development and issues related to the construction industry that formed the sustainable culture among construction participants and stakeholders. Based on the theories of TRA and TPB developed by Azjen (1991) and other theories that dealt with human behaviour, a cultural shift model containing four cultural components comprising awareness, concern, motivation and implementation is developed.

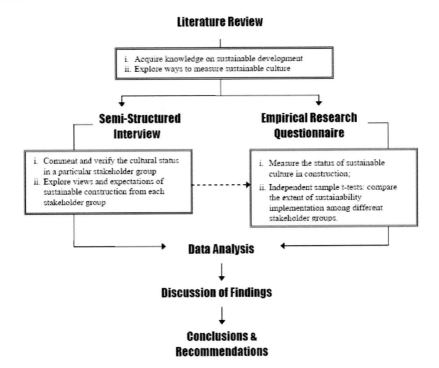

Figure 5. Research Framework.

To measure the status of sustainable culture in construction in general and compare the extent of sustainability implementation among different stakeholder groups in particular, surveys were conducted separately in 2004 and 2006 by using the same questionnaire. The surveys attempt to investigate respondents' awareness, concern, motivation and implementation of sustainability towards the ultimate goal of sustainable development within their respective professions. The survey has been validated through a pilot test which was carried out in year 2004 and was reported by Poon and Yip (2005). The design of the questionnaire is based on studies which are about attitudes and behaviours on sustainability and the implementation of sustainable construction.

It should be noted that construction participants and stakeholders of different discipline exert different influential power due to their distinctive functions and this may result in different extent of achievements in sustainable construction that affect the eventual outcome of the built environment. Based on their functions and duties, these construction participants and stakeholders of different discipline can be fundamentally divided into five groups:

a) the Government;
b) the Developer;
c) architects, structural engineers, electrical and mechanical engineers, quantity surveyors (collectively the Consultant);
d) main contractors, subcontractors, suppliers (the Contractor);
e) site agents, site supervisors, foremen (the Non-professionally recognized frontline participants, or the NPP).

Table 1. Summary of Questionnaire for each Cultural Components in three Generalized Statement

Awareness	I believe sustainable development is important to the world.
	Use of environmentally friendly materials and green construction methods will eventually help to preserve natural resources.
	I am aware that sustainable development is getting more recognition among my colleagues and co-workers.
Concern	Caring about safety and health in the project design and construction process, and an emphasis of quality of products is a sign of recognising the importance of the socio-economic equity of sustainable development.
	I believe that using environmentally friendly materials and green construction methods, caring for safety and health, and emphasising product quality will increase construction cost and time.
	On the contrary, the use of environmentally friendly materials and green construction methods, caring of safety andhealth, and enhancing product quality would reduce construction cost and time.
Motivation	Even if there is an increase in the construction cost and time, I intended to apply sustainable construction methods in Hong Kong.
	Even if there is an increase in the construction cost and time, I have noticed that my colleagues and co-workers intended to apply sustainable construction methods n Hong Kong.
	Even if there is an increase in the construction cost and time, I have noticed that my clients intended to apply sustainable construction methods in Hong Kong.
Implementation	I have implemented sustainable development principles in my work and will practice the same continuously.
	I have found that my colleagues and co-workers have implemented green ideas and added socioeconomic equity elements in their work.
	I have found that my clients have implemented green ideas and added socio-economic equity elements in their proposed projects.

Hence, practitioners within these five groups were the target respondents of the questionnaire. They were asked to evaluate the influential power (in percentage) of each group that influences the output of sustainability in construction. As the questionnaire was designed to measure the four cultural components (awareness, concern, motivation and implementation), the significances of all cultural components were extensively delineated., Table 1 below has simplified and consolidated the lengthy questionnaire of each cultural component into three generalized statements and were rated by the respondents according to a five-point Likert scale delineating different levels of agreement (1 = strongly disagree; 3 = neutral and 5 = strongly agree).

The questionnaires were distributed by emails, post and personal contacts to construction participants in government departments responsible for construction project, as well as private sector organizations including developers, consultant firms, general contractor firms and specialist contractor firms. Valid responses totalling in 446 and 317 were received from two surveys conducted in years 2004 and 2006 respectively.

Table 2. Detail of Interviewees

ID	Sector	Stakeholder	Position of Interviewee	Type of Organisation
1	Private	Developer	General Manager	Leading private property developer
2	Private	Contractor	Senior Manager (Safety and environmental Management)	Leading Contractor firm
3	Private	Contractor	Project Manager	Leading Contractor firm
4	Private	Contractor	Director	Sub-contractor firm for formwork erection
5	Public	Government	Project Architect	Architect Services Department HKSAR
6	Public	Government	Deputy Director	Environmental Protection Department
7	Private	Consultant	Director	Leading Architect and Structural Engineer consulting firm in Hong Kong
8	Private	NPP	Site Agent	Leading construction firm

Descriptive analysis was firstly carried out where means as well as standard derivations of the data were analyzed. By the use of Statistical Package for Social Sciences (SPSS), independent sample t-tests have been done to compare the results among different stakeholder groups in determining which stakeholder groups had achieved more significant sustainable outcome than the others. If the test result was significant at the 5% level ($p < 0.05$), then the null hypothesis that no significant differences in the mean values between the corresponding groups can be rejected (Norusis, 2002).

Supplementing the survey findings, semi-structured interviews were also conducted to explore views and expectations of sustainable construction from each stakeholder group. A total of eight interviewees were invited to participate the interviews (Table 2). All of the selected interviewees have practised in the construction industry in Hong Kong for at least 15 years, some of them have over 25 years post qualification experiences, although some of the experiences were gained abroad. The views and opinions on the research topic are therefore reliable and of a profound depth.

RESULTS AND DISCUSSION

Respondents' Profile

Survey respondents were random samples from the five industry stakeholder groups. Demographic information of these random respondents displayed in Figure 6 showed that the majority of them were affiliated to professional bodies of Hong Kong. Most of the respondents had over 5 years of prior practical working experience in the construction industry in Hong Kong, and nearly three quarters and over half of them had over 10 years of experience in the 2004 and 2006 surveys respectively.

These show that the respondents were experienced in the local construction industry and their views and opinions were therefore valuable and representative.

Demographic Information of Respondents	Survey Result of Year 2004	Survey Result of Year 2006
Number of Qualified Professionals / Total Number of Respondents	323/446 (72%)	192/317 (61%)
Number of Respondents over 20 Years Experience / Total Number of Respondents	151/446 (34%)	74/317 (23%)
Number of Respondents over 10 Years Experience / Total Number of Respondents	326/446 (73%)	183/317 (58%)
Number of Respondents over 5 Years Experience / Total Number of Respondents	410/446 (95%)	285/317 (90%)

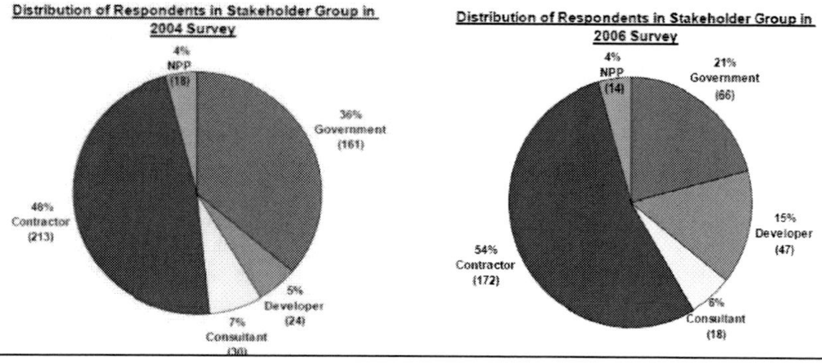

Figure 6. Demographic Information of Survey Respondents.

Influential Power

As stressed earlier, different stakeholder groups bear different influences in both the course of construction and the final built asset on sustainability. The mean values of the influential power as perceived by the survey respondents are computed and presented in Figure 7. The results of the two surveys revealed that the findings of influence power of the same group are close to each other and different groups exerts different influential power due to their hierarchy of functions and has different fundamental impacts on the built assets. The hierarchical order of influence is generally established and represents the sustainable output of the specific stakeholder group.

It is understandable that the government possesses the highest influential power (37-40%), because it not only pioneers the sustainable development of Hong Kong through policy making actions and sustainable initiatives but also takes a leading role enabling sustainable construction (Spence and Mulligan, 1995). The developers are the investors of built assets capturing around 28% of the influential power thus occupying the second highest among all stakeholders. The developer group employs consultants to design and contractors to build, the requirements and the final functions as prescribed by the developer group are therefore imperatively governing the sustainable output of the built assets.

The consultants have less influential power than the developer group due to the nature and the roles of their services, however, they are important players influencing the sustainable output of the built assets by applying considerable sustainable elements in their designs and using substantial amount of environmentally friendly materials.

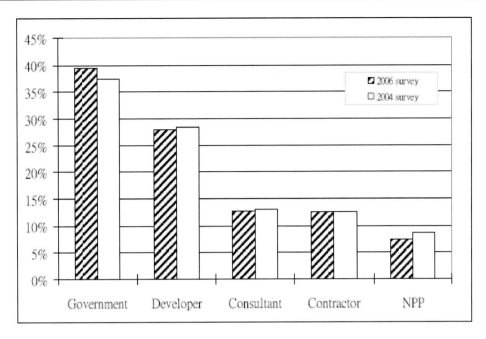

Figure 7. Influential power of each stakeholder group on sustainability.

The contractor's role is to build according to the design and specifications. Although the influential power of the contractor group is not being placed in a high ranking, the method of construction selected, consumption of resources and the methods of handling construction waste are crucial to the sustainable construction result. The NPP group is formed by individuals performing supervisory functions in construction sites as front line managers. Even if they have a very high recognition in sustainable construction and perform diligently in their respective job duties, their output in sustainable construction remains limited and at the level of an individual. Identical hierarchical order of influential power among stakeholder groups is also revealed in the semi-structured interviews after the two questionnaire surveys.

THE CULTURAL SCORE AMONG STAKEHOLDER GROUPS

Based on the cultural values obtained from the surveys in years 2004 and 2006, the extent of cultural shift for each group is identified and shown in Table 3. Positive shifts in sustainable culture were found in most of the stakeholders groups within the study period. However, there were a few negative shifts appeared among the overwhelming amount of positive shifts, including the government in concern, the developer group in motivation and implementation, and the contractor group in implementation. These negative shifts represent decline of sustainable culture in identified cultural components within these groups.

While the government and developers are playing a critical role in policy making and financial support in driving sustainability in construction, the contractor group's execution in sustainable construction input is also a crucial action in achieving sustainable construction. These cultural shifts are important findings for the decision-makers in the government, private sectors and the contractors to initiate corrective actions to transform attitudes and behaviours in the declining portion of the sustainable culture.

Table 3. The Cultural Mean Scores by Industry Stakeholder

			Government	Developer	Contractor	Consultant	NPP
Awareness	2004	Mean	3.62	3.70	3.54	3.74	3.68
		S.D.	0.70	1.06	0.80	0.38	0.76
	2006	Mean	3.94	3.82	3.62	3.98	3.83
		S.D.	0.61	0.81	0.73	0.31	0.80
Concern	2004	Mean	3.36	3.25	3.31	3.19	3.27
		S.D.	0.62	0.76	0.63	0.80	0.24
	2006	Mean	3.28	3.45	3.33	3.30	3.71
		S.D.	0.66	0.68	0.69	0.46	0.83
Motivation	2004	Mean	3.26	3.29	3.18	3.24	3.12
		S.D.	0.75	1.18	0.87	0.82	0.24
	2006	Mean	3.51	3.07	3.39	3.56	3.74
		S.D.	0.79	0.79	1.00	0.43	0.78
Implementation	2004	Mean	3.32	3.23	3.31	3.64	3.33
		S.D.	0.90	0.94	0.73	0.88	0.83
	2006	Mean	3.35	3.13	3.28	3.66	3.69
		S.D.	0.80	0.82	0.92	0.34	1.04

Notes: Items were rated on a five-point Likert scale with 1 = strongly disagree; 3 = neutral and 5 = strongly agree.

Sustainable construction measures need the support of relevant players at various places and times in the decision making process to become effective. All successful sustainable construction measures depend upon the decisions of other players that are being motivated to genuinely implement the concept of sustainability. The unevenness in balancing cultural components would be a barrier to sustainable construction (Van Bueren and Priemus, 2002). Independent sample t-tests have therefore been performed to compare the means between two construction stakeholder groups in order to determine whether the means on the four cultural components are significantly different from one another in a specific point in time. If the test result was significant at 0.05 level, then the null hypothesis that there is no significant difference in the mean values between the stakeholder groups can be rejected. Tables 4 and 5 reveal the results of the independent-sample t-tests based on the surveys conducted in years 2004 and 2006.

Table 4. Summary of Independent-sample t-tests Result (2004 survey)

Government	Developer	Contractor	Consultant	NPP
Government compares with →	A: 0.732 (0.346) C: 0.447 (0.763) M: 0.927 (0.093) I: 0.095 (1.676)	A: 0.292 (1.054) C: 0.482 (0.704) M: 0.301 (1.036) I: 0.132 (1.508)	A: 0.191 (1.323) C: 0.214 (1.246) M: 0.899 (0.127) I: 0.209 (1.259)	A: 0.731 (0.345) C: 0.238 (1.196) M: 0.439 (0.776) I: 0.110 (1.607)
	Developer compares with →	A: 0.373 (0.893) C: 0.667 (0.431) M: 0.658 (0.447) I: 0.195 (1.300)	A: 0.847 (0.195) C: 0.782 (0.279) M: 0.881 (0.150) I: 0.031* (2.220)	A: 0.955 (0.057) C: 0.920 (0.101) M: 0.608 (0.517) I: 0.987 (0.017)
		Contractor compares with →	A: 0.030* (2.215) C: 0.366 (0.906) M: 0.702 (0.383) I: 0.098 (1.354)	A: 0.466 (0.731) C: 0.550 (0.602) M: 0.779 (0.281) I: 0.209 (1.260)
			Consultant compares with →	A: 0.755 (0.316) C: 0.638 (0.474) M: 0.621 (0.498) I: 0.209 (1.259)

Notes:
A – Awareness; C: Concern; M: Motivation; I: Implementation.
** t-statistic significant at .01 level.
* t-statistic significant at .05 level.
Values in parentheses are absolute t-statistics.

In 2004, merely the developer group has a significant lower implementation level than the consultant group (sig. = 0.031), and the contractor group has a significant less awareness than the consultant (sig. = 0.030).

The null hypothesis of "no significant difference on the four cultural components among the stakeholder groups" cannot be rejected in the remaining aspects at the 5% significance level. However, the diverse of sustainable construction raised in 2006 is reflected by the survey results. This reflects that barrier is present in the gaps that arise between the links in promoting sustainable construction in Hong Kong among stakeholders.

Table 5. Summary of Independent-sample t-tests Result (2006 survey)

	Developer	Contractor	Consultant	NPP
Government compares with	A: 0.406 (0.835) C: 0.180 (1.349) M: 0.005** (2.882) I: 0.448 (0.761)	A: 0.001** (3.417) C: 0.624 (0.491) M: 0.386 (0.869) I: 0.848 (0.192)	A: 0.780 (0.281) C: 0.935 (0.082) M: 0.720 (0.361) I: 0.021* (2.371)	A: 0.580 (0.556) C: 0.036* (2.132) M: 0.371 (1.006) I: 0.157 (1.482)
Developer compares with		A: 0.121 (1.568) C: 0.281 (1.080) M: 0.025* (2.278) I: 0.336 (0.803)	A: 0.257 (1.143) C: 0.366 (0.912) M: 0.017* (2.462) I: 0.005** (2.911)	A: 0.966 (0.043) C: 0.234 (1.203) M: 0.007** (2.787) I: 0.083 (1.834)
Contractor compares with			A: 0.000** (3.925) C: 0.772 (0.293) M: 0.187 (1.343) I: 0.012* (2.605)	A: 0.297 (1.046) C: 0.052 (1.959) M: 0.201 (1.283) I: 0.110 (1.606)
Consultant compares with				A: 0.523 (0.653) C: 0.105 (1.704) M: 0.438 (0.792) I: 0.648 (0.465)

Notes:
A – Awareness; C: Concern; M: Motivation; I: Implementation.
** t-statistic significant at .01 level.
* t-statistic significant at .05 level.
Values in parentheses are absolute t-statistics.

Although there has been increasing public demand for developers to pay greater attention to the development of sustainable buildings (Lee and Burnett, 2006), the findings indicated that the developer group had a relatively low apprehension on sustainability in construction, particularly in motivation and implementation of sustainability. While the developers association in Hong Kong launched a voluntary private sector environmental assessment scheme, named the Hong Kong Building Environmental Assessment Method (HK-BEAM), for for new and existing buildings in Hong Kong since 1996, profitability was still the sole concern for most private developers, low motivation and implementation in achieving HK-BEAM criteria may be due to insufficient financial incentive to provoke sustainable construction, as revealed by most of the interviewees.

Nevertheless, Government and public organizations should take leading steps in applying the relevant policies to both new and existing public buildings, and offering incentives for developers to actively promote compliance in implementing policies on sustainable urban development and green buildings (Jallion and Poon, 2009). Exemption of gross floor area for balconies, communal sky gardens, communal could provide a strong incentive to developers. The initiation for sustainability by the developer project team might thereby influence the attitudes and behaviours of other project participants in the implementation of sustainable construction. Interviewees have a consistent view that such a change has propelled other project teams to focus on achieving sustainability in project design and construction.

The survey findings reveal that government group had a remarkably higher score on awareness and motivation of sustainability than the contractor group (sig. = 0.001) and the developer group (sig. = 0.005) respectively in 2006. It is, however, interesting to note that the score in implementation of sustainability is on the low side when compared with other industry stakeholders. This may reflect that genuine execution of sustainable construction is limited to frontline runners including the design consultants and NPP. But their influences to implement sustainable construction are limited as discussed earlier. Development of sustainable culture in construction should therefore be initiated in the upper stream of the construction supply chain.

The remarkable awareness and motivation of sustainability by the Government could be induced from the establishment of the working group to study Hong Kong's role in the Sustainable Development for the 21st Century project (SUSDEV 21), with the mission of developing guiding principles, indicators, and criteria for the sustainable development of Hong Kong (ERM, 2000). However, the preliminary focus of this project is on the promotion of public discussion of sustainable development, rather than the development of an actual strategy for implementation (Lee and Burnett, 2006). The second initiative was started in 2002. A consultant was commissioned by the Government to design a comprehensive environmental performance assessment scheme (CEPAS), which is , an environmental impact assessment tool for various building types covering the pre-design, design, construction and operation stages (BD, 2006). As a green building labelling scheme, the CEPAS endeavours to address both physical and human-related issues amongst the core aspects of sustainability. In addition, a study on life cycle energy analysis (LCEA) of building construction was commissioned by the Electrical and Mechanical services Department (EMSD). An assessment tool was produced along with the required databases to appraise the life cycle performance of commercial building developments in Hong Kong in respects of their environmental and financial impacts (EMSD, 2006). In spite of these initiatives, policy framework and implementation strategies are lacking to promote sustainability and put sustainable construction into action, resulting in ineffectual implementation of sustainable development. Objectives and initiatives of the schemes developed have thus not been explicitly implemented.

Despite that the contractor group exhibited an improvement in sustainable culture throughout the research period, the mean sores of various components of sustainable culture are relatively low as revealed by both survey results. In particular, their awareness of sustainability is significantly inferior to the government in 2006 (sig. = 0.001) and consultant in 2004 (sig. = 0.030) and 2006 (sig. = 0.000), respectively. The contractor was also rated significantly lower than the consultant group (sig. = 0.012) in implementation. Some interviewees explained that contractors may not concern or be motivated on sustainability as they are merely obliged to carry out their duties according to respective contractual requirements despite the fact that they noted the obligations contained requirements to enhance sustainable construction issues; in most case, they performed these obligations in an involuntary manner. Shen et al. (2000) also suggested that the demand for a significant amount of time and cost investment in order to apply advanced measures in sustainable construction decreases contractors' interests in doing so. Lack of support from the industry stakeholders in the upper stream (government and developers) as well as lower stream (sub-contractors and suppliers) on the supply chain also creates barriers for contractors to develop sustainable culture (Chen et al., 2000; Jeljeli and Russel, 1995). An important message was given by one of the interviewees from the contractor group. It was suggested that when sustainable construction is fully implemented as a basic requirement for all construction projects and stakeholders are performing the same autonomously, a moral sentiment can be built up among them as an obligation, all construction participants at this stage are "committed to perform sustainability" as a voluntary action.

In view of practicality in promoting sustainable construction, the interviewees have been asked what they would propose to their superiors on enhancing sustainability. Except the NPP interviewee who suggested improvement on safety and environmental protection on construction sites, the interviewees of all other stakeholder groups suggested partnering approach in construction management to their superiors as a tool to enhance sustainability. The overwhelming recognition of partnering approach as an effective tool to enhance sustainability is a therefore a new topic for further study. Yet studies (e.g. Chidiak, 2002; Rietbergen et al., 2002) are advocating the adoption of a well articulated mix of regulatory and voluntary instruments (Lee and Yik, 2004). Not only setting a minimum standard for all buildings, regulatory controls can also augment co-existing voluntary schemes. The voluntary schemes can benefit from the increased awareness and drive towards improvements triggered by the regulations, use the regulatory requirement as a baseline for defining enhanced performance, and provide an incentive for buildings to achieve a standard above the minimum (Lee and Yik, 2002).

Interviewees have signaled an interaction among different groups, where research activities carried out by academia and professional bodies motivated the change in sustainable culture by the output of their research works, government initiatives in policy making and legislation for sustainable development enabled the process of sustainability. Based on the promulgated ordinances and regulations, stakeholder groups played their respective roles implemented sustainability in accordance with their professional functions. It is therefore logical to conclude that the research results facilitated government policy-making which in turn had a direct impact on practice irrespective of stakeholder grouping. Government is therefore taking an enabling role towards sustainability while sustainable culture among stakeholders is nurtured and developed through the process of implementation.

CONCLUSION

The construction industry plays a leading role to improve the quality of the built environment, Over the years, the stakeholders have established their own ways and means of getting jobs done; their practices became habitual and are accepted by all participants as a culture despite that there are areas where progress is behind the global trend and local demands for sustainability.

The investigation on the change of sustainable culture by means of questionnaire surveys and supplemented with interviews has concluded the followings:

- The influential power of the construction industry in Hong Kong is in a hierarchy of the Government, the Developer, the Consultant, the Contractor and the Non-professional frontline supervisors.
- Positive cultural shift is evident among all categorized stakeholder groups throughout the research period. Other than the overwhelming findings of cultural shifts on the positive side, there were a few negative shifts appeared in the government group in concern, the developer group in motivation and implementation, and the contractor group in implementation. These negative shifts are good reference to implement rectification actions.
- Differences in mean value are found between stakeholder groups in the four cultural components in different timeframe. The developer group has a significant lower implementation level than the consultant group and the contractor group has a significant less awareness than the consultant group. Likewise, the government group and the developer group revealed a higher score on awareness and motivation but their scores in implementation are on the low side, it is a clear indication that the value of sustainable construction is not identical to all stakeholders.
- Genuine execution of sustainable construction is limited to frontline runners and their influential power to bring up sustainable construction is insufficient, therefore, fostering of sustainable culture in the construction industry should be initiated in the upper stream of the construction supply chain.

If the ultimate aim of sustainable development is to balance the social, economic and environmental needs for both present and future generations through the efforts of the community and the government, then this study has shown that all parties listed have a role to play and their concerted efforts hinge upon one core value: how society and future generations may benefit from present day activities.

ACKNOWLEDGMENTS

The authors gratefully acknowledge the professionals and academia who gave valuable comments and advice to the design of the questionnaire. We thank the respondents of the two questionnaire surveys, and the interviewees whom have given their valuable views for this study. The financial support provided by The Hong Kong Polytechnic University for the research activities of this article is also acknowledged.

REFERENCES

Ajzen, I. (1991), The theory of planned behavior, *Organizational Behavior and Human Decision Processes*, 50(2), 179-211.

Allport, G. (1935) *Attitudes: A Handbook of Social Psychology*, 117-138, New York: Guilford.

Bordens, K.S. and Horowitz, I.A. (2001) *Social Psychology*, Lawrence Erlbaum Associates, London, 2nd ed., ISBN 0-8058-3520-2, 6-7.

Blank, L. (1996) *Changing Behaviour in Individuals, Couples, and Groups*, Charles C Thomas Publisher, Springfield, Illinois, USA, ISBN 0-398-06657-4, 9-10.

BD (2006) Comprehensive Environmental Performance Assessment Scheme (CEPAS) for Buildings, Application Guidelines, Buildings Department, Government of the HKSAR, Hong Kong: Government Printer.

Chen, Z., Li, H. and Wong, T.C. (2000) Environmental management of urban construction projects in China, *Journal of Construction Engineering and Management*, 126(4), 320-324.

Chidiak, M. (2002) Lessons from the French experience with voluntary agreements for greenhouse-gas reduction, *Journal of Cleaner Production*, 10(2), 121-128.

Eagle, A.H. and Chaiken, S. (1993) *The Psychology of Attitudes*, Fort Worth, TX: Harcourt, Brace, Javanovich.

EMSD (2006), Consultancy Study on Life Cycle Energy Analysis of Building Construction, Final Report submitted by the Ove Arup and Partners, Consultancy Agreement No. CAO L013, Electrical and Mechanical Services Department, Government of the HKSAR, Hong Kong: Government Printer.

ERM (2000) *Sustainable Development for the 21st Century*. Final report 2000, August 2000.

Fazio, R.H. (1990) Multiple processes by which attitudes guide behavour: The MODE model as an integrative framework, *Advances in Experimental Social Psychology*, 23, 75-109, San Diego, CA: Academic Press.

Fazio, R.H., Roskos-Ewoldsen, D.R. (1992) On the orienting value of attitudes: attitude accessibility as a determinant of an object's attraction of visual attention, *Journal of Personality and Social Psychology*, 63(2), 198-211.

Hungerford, H.R. and Volk, T. (1990) Changing learner behaviour through environmental education, *Journal of Environmental Education*, 21(3), 8-12.

Jallion, L. and Poon, C.S. (2009) The evolution of prefabricated residential building systems in Hong Kong: A review of the public and the private sector, *Automation in Construction*, 18(3), 239-248.

Jeljeli, M.N. and Russell. J.S. (1995) Coping with uncertainty in environmental construction: decision-analysis approach, *Journal of Construction Engineering and Management*, 121(4), 370-379.

Kibert, C.J. (1994), Establishing principles and a model for sustainable construction, in *Proceedings of the First International Conference of CIB TG16*, 6-9 November 1994, Tempa, FL. 3-12.

King, S.B. (2004), *Sustainable Development and Civil Society, Sustainable Development in Hong Kong*, Hong Kong University Press, 2004, ISBN962-209-491-0, 252-253.

Lee, W.L. and Burnett, J. (2006) Customization of GBTool in Hong Kong, *Building and Environment*, 41(12), 1831-1846.

Lee, W.L. and Yik, F.W.H. (2002) Regulatory and voluntary approaches for enhancing energy efficiency of buildings in Hong Kong, *Applied Energy*, 71(4), 251-274.

Lee, W.L. and Yik, F.W.H. (2004) Regulatory and voluntary approaches for enhancing building energy efficiency, *Progress in Energy and Combustion Science*, 30(5), 477-499.

Lewin, K. (1936) *A Dynamic Theory of Personality*, New York: McGraw Hill.

Lozar, C. (1994), Concepts for recycling construction /demolition materials: towards a predictive model, *Proceedings of the First International Conference of CIB TG 16*, November 1994, Tampa, Florida, U.S.A. 269-278.

Norusis, M.J. (2002) SPSS 11.0 Guide to Data Analysis, Prentice Hall, Upper Saddle River, NJ.

Poon, C.S., Yu, A.T.W. and Ng, L.H. (2003), Comparison of low-waste building technologies adopted in public and private housing projects in Hong Kong, *Engineering, Construction and Architectural Management*, 10(2), 88-98.

Poon, C.S. and Yip, R.C.P. (2005), Culture shift of the construction industry of Hong Kong under the influence of sustainable development, *Proceedings of the 6th International Conference on Tall Buildings*, December 2005, Hong Kong, 987-992.

Rietbergen, M.G., Farla, J.C.M. and Blok, K. (2002) Do agreements enhance energy efficiency improvement? Analysing the actual outcome of long-term agreements on industrial energy efficiency improvement in The Netherlands, *Journal of Cleaner Production*, 10(2), 153-163.

Shen, L.Y., Bao, Q. and Yip, S.L. (2000) Implementing innovative functions in construction project management towards the mission of sustainable environment. Proceedings of the millennium conference on construction project management, *Hong Kong Institution of Engineers*, 24 October 2000, 77-84.

Spence, R. and Mulligan, H. (1995) Sustainable development and the construction industry, *Habitat International*, 19(3), 279-292.

Tam, V.W.Y. (2009) Comparing the implementation of concrete recycling in the Australian and Japanese construction industries, *Journal of Cleaner Production*, 17(7), 688-702.

Teo, M. and Loosemore, M. (2001) A theory of waste behaviour in the construction Industry, *Construction Management and Economics*, 19(7), 741-751.

UNESCO (2002) *UNESCO Universal Declaration on Cultural Diversity*, The United Nations Education, Scientific and Cultural Organization.

Van Bueren, E.M. and Priemus, H. (2002) Institutional barriers to sustainable construction, *Environment and Planning B: Planning and Design*, 29(1), 75-86.

WCED (1987), Our Common Future, The World Commission on Environment and Development, Oxford and New York: Oxford University Press

Wyatt, D.P., Sobotka, A. and Rogalska, M. (2000) Towards a Sustainable Practice, *Facilities*, 18(1/2), 76-82.

Yip, R.C.P. and Wong, E.O.W. (2004) Promoting sustainable construction waste management in Hong Kong, *Construction Management and Economics*, 22(6), 563-566.

Yip, R.C.P. and Poon, C.S. (2009) Cultural shift towards sustainability in the construction industry of Hong Kong, *Journal of Environmental Management,* 90(11), 3616-3628.

In: Environmental Management
Editor: Henry C. Dupont

ISBN: 978-1-61324-733-4
© 2012 Nova Science Publishers, Inc.

Chapter 7

THE PROS AND CONS OF ISO 14000 ENVIRONMENTAL MANAGEMENT SYSTEMS (EMS) FOR TURKISH CONSTRUCTION FIRMS

Ahmet Murat Turk[*]

Department of Civil Engineering, Istanbul Kultur University,
Atakoy Kampus, Yanyol, Bakirkoy, Istanbul, Turkey

ABSTRACT

Recently, the ISO 14000 environmental management system (EMS) has been widely utilized by all sectors throughout the world prepared by ISO (International Standard Organization). With the purpose of keep up and develop the environmental performance within all sectors, including the construction sector, some methods exist for the protection of sustainable development and the environment worldwide. ISO 14000 EMS originated from such necessity. Lately, an increased awareness has emerged related to the use of this system in the construction industry. Despite such interest, the research related to the implementation practice of the ISO 14000 EMS by the construction firms has not reached the desired level. In this study, the major motives for seeking ISO 14001 certification for Turkish construction firms is being examined by using the questionnaire survey method. Questionnaire survey was conducted with 68 individual construction firms, which represent the top firms in Turkey and operate in national and international markets as they are members of the Turkish Contractors Association (TCA). Descriptive and factor analyses were used with the obtained data from questionnaires of advantages and disadvantages related to ISO 14000 EMS. In the study, factor analysis was used to summarize many variables with a few factors. Each of the factors acquired from these analyses presents advantages and disadvantages of ISO 14000 EMS factors. As the advantages of ISO 14000 EMS, mainly two factor dimensions are found in the analysis. First dimension is entitled as "related to environment" and second one is called as dimension "related to company". As the disadvantage of ISO 14000 EMS, mainly three factor dimensions have found in the analysis. First one is named as "related to lack of the knowledge and personnel". Second one is called as "related to the cost and implementation" and third dimension is entitled as "related to no apparent benefits".

---

[*] E-mail: murat.turk@iku.edu.tr phone: +90-212-4984257.

This study shows that there is a positive approach to the ISO 14000 EMS within the construction sector in Turkey as well as indicating that the utilization of the ISO 14000 EMS is not yet at the ideal level. In particular, the problems of lack of information and qualified personnel revealed in the results of the analysis should be overcome. Personnel should be qualified in the concept of EMS and on the technical details. In the global construction market, an increase in the number of firms having EMS will both reduce environmental impact and develop the potential of awarding contracts to the construction firms from underdeveloped and/or developing countries.

Keywords: ISO 14000, environmental management system, environment, standards, construction.

1. INTRODUCTION

The threat caused by the seriously increasing environmental pollution all over the world on natural resources has forced nations to take safety measures on environment. Mainly, the basic principles agreed on by many of the nations are "sustainable development" and "the protection of the environment" following the Rio Conference which was held in 1992. This situation has introduced a significant pressure on all of the industrial sectors within countries for the upgrading of the environmental performance and the use of the essential processes. According to Hawken (Hawken 1993), such regulatory and competitive pressures have caused the companies to face the requirement of overrating the environmental issues within their own productions and market plans. As a result, many sectors within developed countries have made some revisions in their policies and implementations (Stingson 1998). These pressures are valid for the construction industry as well. While the construction industry produces the essential infrastructure for human life, it causes serious adverse effects to the environment at the same time. Construction activities cause to different types of pollution such as land deterioration, resource depletion, waste generation, air pollution, noise pollution and ground water pollution (UNCHS 1990; Ofori 1992; Shen and Tam 2002). For example, the amount of annual solid waste arising from construction activities in China is approximately 30-40 % of the total municipal waste quantity (Zeng et al. 2003). It is known that the sum of construction demolition waste in Hong Kong in 1995 covers 65 % of the landfill areas. The amount of annual construction and demolition waste per capita in the European Union is estimated to be 0.5–1 ton (Ekanayake and Ofori 2004). This data show that the construction waste is a significant problem for the countries. According to Hendrickson and Horvarth, the sulphur dioxide (SO_2), nitric dioxide (NO_2), explosive organic compounds, toxics, dangerous wastes formed as a result of construction activities are causing air pollution (Hendrickson and Horvarth 2000). The impact of the construction activities on the environment is also important, as they are resource concentrated. It is estimated that the construction activities are responsible for approximate 40% of the world's material and energy flow (Lenssen and Roodman 1995).

In addition, most of the materials utilized in the construction sector contain irrecoverable raw materials (CIRIA 1995). These and similar problems have revealed the requirement of environmental management systems in order to minimize the impacts of the construction activities on the environment within various countries. The ISO 14000 environmental management system (EMS) presents a framework to maintain an environmental management system for the construction firms as well.

In Turkey, significant motives exist which are requiring the expansion and implementation of the ISO 14000 EMS for the construction sector as well as ISO 9000 QMS (Turk 2006 and 2009 a,b). First one is that the construction sector is an essential economic activity area in terms of the Turkish economy, as in the case for all countries all over the world. The construction is considered as one of the leading sectors of the Turkish economy because the sector depends on the local workforce to a great extent by providing large employment opportunities. In terms of production, it has been calculated that the construction sector takes 6% share of the gross national product in the Turkish economy. Furthermore this share reaches up to 1/3 with the contribution of the other 200 different sub-sectors that carry on their activities dependent to the construction sector. As this sector is directly related with the fixed capital investment, the growth of the construction sector also affects the growth rate of the country's economy. In general, the construction investments in Turkey account for approximately 60% of the realized total investments (DPT 2007). Due to the rapid population growth and urbanization rate in Turkey, the construction investments within the country's economy are considered to carry on for long years. In a country where the share of construction investments is so high, it is inevitable to think of the impacts of such investments on the environment.

The second is that there is no exact data about the direct impacts of the construction sector in Turkey on the environment. However, just like the examples in other countries, the construction waste is also a significant problem for Turkey. In order to eliminate such problem, the "bylaw on the control of excavation, construction, and demolition wastes" has been in effect since 2004. However, such problems could not be eliminated despite such a regulation. The best example is the dumping of tons of construction waste on to the Turkish coast in August 2005 which was removed from a construction site of a tourism facility which was constructed in Belek, Antalya, whose beach is one of the most important egg-laying and sheltering areas of the Caretta Caretta sea turtles. Egg-laying areas of sea turtle species have been put under strict protection by Environmental Ministry for their actual protection and for raising environmental consciousness of the public as well. Approximately, 600 baby Caretta carettas were killed before birth (Milliyet 2005). This particular example shows that the importance of the situation what if an adequate environmental awareness does not exist in the construction sector.

The third one is Turkey has accepted the principles of sustainable development, protection of the environment and the related international agreements. Under this framework, laws and various regulations have been put into force for the protection of the environment. Furthermore, Turkey's adaptation process for accession to the European Union, the intensive utilization of the ISO 14000 EMS within the construction sector by the member states of the European Union and its standardization introduces an obligation for Turkey, especially on the subject of utilization of the EMS by the construction firms (ISO 2004). With the Regulation on Evaluation and Management of Environmental Noise put into force in 2005 regarding the protection of the environment, some sanctions are applied in relation with construction site noise and the noise created by the equipment used in the site. It is evident that these kinds of legal arrangements put pressure on the construction sector. These pressures tend to grow stronger.

The fourth is the actions of Turkish construction firms in international markets. There are 5 Turkish firms among the leading 225 construction firms throughout the world (ENR 2008). The amount of international work contracted by Turkish firms in 2005 reached to USD 9.3 billion (Milliyet 2006). The ISO 14000 EMS is important for the Turkish construction firms to increase their chance to get new projects or increase their prestige in international markets. In addition, ISO 14000 EMS is important for foreign investment firms also within the construction sector to undertake work in Turkey. In particular, with the legal arrangements carried out in 2003, foreign direct investment of capital in Turkey has become easier. As a result of these legal arrangements, a significant foreign investment flow has been achieved (Berkoz and Turk, 2008). ISO 14000 EMS certificates for foreign companies planning to make investments may provide benefits for the Turkish construction sector in two respects. First, it facilitates the adoption of environmental management systems by the construction firms in the internal market; secondly, it means a guarantee for Turkey in the protection of the environment just as in other developing countries.

A detailed evaluation of ISO 1400 EMS system for Turkish construction sector is discussed here for the determination of what types of advantages and disadvantages are significant to obtain ISO 14000 certification in detail. Initially, brief information on ISO 14000 EMS is provided. Then, the theoretical background regarding the previous studies is established. After that, information on use of the ISO 14000 EMS within the construction sector in Turkey is given. The next part contains the questionnaire survey and its results developed for the purpose of analyzing the impact of the ISO 14000 EMS on Turkish construction firms. After the discussion of findings, the conclusion and suggestions are presented.

2. ENVIRONMENTAL MANAGEMENT SYSTEMS

First environmental management standard, BS 7750, is prepared in 1992. In 1993, the start of "Eco-management Audit Scheme" EMAS, prepared by the European Union, has been given start for implementation. After then, different countries have developed their own EMS (Kein et al. 1999). Later, standards have been introduced for the environmental management known as ISO 14000 EMS series developed by the International Standards Organization 1996 (ISO 2004).

With the use of environmental management standards, a new approach is adopted which made its way from a passive construction method on the issue of minimizing pollution to the active EMS (Chen et al. 2000). Firms can maintain the stability between the economical development and environmental sustainability by implementing the EMS. The firms can overcome the pressures put by the clients on the subject of environment, reduce their costs and reduce the violation of the environmental laws by the use of environmental management implementations (Ofori et al. 2000). The EMS is considered to provide an essential role for the improvement of the manufactured goods, efficient waste management, avoidance of in-house accidents and advance in a continually improving performance. The firms apply the EMS in order to carry on their activities appropriate to the environmental regulations, maintain lower environmental costs, minimize risks, train employees and improve environmental performance (Christini et al. 2004).

The ISO 14000 EMS series lead the firms to minimize all the waste materials which are dangerous for the environment by implementing the production process in a method that minimizes the waste amount and by developing waste management in a way that directs the respective management strategies related to the environment (Bolat and Gozlu 2003, Bolat 2003).

Today, all over the world, many firms are seeking ISO 14001 certification. According to statistics published by ISO, by the end of 2006, 129.199 certificated have been issued in 140 countries, an increase of 18.037 certificates since the end of 2005 when the total number of was 21.225 in 138 countries (ISO Survey, 2006). According to statistics published by ISO, the top five industrial sectors for ISO 14001 certifications are electrical and optical equipment (9423), construction (9095), basic metal and fabricated metal products (7521), chemicals, chemical products and fiber (5041), and machinery and equipment (4554), respectively. The share of construction certificates in industrial sectors is quite high. While 4660 firms were registered at the end of 2006. In this sector, the share all over the world has increased by 51.2% in one year (ISO Survey, 2006). The more and more increasing interest of construction firms to obtain the ISO 14001 certificate depends on benefits associated with it (Turk, 2009b).

3. LITERATURE REVIEW

Limited numbers of research have been carried out in different countries about EMS implementations in the construction sector. In these studies, by carrying out a questionnaire survey with the construction firms, the perspective of the construction firms on the EMS, the benefits of the ISO 14000 EMS, the problems and difficulties faced within its implementations are determined. In addition, case studies based on the implementation of ISO 14000 EMS for a single construction firms were summarized in Table 1.

In a survey carried out in Singapore (Kein, Ofori and Briffett 1999), 24 construction firm's involved-questionnaire, the most important priorities for having ISO 14000 certification are determined to obtain the client satisfaction, to minimize the costs, and to complete the work within the specified period.

According to the firms involved in the study, the environmental protection is important, but it is not the main concern for implementation. Although the necessity for environmental protection is agreed on in the study, the construction firms in Singapore were not found ready for the implementation of EMS such as ISO 14000. According to the survey results of other study carried out in Singapore (Ofori et al. 2000), the major problems for the construction firms in Singapore in the implementation of the ISO 14000 EMS were identified as lack of personnel and knowledge within the construction sector, high costs and the discrepancies due to the change of traditional implementations.

According to the findings of the same study, the reasons for which the construction firms in Singapore are not ready for the ISO 14000 EMS, were high costs in implementation, the dissatisfaction of the ISO 14000 EMS to cover the costs and the lack of client support. In the study, the reasons for obtaining ISO 14001 certificates were determined as the decrease of waste materials and prevention of noncompliant behavior against legal arrangements.

Table 1. The summary of previous studies based on the EMS within the construction industry

Field Study	Country and Targeted group	Method	No of Samples in the Study	Findings of Research
Kein et al. (1999)	Singapore, Construction firms	Questionnaire Survey	24	* Environmental protection is important for the construction firms but it does not have a priority. The construction firms in Singapore are not found to be ready for ISO 14000 although environmental awareness exists among firms.
Ofori et al. (2000)	Singapore, Clients, consultants and contractors	Questionnaire Survey	33 (contractors)	* The need of qualified personnel, lack of knowledge, high implementation costs and changes in traditional applications are identified as main problems. High costs of ISO 14000, common belief about ISO 14000 such as it does not have any advantage for construction firms, benefits of ISO 14000 does not compensate the costs of implementing it, lack of client support impeding the use of ISO 14000. The main reasons for seeking ISO 14001 are minimizing waste production, stopping illegal behavior and to stand behind the legal barriers in terms of environment.
Tse (2001)	Hong Kong, Construction firms	Questionnaire Survey		* The obstacles in the use of EMS in construction are lack of governmental pressures, lack of client support, high costs of implementation of EMS, the problems related to subcontractors in terms of EMS.
Shen and Tam (2002)	Hong Kong, Construction firms	Questionnaire Survey	72	* The main benefits of EMS are summarized as the contribution to environmental protection, minimization of environmental risk, improving the environmental image and cost savings due to reductions of environmental pollution. Increasing management costs, lack of qualified personnel, lack of subcontractor cooperation, lack of client support, time consuming feature of improving environmental performance are main obstacles for implementation of EMS in construction sector.

	Country and Targeted group	Method	No of Samples in the Study	Findings of Research
Valdez and Chini (2002)	USA, Construction firms	Literature review and case study of a ISO 14001 certified construction firm	1 (Case study on Beers Skanska)	* The positive aspects of ISO 14000 is identified such as making commitments to environmental responsibility, improvement in environmental performance, increasing employee awareness, generating benefits like monetary savings and creating new marketing opportunities.
Zeng et al.(2003)	China, Construction firms	Questionnaire Survey	60	* The major motivation for Chinese construction firms to have ISO 14001 certification is to entry to international markets. The others can be summarized as standardization of environmental management procedures, to get social recognition and client confidence, to enhance company image, to increase the environmental awareness of subcontractors and to get cleaner construction sites for better housekeeping. Main obstacles are found such as financial burden for firms, imbalance between costs and benefits, low environmental awareness and lack of governmental pressure and enforcement.
Field Study				
Chen, et al. (2004)	China, Construction firms	Questionnaire Survey	72	* A decision making model is developed for construction companies by comparing the critical factors such as governmental regulations, technology conditions, competitive pressures, cooperative attitude, cost-benefit efficiency whether to pursue ISO 14001 certification.
Christini et al. (2004)	USA, A construction firms	Case study, detailed examination and interview with firms employee	1 (Case study on Beers Skanska)	* The construction firms are realizing that the elimination or minimization of harmful environmental impacts from construction site is crucial. Although ISO 14001 does not have specific environmental performance criteria for construction firms, firms can seek the balance between costs and benefits with the implementation of EMS. Beers Skanska did not receive the ISO 14001 certification with governmental or other pressures. On the contrary, the firm is enthusiastic to fulfill the environmental commitment and get competitive advantage against competitors in the construction industry.

Table 1. (Continued)

Turk (2009a)	Turkey, Construction firms	Questionnaire Survey	68	* This study shows that there is a positive approach to the ISO 14000 EMS within the construction sector in Turkey as well as indicating that the utilization of the ISO 14000 EMS is not yet at the preferred level. *In the Turkish construction sector, the most significant reason for the certified firms obtaining ISO 14001 certificates is their desire to access the international market to get a competitive edge. *Among the difficulties encountered by the ISO 14001 certified firms in obtaining ISO 14001 certificates are found: company management is not open to research and criticisms, the registration process is too lengthy, the volume of documentation and paperwork has increased and ISO 14000 EMS has increased expenses.
Turk (2009b)	Turkey, Constructions firms	Questionnaire Survey	68	*There is not any difference in perceptions on ISO 14001 certification in terms of firm characteristics and being as certificated-non-certificated and their both positive opinions about ISO 14001 certification. There is a relation between firms characteristics and having ISO 14001 certification. ISO 14001 certification contributes to construction firms not only in terms of environmental benefits but also with corporate management and marketing effects, thus verifying that the ISO 14001 has a positive impact on Turkish construction sector.

Notes: The table is improved from the tables conducted by Turk (2009a, b).

In a study carried out in Hong Kong (Tse 2001), the obstacles in the implementation of the EMS within the construction industry were defined as the lack of government demand, the lack of client support, the high implementation costs, the existing sub-contractor system and the difficulties in the management of the system. Again, according to the results of another study conducted in Hong Kong (Shen and Tam 2002), the most important benefits of the environmental management implementation in construction were determined as the contributions to the environmental protection, minimization of environmental risk, the development of positive environmental image and cost savings due to compliance with environmental guidelines. On the other hand, existence of various obstacles in the implementation of EMS were identified and determined as increasing managerial costs, lack of qualified personnel and experts, lack of sub-contractor cooperation, lack of client support and the time consuming characteristic of the environmental performance development.

In the study conducted by Valdez and Chini, the key benefits of the implementation of ISO 14001 within US construction industry are outlined. These benefits were determined as improvement of public image, providing the compliance with the regulations and competitive benefit. Further, the examples of benefits such as process development and cost savings were discussed (Valdez and Chini 2002).

In a research carried out in China, the benefits of the ISO 14000 EMS were analyzed under five subtitles (Zeng et al. 2003). These were internal operations, corporate management, marketing effects, subcontractor relations and site cleanliness. In conclusion, the benefit of the ISO 14000 EMS was determined as the achievement of standardization for the management under the title of internal operations. Under corporate management, the benefit of the ISO 14000 EMS was determined as protection of the resources and minimization of waste. In terms of market effects, the benefits were determined as being sensitive to environment and getting the confidence of the clients. In terms of sub-contractor relations, the benefit was determined as emphasis on the importance given to environmental issues by the sub-contractors. In terms of site cleanliness, it was affirmed that the ISO 14000 EMS has provided a significant development. In the same study, the obstacles in the implementation of ISO 14000 EMS were determined as the high costs of the implementation and the increase in the paperwork.

According to other study conducted in China (Chen and Hong 2004), the critical factors in implementation and acceptance of the ISO 14000 EMS were examined under five titles. The first factor was governmental regulations whereas the second factor was the state of the technology. According to the results of the analysis, technology was found important for the implementation and acceptance of the ISO 14000 EMS and it was concluded that if the construction firms have adequate technology in order to minimize and control the adverse effects of the construction on the environment, then they prefer to accept the ISO 14000 series. The third factor was competitive demand, it was essential for acceptance and implementation of the ISO 14000 EMS for the construction firms. If there was sufficient competitive pressure, construction firms accept the ISO 14000 EMS. The fourth factor was the cooperation behavior then the construction firms are determined to accept the ISO 14000 EMS if there is satisfactory cooperation with local and foreign firms on the subject of environmental management. The fifth factor was the cost-benefit efficiency. It was determined that indefinite cost-benefit efficiency prevents the construction firms from accepting the ISO EMS series.

In a case study carried out by Christini and others, the experience of Beers Skanska in relation with the ISO 14001 was analyzed (Christini et al. 2004). According to the findings, Beers Skanska did not obtain ISO 14001 certificate because of government or client pressure. The firm obtained the ISO 14001 certificate in order to fulfill its environmental commitments and provide a competitive advantage. Furthermore, the sub-contractors of Beers Skanska were not required having their own EMS. The obligation of applying the firm's own EMS on the operation areas of the firm is introduced and the sub-contractor firms were determined to comply with such an obligation.

In a study carried out by Turk (2009a), the impact of the ISO 14000 EMS on construction firms within Turkey is examined by using the questionnaire survey method. According to the findings of the study shows that there is a positive approach to the ISO 14000 EMS within the construction sector in Turkey as well as indicating that the utilization of the ISO 14000 EMS is not yet at the preferred level. In the Turkish construction sector, the most significant reason for the certified firms obtaining ISO 14001 certificates is their desire to access the international market to get a competitive edge. Among the difficulties encountered by the ISO 14001 certified firms in obtaining ISO 14001 certificates are found: company management is not open to research and criticisms, the registration process is too lengthy, the volume of documentation and paperwork has increased and ISO 14000 EMS has increased expenses.

Again, a study carried out by Turk (2009b) investigates whether there is any dependence or relation between construction firm characteristics and having ISO 14001 certification and any difference in the perceptions related to ISO 14001 by considering both firm characteristics and two different groups as certificated and non-certified firms. There is not any difference in perceptions on ISO 14001 certification in terms of firm characteristics and being as certificated-non-certificated and their both positive opinions about ISO 14001 certification. There is a relation between firms characteristics and having ISO 14001 certification. ISO 14001 certification contributes to construction firms not only in terms of environmental benefits but also with corporate management and marketing effects, thus verifying that the ISO 14001 has a positive impact on Turkish construction sector.

4. CASE STUDY FOR TURKISH CONSTRUCTION FIRMS

4.1. Survey

A questionnaire survey was conducted in order to collect the information depending on the utilization of the ISO 14000 EMS by construction firms in Turkey. Literature review was utilized for the preparation of the questionnaire survey form. After the preparation of questionnaire, pilot surveys were carried out with 12 construction firms. According to the results of this survey, the questionnaire survey questions were modified. The forms were sent the construction firms which were the members of Turkish Contractors Association with return envelopes and 42 out of 138 construction firms returned the questionnaire survey forms (TCA 2006). Then, 68 forms were returned after making one-on-one telephone calls with the non-responding construction firms. The final rate of return of the conducted questionnaire survey was 49%. After completion of the questionnaire survey, the data obtained from the results of such questionnaire survey were processed in SPSS 9.0 software. In the analysis of the obtained data, one sample t-test and factor analyses are used.

4.2. The General Profile of the Construction Firms in the Survey

The evaluations were conducted according to the 68 responding construction firms. Among the construction firms involved in the survey, it is determined that 41.2% (28) of construction firms has ISO 14001 certificates and 57.3% (40) of them has not.

Out of 68 questionnaires returned by the firms, 67.6% (46) were answered by the quality department managers, 10.3% (7) by the general managers, 5.9% (4) by the administrative managers, 5.9% (4) by the engineering department managers and 10.3% (7) by other personnel. 95.6% (65) of the firms involved in the survey were contracting firms whereas 4.4% (3) was consultant firms. The percentage of the average amount of work undertaken by the 68 firms for the last five years was as follows: 37.8% (25) USD 0-25 million, 13.6% (9) USD 25- 50 million, 18,1% (12) USD 50-100 million, 19.7% (13) USD 100-250 million, and 10.6% (7) USD 250 million. 33.3% (22), 13.6% (9), 10.6% (7), and 9.1% (6) of the 68 surveyed firms have undertaken 0-25%, 25-50%, 50-70%, and 100% of their volume of work from international markets for the last five years respectively. 33.3% (22) of the surveyed construction firms have undertaken their entire volume of work from national markets for the last five years. The average personnel numbers of the surveyed construction firms are as follows: 53.0% (35) have over 200 personnel, 16.6% (11) have between 100-200 personnel, 12.1% (8) have between 50-100 personnel, 15.1% (10) have between 20-50 personnel, 1.5% (1) has between 10-20 personnel and 1.5% (1) has between 1-10 personnel.

When the profiles of the firms holding ISO 14001 certificates were examined, the percentages of the staff answering the questionnaire survey are as follows: 71.4% (20) were quality department managers of the firms, 10.7% (3) were general managers, 7.1% (2) were administrative managers, and 10.7% (3) were other staff.

The operation field of 85.7% (24) of the firms holding ISO 14001 certificates involved in the survey was construction services. The percentages of average amount of work undertaken by the firms holding ISO 14001 certificates for the last five years was as follows: 17.9% (5) USD 0-25 million, 10.7% (3) USD 25- 50 million, 28.6% (8) USD 50-100 million, 25.0% (7) USD 100-250 million, and 17.9% (5) over USD 250 million. 46.4% (13), 17.9% (5), 10.7% (3), and 3.6% (1) of the surveyed firms holding ISO 14001 certificates has undertaken 0-25%, 25-50%, 50-70%, and 100% of their volume of work from international markets for the last five years, respectively. 21.5% (6) of the surveyed construction firms has undertaken their entire volume of work from national markets for the last five years. The average personnel numbers of the surveyed construction firms were as follows: 64.3% (18) over 200 personnel, 14.3% (4) between 50-100 personnel, and 14.3% (4) between 20-50 personnel.

Among the firms surveyed, 71.8% (28) that do not have ISO 14001 certificates, were considering obtaining them in the near future. Among the firms holding ISO 14001 certificates surveyed, 42.9% (12), 50% (14) and 7.1 % (2) of them have been holding ISO 14001 certificates between 0-1 years, 1-3 years and 3-5 years, respectively. The utilization of the ISO 14000 EMS by construction firms in Turkey is rather new.

Table 2. The Advantages of ISO 14000 EMS

Advantages	Mean	Standard deviation	t-statistic	p-value
It improves the environmental awareness of company.	1.4697	0.7888	15.137	0.000
It improves the standardization in environmental management	1.2879	0.8729	11.986	0.000
It decreases the possible environmental impacts	1.2576	0.8649	11.813	0.000
It provides sustainable development in environment	1.2424	0.8604	11.731	0.000
It enhances the company's image	1.1364	0.8751	10.550	0.000
It decreases the complaints against the company due to the environmental problems	0.9848	1.0883	7.352	0.000
It increases the self-confidence of the company	0.8939	0.9943	7.304	0.000
It improves client satisfaction	0.8636	1.0653	6.586	0.000
It improves the communication with subcontractors	0.5152	1.0988	3.809	0.000
It increases the social recognition of the company	0.5303	1.1925	3.613	0.001*

Notes: (N: 66), * Do not reject null hypothesis at 95 % level of significance.

Table 3. The Disadvantages Related to ISO 14000 EMS

Disadvantages	Mean	Standard deviation	t-statistic	p-value
High initial and operating costs of ISO 14001 certification	0.5645	1.1959	3.717	0.000
Lack of qualified personnel about ISO 14000 EMS	0.4194	1.1095	2.979	0.004
Lack of knowledge about ISO 14000 EMS and certification	0.1774	1.2351	1.131	0.262*
Lack of knowledge of clients about ISO 14000 EMS and certification	-0.1290	1.1802	-0.861	0.393*
Conflict of purposes of ISO 14001 certificate and the firm	-0.5806	1.1242	-4.067	0.000
No apparent and practical benefits of ISO 14000 EMS exist	-0.5484	1.0029	-4.305	0.000
Some parts of ISO 9000 QMS and ISO 14000 EMS overlap	-0.6333	1.0571	-4.641	0.000

Notes: (N: 66) * Do not reject null hypothesis at 95 % level of significance.

4.3. Methodology of the Research

When the survey had been completed, the data obtained from these questionnaires were transferred into the SPSS software. One sample t test and factor analyses were used with the obtained data from questionnaires of advantages and disadvantages related to ISO 14000 EMS. In this study, factor analysis was used to summarize many variables with a few factors. Each of the factors acquired from these analyses presents advantages and disadvantages of ISO 14000 EMS in detail. (Tables 2 and 3).

4.4. Empirical Results

The evaluations of the firms on the subject of the benefits obtaining ISO 14001 certificates are analyzed using t-tests. According to the results of the analysis, the most essential benefit of obtaining ISO 14001 is the achievement of environmental awareness for

Table 4. Factor Analysis Results of the Advantages of ISO 14000 EMS

Factors	Factor loading	Eigen-values	Percentage of variance	KMO	Barlett Test
				0.674	96.440***
Dimension 1: Related to the environment		5.824	58.240		
It improves the environmental awareness of company.	0.893				
It improves the standardization in environmental management	0.850				
It decreases the possible environmental impacts	0.845				
It provides sustainable development in environment	0.773				
It enhances the company's image	0.670				
It decreases the complaints against the company due to the environmental problems	0.614				
Dimension 2: Related to the company		1.202	12.023		
It increases the social recognition of the company	0.79				
It improves the communication with subcontractors	0.772				
It increases the self-confidence of the company	0.682				
It improves client satisfaction	0.636				

Notes: *** Significant at the 0.1 level.

Table 5. Factor Analysis Results of the Disadvantages of ISO 14000 EMS

Factors	Factor loading	Eigen-values	Percentage of variance	KMO	Barlett Test
				0.645	64.123***
Dimension 1: Related to lack of knowledge and personnel		2.062	29.462		
Lack of knowledge of clients about ISO 14000 EMS and certification	0.805				
Lack of knowledge about ISO 14000 EMS and certification	0.715				
Lack of qualified personnel about ISO 14000 EMS	0.707				
Dimension 2: Related to the cost and implementation		1.506	21.520		
High initial and operating costs of ISO 14000 certification	0.812				
Some parts of ISO 9000 QMS and ISO 14000 EMS overlap	0.642				
Conflict of purposes of ISO 14001 certificate and the firm	0.623				
Dimension 3: Related to no apparent benefits		1.149	16.419		
No apparent and practical benefits of ISO 14000 EMS exist	0.879				

Notes: *** Significant at the 0.1 level.

the firm. Other important benefits of obtaining ISO 14001 EMS according to the surveyed firms are: providing standardization in environmental management, minimizing the adverse impacts on environment and its support to establish a system for sustainable environmental development (Table 2)

The results of factor analysis about the advantages of ISO 14000 EMS are statistically significant (KMO 0.674, Barlett Test: 96.440, Sig. 0.00). According to the results, mainly two factor dimensions have arisen. First dimension is entitled as "related to environment" and it explains 58.24 % of total variance. Second one is called as dimension "related to company". Second dimension explains 12.023 % of total variance (Table 4). According to construction firms the dimension related to environment to obtain ISO 14001 certification is the more important than the dimension related to company. This demonstrates that the construction firms are quite conscious in Turkey.

The evaluations on disadvantages of obtaining ISO 14001 certificates are analyzed by using t-tests in Table 3. According to the results of the analysis, the greatest disadvantage of obtaining ISO 14000 for the firms is 'high initial and operation costs of ISO 14001 certificates'. Other disadvantages include 'lack of qualified personnel' and 'lack of sufficient information regarding the certificates'. 'No apparent benefits of the ISO 14000 certificates' and 'conflict of purposes of ISO 14000 certificate and the firm' are not seen as disadvantages by the surveyed firms (Table 3).

Likewise, the results of factor analysis about the disadvantages of ISO 14000 EMS are statistically significant as well. (KMO 0.645, Barlett Test:64.123, Sig. 0.00). According to the results of factor analysis, mainly three factor dimensions are obtained. First one is named as "related to lack of the knowledge and personnel" and it explains 29.462 % of total variance. Second one is called as "related to the cost and implementation". Second dimension explains 21.520 % of total variance. Third dimension is entitled as "related to no apparent the benefits" and it explains 16.419 % of total variance (Table 5). According to construction firms the dimension related to lack of the knowledge and personnel to obtain ISO 14001 certification is the more important than the dimension related to cost and implementation. Especially, lack of knowledge of clients is important disadvantage for construction firms. According to Turkish construction firms, if clients pay attention to ISO 14001 certification, the construction firms prefer to obtain ISO 14001.

CONCLUSION

Turkish contractors in general have very positive perceptions regarding ISO 14001 standards. The firms believe that implementation of ISO 14001 standards not only have a positive impact on environmental protection but are also crucial for being competitive in the local and international markets (Turk, 2009 a; Turk, 2009b). The firms without any exception indicated that the advantages of the standards were more than the disadvantages.

The most significant advantages of obtaining ISO 14001 certificates for Turkish construction firms are determined as: the improvement of environmental awareness of the firm, improvement of standardization in environmental management, minimizing the adverse impacts on environment and its assistance to establish a system for sustainable environmental development. These results have similarities with the findings of Shen and Tam (2002) and Valdez and Chini (2002).

The most significant disadvantages of obtaining ISO 14001 certificates are determined as; the high initial and implementation costs of certification, lack of qualified personnel and lack of information about ISO 14000 EMS. These results have similarities with the findings of Shen and Tam (2002). According to the findings of Tse (2001) and Ofori (2000) high costs of implementation, the need of qualified personnel, and lack of knowledge are defined as disadvantages. Turkish case supports these findings.

According to the results of factor analyses, factors that affecting both advantages and disadvantages of ISO 14000 are examined in detail. Mainly two factor dimensions are effective in determining of the advantages of ISO 14000 EMS. First dimension is entitled as "related to environment" and second one is called as dimension "related to company". First dimension related to environment to obtain ISO 14001 certification is the more important than the second dimension related to company. The construction firms see ISO 14001 certification as a tool for improvement of environmental management more than as a tool for commercial development of the company. This can be interpreted that the construction firms are quite conscious in Turkey.

Mainly three factor dimensions that affecting of the disadvantages of ISO 14000 EMS are found in detail. First one is the dimension "related to lack of the knowledge and personnel". Second one is the dimension "related to the cost and implementation" and third dimension is entitled as "related to no apparent benefits". According to construction firms the first and second dimensions are the more important than third dimension. That is, an increase in the number of ISO 14000 certification is closely related to the removal of lack of knowledge and personnel and the reduction in costs and removal of implementation difficulties.

The number of ISO 14001 certificates held by construction firms in Turkey is observed to be low when compared to other countries, particularly European and Asian countries. An increase in this number can provide a reduction of potential impacts from construction investments to the environment as an important activity area in Turkey and find solutions for construction wastes, facilitate compliance with legal regulations regarding the environment, provide a competitive edge for Turkish firms operating internationally and serve as a guarantee for the protection of the environment.

This study shows that there is a positive approach to the ISO 14000 EMS within the construction sector in Turkey as well as indicating that the use of the ISO 14000 EMS is not yet at the preferred level. In particular, the problems of lack of information and qualified staff revealed in the results of the analysis should be overcome. Staff should be qualified in the concept of EMS and on the technical details. There are important roles for government authorities, trade associations, companies and certification bodies in order to overcome the lack of information in this field.

REFERENCES

Berkoz, L.; Turk, S.S. Determination of location-specific factors at the intrametropolitan level: Istanbul case. Tijdschrift voor Economische en Sociale Geografie TESG (Journal of Economic and Social Geography), 2008, Vol. 99 (1), 94–114.

Bolat HB. The evaluation of the applications of ISO 9000 and ISO 14000: Turkey case. Ph.D. Thesis; Istanbul Technical University, 2003.

Bolat, HB.; Gozlu, S. ISO 14000 Effective factors for application of environmental management system, itudergisi/d, 2003, Vol. 2, 2, p. 39-48 (in Turkish).

Chen, Z.; Li, H.; Hong, J. An integrative methodology for environmental management in construction. *Automation in Construction*, 2004, Vol. 13, 621-628.

Chen, Z.; Li, H.; Wong, C.T.C. Environmental management of urban construction projects in China. *Journal of Construction Engineering and Management*, 2000, Vol. 126, No.4, 320-324.

Christini, G.; Fetsko, M.; Hendrickson, C. Environmental management systems and ISO 14001 certification for construction firms. *Journal of Construction Engineering and Management,* 2004, Vol. 130, No.3, 330-336.

Christini, G.; Fetsko, M.; Hendrickson, C. Environmental Management Systems and ISO 14001 Certification for Construction Firms." *ASCE J. Construction Eng. And Mgmt*, 2004, Vol. 130(3): 330-336.

CIRIA, A client's guide to greener construction: a guide to help clients address the environmental issues to be faced on building and civil engineering projects, Publication code: PR74, London, 1995.

DPT, Report of Special Commission of Construction, Contracting, Engineering and Consultancy Services, *DPT-Report of 9th Development Plan, State Planning Agency*, Ankara, 2007 (in Turkish).

Ekanayake LL.; Ofori G. Building waste assessment score: design-based tool. *Building and Environment,* 2004, Vol. 39, 851-861.

Engineering News Record, The Top 225 International Contractors, www.enr.com, The McGraw-Hill Companies, 2008.

Hawken, P.; *The ecology of commerce*; Harper Business: New York, NY, 1993.

Hendrickson, C.T.; Horvath, A. Resource use and environmental emissions of U.S construction sectors, *Journal of Construction Engineering and Management*, 2000, Vol. 126 (1), 38-44.

Kein, A.T.T.; Ofori, G.; Briffett, C. ISO 14000: Its relevance to the construction industry of Singapore and its potential as the next industry milestone, *Construction Management and Economics,* 1999, Vol. 17, 449-461.

Kein, A.T.T.; Ofori, G.; Briffett, C., ISO 14000: Its relevance to the construction industry of Singapore and its potential as the next industry milestone. *Construction Management and Economics,* 1999, Vol. 17, 449-461.

Lenssen, N.; Roodman, D.M. *Making better buildings in State of the World 1995*, World Watch Institute (ed.) Norton, New York, 1995, pp 95-112.

Milliyet, *Daily Newspaper*, 13 August 2005 (in Turkish)

Milliyet, *Daily Newspaper*, 22 February 2006 (in Turkish)

Ofori G.; Briffett, C.; Gang, G.; Ranasinghe, M. Impact of ISO 14000 on construction enterprises in Singapore. *Construction Management and Economics*, 2000, Vol. 8, 935-947.

Ofori, G.; The environment: the fourth construction objective. *Construction Management and Economics,* 1992, Vol. 10 (5), 369-395.

Shen LY.; Tam VWY.; Implementation of environmental management in the Hong Kong construction industry. *International Journal of Project Management*, 2002, Vol. 20, 535-543.

SPSS for Windows, *Statistical Package for the Social Sciences*, Release 9.0.0.

Stingson, B.; Sustainability in an era of globalization: the business response. In OECD (Ed.) *Globalization and Environment: Perspectives from OECD and Dynamic Non- member countries,* OECD, Paris, 1998, pp 54-64.

TCA, Turkish Constractors Association, www.tmb.org.tr, 2006.

The ISO Survey, The ISO survey of ISO 9001:2000 and ISO 14001 certificates, International Organisation for Standardization, ISO International Organization for Standardization CD-Rom, Geneva, 2006.

Tse, Y.C.R. The implementation of EMS in construction firms: case study in Hong Kong. *Journal of Environmental Assessment Policy and Management*, 2001, Vol. 3 (2), 177–194.

Turk, A.M. ISO 14000 environmental management system in construction: An examination of its application in Turkey, *Total Quality Management and Business Excellence*, 2009, Vol. 20, No. 7, 713–733.

Turk, A.M. ISO 9000 in construction: An examination of its application in Turkey, *Building and Environment*, 2006, Vol. 41, Issue 4, 501-511.

Turk, A.M. The benefits associated with ISO 14001 certification for construction firms: Turkish case, *Journal of Cleaner Production*, 2009, Vol. 17, Issue: 5, 559-569.

UNCHS (United Nations Centre for Human Settlements); People, Settlements and Sustainable Developments, intergovernmental meeting on human settlements and sustainable development, Nairobi, 1990.

Valdez, H.E.; Chini, A.R. ISO 14000 standards and the US construction industry, *Environmental Practice*, 2002, Vol. 4 (4), 210-219.

Zeng SX.; Tam CM.; Deng ZM.; Tam VWY. ISO 14000 and the construction industry: survey in China. *Journal of Management in Engineering*, 2003, Vol.19, No.3, 107-115.

In: Environmental Management
Editor: Henry C. Dupont

ISBN: 978-1-61324-733-4
© 2012 Nova Science Publishers, Inc.

Chapter 8

USERS' PREFERENCES AND CHOICES IN ARGENTINEAN BEACHES

*A. Faggi[1], N. Madanes[2], M. Rodriguez[3], J. Solanas[3], A. Saenz[3] and I. Espejel[4],**

[1]CONICET-MACN. A. Gallardo, Buenos Aires, Argentina
[2]FCEyN, Buenos Aires, Argentina
[3]Universidad de Flores, Buenos Aires, Argentina
[4]Facultad de Ciencias, Universidad de Baja California, Ensenada, Mexico

ABSTRACT

This study analysed the profile and perception – composed of opinions and attitudes – of beach users in Argentina, using data from nine sandy beaches, each with unique environmental and socioeconomic features, located in two coastal municipalities. We distributed 329 surveys composed of 42 questions to Argentinean residents and tourists visiting the beach. Data on the profile were analysed by cluster analysis, and data on perception were explored by Principal Components Analysis. These results allowed grouping all beaches according to variables such as cleanliness, accommodation, infrastructure and services. Users' age and marital status were found to be associated only with certain beaches; married people visited urban and rural beaches, preferring those without infrastructure, while single and young people chose urban beaches with facilities. Contrasting answers regarding environmental beach features were recognised between both municipalities, indicating the success of awareness programs that enhance the beaches' natural values.

Keywords: beach users profile, opinion, attitude, beach, environmental education.

---

* Corresponding Author: Tel.: +52 646 174 5925; fax: +52 646 174 4560. E-mail: ileana.espejel@uabc.edu.mx (I. Espejel). noram@ege.fcen.uba.ar (N Madanes).

INTRODUCTION

Sandy beaches are among the most valued environmental managed ecosystems used for popular recreation, as they offer various activities such as leisure, relaxation, landscape and wildlife observation and sports associated with water and sand. Beaches are multifunctional systems that involve natural, socioeconomic and administrative components (James, 2000; Valdemoro and Jimenez, 2006; Ariza 2008) and are interesting study sites, where environmental information may coincide with human and economic aspects of society (Van der Weide, 1993; James, 2000; Dadon, 2002). Regarding beach management, some authors have established the relevance of considering not only the physical features of the beach but also social factors such as visitors' behaviour, preferences and choices (Morgan, 1999; Enriquez Hernandez 2007; Cervantes et al., 2008; Espejel at al., 2008; Cervantes and Espejel, 2008; Roca et al. 2009).

Beaches are perceived as natural recreational spaces with structural, semantic and connotative values. They vary structurally in terms of their environmental signatures, which influence the quality of the beach and include width, length, gentle slope, the fineness and colour of the sand, the presence or absence of stones, the cleanliness of the water and sand, fauna and vegetation cover. Semantically, they provide insight into the needs and values of their users, as beach users respond to their feelings and preferences by choosing to attend specific beaches, as described by Ritterfeld and Cupchik, (1996) for interior spaces. Users' opinions and attitudes can mirror these connotative values.

Since 1950, coasts worldwide have experienced remarkable changes due to the accelerated spread of beach resorts devoted to sun-and-sand tourism. Massive tourism associated with urban centres has generated a profitable market that demands infrastructure, hotels and beachside residences, commerce, road networks and complementary attractions (Lencek and Bosker, 1998). The current appeal of beaches to society has been associated with the public's conception of an "ideal beach", which, according to Williams et al. (1993) and Micallef and Williams (2003), is expected to be sandy, clean and not too deep and to have water with a pleasant temperature. An "ideal beach" must also have basic infrastructure, a small commercial area, easy accessibility, restrooms, security, lifeguards and shade (Cervantes and Espejel, 2008).

During the last century, massive tourism determined by fast and short-term profit has led to adverse impacts on natural resources due to man-made erosion and the contamination of beach resorts (Dadon, 2002). New threats have emerged regarding the sea level's increase associated with climate change and the continuous spread of urbanisation, forcing a fundamental reassessment of beach management. Additionally, in the last decades of the 20th century, stronger consumer demands, new habits, and a new environmental outlook, together with changes in social structures, prompted a new manner of spending holidays (Aguiló et al., 2005). The "new tourist" looks for novel experiences of short duration, in attractive places and featuring sustainable landscapes. This new paradigm constitutes a challenge regarding the design of beach management plans that seek to construct a balance between the infrastructures offered at the beach and the conservation of the natural as well as cultural environment.

Perception studies have been conducted in coastal tourism sites such as coral reefs (Uyarra et al., 2009) as well as marine protected areas (Petrosillo et al., 2007) and have sought to investigate users' perceptions of beach quality (Ballance et al., 2000; Tudor and Williams, 2003 ; Priskin, 2003 ; Micallef and Williams 2003).

It has been argued that strategies planned to manage natural areas of importance for tourism and recreation should consider users' preferences for specific natural features (Uyarra et al., 2009) and include programs of environmental awareness (Debrot and Nagelkerken, 2000).

Therefore, to understand beach users in a country with multicultural roots such as Argentina, we compared the nine most popular beaches in two coastal municipalities.

We had two main inquiries:

1. Which features of beaches are influencing beach users' preferences and choices?
2. Which features are the most important values used to select a beach: nature-based and scenic values, or the beach's infrastructure and services?

METHODS

Study Area

Sun-and-sand tourism is a principal activity on Argentina's Atlantic coast. Tourism began in the 30s and, with the economic expansion of the middle classes, became massive in the 50s. Initially, many families, having second residences on the coast, would spend their summer holidays there from December to February. This method of spending holidays, paired with European tradition stemming from immigration from Europe, which began during the end of the 19th century led, to a clear predominance of Italians and Spaniards along the coast. As in Europe, the "beach" was considered a social environment in which residents would take long walks and visit small coffee shops or restaurants by the sea or with an ocean view (Lencek and Bosker, 1998).

The study was performed in nine public sandy beaches with similar physiographic characteristics but with differences in some environmental and socioeconomic features. They are very popular beaches located in two municipalities: Necochea (four beaches) and Puerto Madryn (five beaches) (Figure 1). All have scenic views, and, of the nine selected beaches, only one, "Doradillo", is rural, located 17 km from Puerto Madryn (Table 1). The other eight are urban beaches. Necochea (38° 44 ' S - 58° 44 ' W; 79,983 inhabitants) is located in the province of Buenos Aires, 800 km south of the Buenos Aires metropolitan area. Necochea is a favoured holiday spot among the metropolitan's residents. In 2007, local authorities launched an integral coastal plan including activities to raise environmental awareness about beach resources among local residents and tourists http://www.necochea.gov.ar/gestion.

Puerto Madryn (42° 47 ' S - 65° 2 ' W; 57,800 inhabitants), located on the Atlantic coast of the province of Chubut, is placed 1400 km south of Buenos Aires. Puerto Madryn attracts international tourism for marine fauna observation and is on the route to mountain resorts in the Andes. Puerto Madryn launched an environmental education program during the 2004-2005 seasons.

Table 1. Characteristics of the nine studied beaches of two Argentinean municipalities (Necochea and Puerto Madryn)

Provinces	Buenos Aires					Chubut			
Municipalities	Necochea					Puerto Madryn			
Shoreline length	8.5 km					8 km			
Beach name	Quequén	Centro	Escollera	Lillo	Rancho	Mimosa	Muelle	Indio	Doradillo
Abreviation	QU	CE	ES	LI	RA	MI	MU	IN	DO
Beach type	URBAN	URBAN	URBAN	URBAN	URBAN	URBAN	URBAN	URBAN	RURAL
Nr. interviews	32	49	23	25	40	40	40	40	40
Infrastructure	YES	YES	NO	NO	YES	YES	NO	YES	NO
Beach resorts	4	7	0	0	3	4	0	0	0
Lifeguards	YES	YES	YES	YES	YES	YES	YES	YES	NO
Permission to enter domestic animals	NO	NO	YES	YES	NO	NO	NO	NO	NO
Permission to drive off-shore vehicles	NO	NO	YES	YES	NO	NO	NO	NO	NO
Booklets	YES	YES	YES	YES	NO	NO	NO	NO	NO
Guided tours	YES	YES	YES	YES	NO	NO	NO	YES	NO
Cleaning campaigns	YES	YES	YES	YES	YES	YES	YES	YES	YES

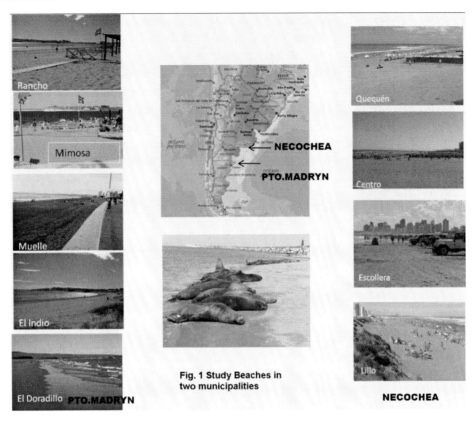

Figure 1. Nine beaches of two Argentinean municipalities (Necochea and Puerto Madryn).

Unlike the holistic awareness campaign of Necochea, Puerto Madryn placed emphasis on reducing garbage. Action was also taken to ban dogs and vehicle circulation along the beach. http://www.madryn.gov.ar/areas/ecologia/index.php.

The presently studied beaches are dissipative and wide as well as have a gentle slope. The beaches are fringed by dunes 6 to 12 m in height and are partially covered by scrub vegetation. This natural vegetation decreases towards the centres. Madanes et al. (2010) made an assessment of these beaches considering physical, biological, environmental features as well as the existence of infrastructure and services, concluding that all of them are acceptable for visiting, with the exception of Centro Beach, which was found to be recommendable. In Puerto Madryn, native and exotic seaweeds notoriously increased over the last decade (Piriz et al., 2003) producing unpleasant odours in the upper beach.

Survey

Survey based on visitors' interviews was selected as the optimum method to describe and compare the profiles of users of the nine beaches and their preferences and choices among beaches. Additionally, the questionnaire contained questions to identify the most important values used to select a beach: nature and scenic values or beaches' infrastructure and services.

A total of 329 surveys were distributed to users of the nine beaches during the summer holidays of 2008. Isovariance curves were used to determine the optimal number of surveys (Cochran and Cox, 1965).

The survey included 40 questions. Overall, six questions collected personal and background data (Figure 2), and 32 questions sampled users' perceptions, that is, their attitude and opinions regarding geomorphological and physical features, environmental quality and issues related to the beaches' landscape, comfort, services and facilities (Table 3 a and b in Appendix). Other issues included a comparison with precedent features of the same beach and users' suggestions for improving beach conditions. The questions were either a) *fixed* (yes or no answers, or a choice among fixed options), b) *open ended* (user expressed his/her opinion) or c) *mixed* (many options were offered and the user explained the reasoning informing his/her choice).

Analyses

Frequencies of respondents' data were calculated in percentages (Figure 2 and Table 3 a and b in Appendix). To avoid redundancy in *fixed* questions, only affirmative answers were considered.

In this study, cluster analysis and principal component analysis were used to explore the data sets. Multivariate analyses were carried out using the STATISTICA software 6.1.

A cluster analysis (distance measure: 1-Pearson, agglomerative linkage: unweighted pair group average) was used to elucidate the potential of user profiles to differentiate beaches. Cluster analysis was performed using a matrix of 120 rows (answers) and nine columns (beaches).

A principal components analysis was used to analyse which of the users' opinions and attitudes could be used to segregate beaches. A data matrix was built for the 9 beaches with 65 answers to 16 questions regarding attitude and 17 regarding opinion (9 rows= beaches x 65 columns = answers) considering only the positive answers for the fixed questions and all of the answers for the rest of the questions (Table 3 a and b; answers used for this analysis are in grey).

Opinion and attitude variables with factor loadings higher than ± 0.6 were analysed and were classified for discussion in categories such as "Beach quality", "Motivations and choices", "Infrastructure and services" and "Knowledge and sense of place" (Table 2).

RESULTS

Beaches and Users' Profile

To answer which features of the profile are influencing beach users' preferences and choices, description of the nine beach users is provided in Figure 2 (profile of 329 interviewees).

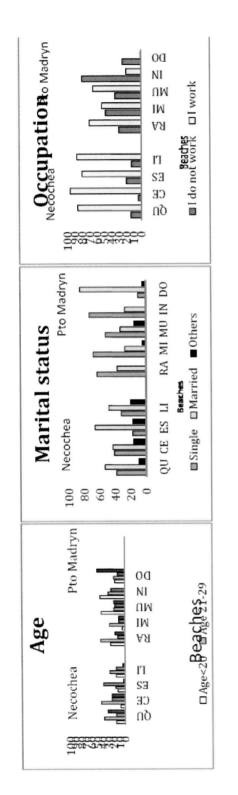
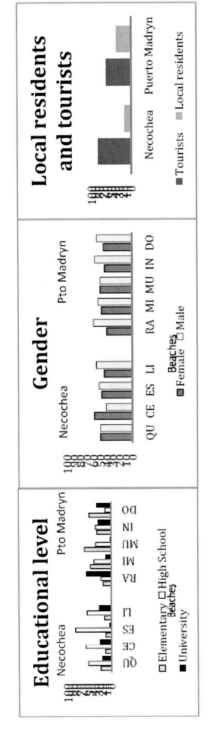

Figure 2. Users's profile in percentage of the nine studied beaches of two Argentinean municipalities (Necochea and Puerto Madryn).

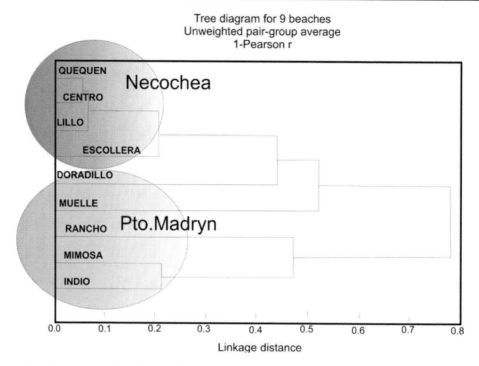

Figure 3. Dendrogram resulting from a cluster analysis of profiles of nine Argentinean beachgoers interviewed.

Men and women in equal proportion visited beaches. Many of them accompanied by their families, most of who went to the beach for recreation as well as relaxation and some of who did so to play sports. Most of the interviewees are young adults (21-29 and 30- 45 years old). Visitors of the nine beaches are mainly Argentinean tourists, followed by local residents, although the beaches in Puerto Madryn are more frequented by local residents than the beaches in Necochea (Figure 3). In the former, most of the respondents had a high school-level education; in Puerto Madryn, the education level was more evenly distributed, although most users have an elementary school education (Figure 2).

Cluster analysis (Figure 3) based on users' profile enabled a distinction to be made between the two main groups of beaches. The first group comprised all Necochea beaches (Quequén, Lillo, Centro and Escollera) together with two beaches in Puerto Madryn: the rural beach (Doradillo) and the central beach (Muelle). This group was mostly visited by employees, most of who have a high school-level education: most were married, older than 30 years and accompanied by their children. The second group, comprised three beaches of Puerto Madryn (Rancho, Mimosa and Indio), was mostly visited by unmarried, young and adolescent men with an elementary school education.

BEACHES AND USERS' PERCEPTION

The Principal Components Analysis (Figure 4) resulted in two components accounting for 53.58 % of the total variance. Axis 1 explained 35.73%, and axis 2 explained 17.81 % of the variance. We observed four groups.

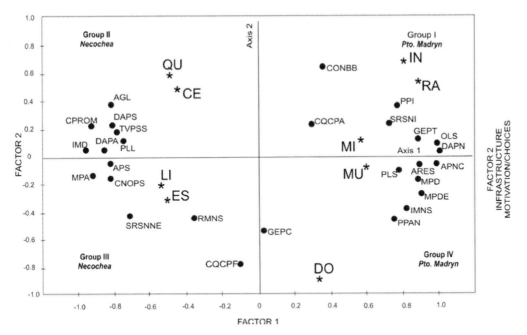

Figure 4. Ordination of beach user's perception in nine Argentinean beaches using principal components analysis. Opinion and attitude variables with factor loadings higher than ±0.6 classed in beach quality, infrastructure and services, motivation and choices, knowledge and sense of place. See data base in Table 3 a and b in Appendix. Group I (INdio, RAncho and MImosa beaches); CONBB=good beach services, PPI=beach lot infrastructure, CQCPA=interviewees go to the beach with friends, SRSNI= inadequate public services, GEPT=would like tranquility, OLS=interviewees smell odors, DAPN=dislike wind and stones. Group II (QUequen and CEntro beaches); AGL= water is clean; CPROM=beach condition better than others, DAPS=dislike crowding, music, offshore vehicle, dogs, sports practice, garbage, TVSS=visit in Easter holidays, DAPA=dislike nothing, IMD=dunes are important for sports and PLL=beach is clean. Group III (LIllo and EScollera beaches); APS=interviewees like music, offshore vehicle, dogs, sports, MPA=go to the beach because they like it, CNOPS=interviewees know other beaches, RMNS=yes, interviewees care about the environment, SRSNNE=there are no public services, CQCPF=interviewees usually go to the beach with family. Group IV (MUelle and DOradillo beaches); PLS=beaches are dirty, ARES=interviewees like the sand of the beach, APNC=interviewees do not know what they like, MPD=interviewees go to the beach for fun, MPDE=interviewees go to the beach to practice sports, IMNS=interviewees do not know the importance of sand dunes, PPAN=beaches loose sand with wind and storms.

The first group (I) is situated to the right of Figure 4, towards the positive values of Axis 2, and is made up of the beaches Indio (IN), Rancho (RA) and Mimosa (MI). This group was associated with some variables addressing what users like or dislike. For example they disliked the features of the environment and bad odours (DN and DO in Table 1) and considered that the public services were inadequate (IS). Additionally, visitors were satisfied with the services and facilities of the beach resorts (PO) and found that the beach had more infrastructure than before (MI). These beaches gathered people who went with friends (FC); some of them wished for more tranquillity (WT).

The second group (II) comprised two beaches (Quequén and Centro) and were defined by the cleanliness of water and of the beach, (WC, BC). Respondents were satisfied with the condition of the beach (BB and ND); they disliked the presence of off-shore traffic, dogs, overcrowding, and the playing of sports, music, the lack of security and the presence of ambulant salesmen (DS). Some visitors came during Easter (EC), and thought that sand dunes were important for the practice of sports (SP)

The third group (III) of the Principal Components Analysis was built by two beaches of the Necochea Municipality: Lillo and Escollera. In this group respondents liked everything at the beach, while others preferred the social life of the beach (BL, SL). Some users disliked the comfort at the beach and the lack of public services (MC, NP). Respondents came with their families (FY), knew other beaches in the region (VK) and thought that people cared about the environment (EN).

The last group (IV) comprises the beaches Doradillo and Muelle. In this case, the respondents enjoyed the sand (LS), disliked the presence of litter and seaweeds (DL) and found the beach to be entertaining (BE). They expressed that beaches normally lose sand in a natural way because of winds and storms (WS); others did not know what the beach has lost (NK) and did not perceive the importance of the dunes (DU).

DISCUSSION

Results allowed us to delineate a descriptive model defined by users' perceptions of the beach quality. The model of the nine beaches' users' perceptions (Figure 4) showed two main principal components: first, "beach quality and motivation and choices" were associated with Axis 1 of the Principal Component Analysis, and "infrastructure and services" together with "company" were associated with Axis 2. In some cases, respondents were satisfied by beaches' characteristics in terms of width, waves for surfing, cleanliness and social life. In some others, they disagreed with the sanitary conditions and disliked off-shore traffic, dogs, overcrowding, people playing sports, music, the lack of security and the presence of ambulant salesmen (Table 3b, DAPS).

Results indicated the degree of the component "water and beach dirtiness" (Table 3a PLS and AGL) as a primary reason for dissatisfaction. As argued by Defeo et al. (2009), pollution is a subject of great sensitivity in areas devoted to recreation and tourism. Many authors studying beach users' perceptions in different countries, such as Eiser et al. (1993); Ballance et al. (2000); Tudor and Williams (2003); Marin et al. (2009), found that the cleanliness of the sea and the beach were considered to be far more important than other features. In our case, these variables related to beach quality enabled differentiating beaches at a regional level. Puerto Madryn beaches were segregated from Necochea beaches through the decline in aesthetic quality and the presence of bad odours (decomposition of accumulated seaweeds on the sand) (DO quotation in Table 2).

The majority of Puerto Madryn's users (68% mention DL commentary in Table 2 especially in Mimosa beach) – predominantly residents were dissatisfied with the beach's state of "cleanliness". Users' olfactory perceptions associated dirtiness with an aesthetic visual appreciation, a trend also discussed by Tudor and Williams (2003).

Table 2. Quotations from interviewees representing opinions and attitudes variables with factor loadings higher than ±0.6, related to Figure 4 axes

BEACH QUALITY

Opinion: DN: "I dislike things from the natural environment", DO: "I dislike the presence of bad odors (seaweeds)", DS: "I dislike off shore traffic, dogs, overcrowding, the practice of sports, music, insecurity and the presence of ambulant salesmen", ND: "nothing dislike to me in this beach", WC: "The water is clean", BC: "The beach is clean", DL: "I dislike the presence of litter and seaweeds".

Attitude: BL: "I like this beach very much", BB: "comparatively to other beaches the conditions of this beach are better": "because it is wider, has good waves for surfing, has a familiar atmosphere and vehicular traffic is allowed.

INFRAESTRUCTURE/SERVICES

Opinion: PO: "I am pleased with the offer of services and facilities, i.e. showers, restrooms", IS: "public services are inadequate", MI: "The beach has more infrastructure than before", NP: "has not public services", MC: "I would like more comfort at the beach".

MOTIVATION AND CHOICES

Opinion: LS: "I like the sand of the beach", SL: "I like the social life of the beach", WT: "I wish more tranquility", SP: "Dunes are important for sports".

Attitude: ES: "I enjoy the practice of sports", BE: "I find the beach entertaining", EC: "I come during Eastern", FC: "I come with friends", FY: "I come with my family".

KNOWLEDGE AND SENSE OF PLACE

Opinion: DU: "Don't know the significance dunes have", NK: "Don't know what the beach has lost", WS: "winds and storms are responsible for losing sand", EN: "people care about the environment".

Attitude: VK: "Visitors knew other beaches in the region".

When considering for Puerto Madryn the mean value of the variable that compares the quality of the elected beach with all beaches (Table 3 a, Answer 16), we saw that, surprisingly, 40% of the respondents thought the presence of seaweeds as a source of dirtiness to be condition common to all beaches. This emergent perception in the beaches of Puerto Madryn can be explained, according to Roca et al. (2008), because residents tend to perceive their local beaches as being less polluted than they really are as an emotional way to protect their place identity. Residents in Puerto Madryn represented 37% of the visitors coming to the beach (Figure 2). In contrast, respondents seemed to perceive Necochea beaches near the standards of an "ideal beach" according to the definition given by Williams et al. (1993) and

Micallef and Williams (2003). A higher satisfaction of Necochea respondents, who were predominantly tourists, could be explained because users were on holiday, the moment of the year when people have a more positive disposition (Roca et al. 2008).

For the most part, people chose to visit a beach because of the cleanliness of the sand and water and as a social stage for enjoying and meeting friends and playing sports in beaches with the appropriate infrastructure and services (Figure 4). However, surveyed beach users did not consider the beautiful and calm of the pristine beaches to be the main drivers of their choices. These findings were in accordance with results given by Marin (2009) in Italy and mirrored the Italian and Spanish heritage within Argentina.

The study also found that the existence of infrastructure and services, especially restrooms (PO commentary in Table 2), were highly valued components and were reasons for visitors' beach choices. These results, which enabled segregating beaches at the resort level, were coincident with findings from other regions (De Ruyck et al., 1995; Ariza et al., 2008; Cervantes et al., 2008; Espejel et al., 2008; Roca et al., 2009).

Users' opinions and attitudes from the surveyed beaches confirmed findings from other parts of the world that sun-and-sand recreation plays a key role, especially for tourists and residents seeking relaxation, amusement and the common coast attractions, including services. Among the studied beaches, we recognised a general trend described by Roca et al. (2008) for Spain, that "beach going" is a social activity that involves families and friends who visit beaches to enjoy themselves with passive activities but also to play sports, especially during the summer months and holidays (Eastern).

The survey data suggested that infrastructure and services were associated with the users' characteristics, which include age, social status and family characteristics (Fig. 2), a conclusion also demonstrated by De Ruyck et al. (1995) and Chapman (1989) for South African and Australian beaches, respectively. Beaches without facilities located far from the centres were preferred by parents accompanied by their children who reached the beach with their own cars. If they were residents, then they visited the beach on the weekends (Answer 11 in Table 3a in Appendix). Although these users chose more naturalistic beaches, they claimed to pursue more comfort (Table 3b in Appendix, Answer 13 -MC,). This finding is somehow paradoxical and opposite to the behaviour observed among users of natural beaches in South Africa (De Ruyck et al., 1995), where the landscape was the main reason driving the choices made.

Age was an important factor driving preferences for urbanised beaches and choices of recreational services. A higher percentage of users between 21 and 45 indicated that they valued the recreational attributes of the beach – including the ability to play sports - more highly than respondents aged $\geq$ A did (Figure 2). This was coincident with claims made by Crow et al., (2006), but was somehow different from an assessment made by Marin (2009) in Argentina where the beach was significant among young people as a space for playing sports.

In addition, differences in environmental knowledge between Necochea and Puerto Madryn users could be stated, and their answers somehow mirrored the environmental programs of both municipalities. It was already mentioned that Necochea in 2004-2005 launched an integrated coastal management program that included thematic guided tours and the distribution of booklets, making people aware of the significance of natural coastal resources. This could explain why people visiting beaches in this municipality revealed greater concern about beach erosion and the decrease of fauna populations and ecological services on the coast (Figure 5).

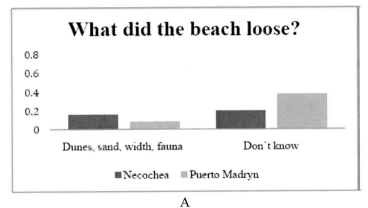

A

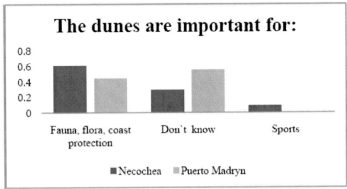

B

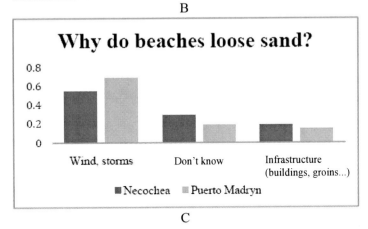

C

Figure 5. Frequencies of answers related to environmental knowledge given by users of the studied beaches in the municipalities of Necochea and Puerto Madryn. A. Responses to question regarding "loss"; B. Responses to question regarding important issues to do; C. Responses to question regarding erosion reasons.

Beach users recognised that dunes are important as a natural resource for the conservation of fauna and flora. They also linked dunes to coastal protection and to natural sand nourishment. Interviewed visitors to Necochea knew that fixed infrastructure like buildings, groins and jetties, on the beach could be responsible for beach erosion. These answers confirmed that in Necochea, some beach users perceived the negative effect of the harbour jetty on the sediment dynamics and more deeply recognised the beach as a multidimensional system (Ariza 2008) than respondents in Puerto Madryn did.

In contrast, visitors to Puerto Madryn, where there is no integral management plan (Table 1) found the beach to be a natural system dominated by natural forces. They believed natural factors such as winds and storms to be responsible for beach erosion (Figure 5c). Additionally, Puerto Madryn respondents emphasised their concerns regarding some issues such as litter (DL commentary in Table 2) and the presence of domestic animals (answer 11 in Table 3 b).

Although most of Necochea's beaches were in good condition (Madanes et al., 2011), some respondents mentioned they were worried about the environment and possible inappropriate future scenarios concerning erosion and increasing urbanisation. It is interesting that several adults at Quequén criticised the beach's condition in terms of its width, the decrease of shells and fish and the increase of stones, as Kokot and Otero (1999) documented. Criticism from users, who traditionally and recurrently spend their holidays at a certain resort, as the users of Necochea did, showed that those respondents are more locally bounded, demonstrated a feeling of identity with those beaches (Roca et al., 2009). Like these authors, we found that parents preferring rural beaches or urban beaches without facilities such as Lillo and Escollera were more receptive to environmental issues and believed that visitors of those beaches care about environmental issues Figure 4, left down). Crow et al. (2005) found that interviewees over the age of 55 gave higher ratings to nature-related benefits.

Conclusion

Argentinean beach users' profiles and values influenced their preferences and choices among beaches. The results indicated that different beaches satisfied diverse beach users. Our results agreed with other beach users' preferences reported in other studies. For instance, according to Nelson et al. (2000), scenic quality and infrastructure drove visitors' choices among Welsh beaches. These two features were even rated more highly than environmental issues such as water quality, security, and the presence of algae, odours, waste, gas and noise. Instead, Morgan (1999) mentioned that visitors to Aegean beaches preferred organised resorts with services and infrastructure rather than attractive landscapes or environmental features.

Among our nine Argentinean beaches, we found both types of beach users, those preferring beaches without infrastructure and focused on environmental quality and the protection of high natural values and those electing beaches with high user density, strong infrastructure and facilities, and diverse tourist services (Bottero and Hurtado, 2009).

The diversity of beaches can help bring about a more even distribution of visitors along the shore, which, from an environmental management systems point of view, is very advisable for ensuring the better distribution of beach use intensities. To summarise, our results can guide some beach management priorities; for instance, urban beaches should solve pollution by controlling pluvial drainage and the cleaning of decaying fish and seaweeds, while beaches without infrastructure need to enhance services such as lifeguarding, restrooms and showers as well as maintain a minimum of restaurants and cafeterias.

APPENDIX

Table 3a. Questions to measure attitude of the nine studied beaches of two Argentinean municipalities (Necochea and Puerto Madryn)

Attitude	Variable	QU	CE	ES	LI	RA	MI	MU	IN	DO
1 Accomodation										
Tourist	LRT	0.89	0.86	0.74	0.83	0.40	0.75	0.83	0.65	0.55
Resident	LRR	0.11	0.14	0.26	0.17	0.60	0.25	0.18	0.35	0.45
2 Have you visited this city before?										
Yes	PVS	0.78	0.76	0.83	0.72	0.63	0.95	1.00	0.78	0.95
No	PVN	0.22	0.24	0.17	0.28	0.38	0.05	0.00	0.23	0.05
3 How often do you come here?										
Seldom	FVP	0.34	0.45	0.26	0.48	0.80	0.20	0.43	1.00	0.78
Frequently	FVF	0.66	0.55	0.74	0.52	0.20	0.80	0.57	0.00	0.22
4 Have you visited this beach before?										
No	PVPN	0.69	0.65	0.78	0.68	0.98	0.93	0.00	0.00	0.95
Yes	PVPS	0.31	0.35	0.22	0.32	0.03	0.08	1.00	1.00	0.05
5 How often do you visit the beach?										
Seldom	FCPP	0.16	0.21	0.10	0.00	0.13	0.25	0.60	0.88	0.25
Frequently	FCPM	0.84	0.79	0.90	1.00	0.88	0.75	0.40	0.13	0.75
6 If you are resident, do you live near the beach?										
No	RVCPN	0.25	0.13	0.38	0.60	0.25	0.55	0.19	0.58	0.83
Yes	RVCPS	0.75	0.88	0.63	0.40	0.75	0.47	0.82	0.42	0.17
7 If you are tourist why do you choose this beach?										

Table 3a. (Continued)

Attitude						Beaches						
Because I have a house, family, friends, by recommendation	TEDI	0.39	0.30	0.35	0.30	0.64	0.08	0.13	0.41	0.18
Because I wanted to know the place. because I like it	TEDE	0.61	0.70	0.65	0.70	0.36	0.93	0.88	0.59	0.82
8 Why do you come to the beach?										
It is near	MPP	0.12	0.07	0.06	0.03	0.12	0.00	0.02	0.09	0.03
I like it	MPA	0.44	0.51	0.44	0.54	0.21	0.35	0.28	0.27	0.22
to practice sports	MPDE	0.05	0.05	0.06	0.03	0.18	0.12	0.11	0.09	0.11
for fun	MPD	0.39	0.37	0.47	0.43	0.48	0.53	0.59	0.56	0.65
9 With whom do you usually go to the beach?										
Family	CQCPF	0.54	0.66	0.71	0.55	0.40	0.50	0.44	0.50	0.91
Friends	CQCPA	0.27	0.23	0.23	0.30	0.50	0.44	0.44	0.33	0.05
Partners	CQCPP	0.10	0.07	0.06	0.10	0.03	0.06	0.10	0.15	0.05
Other	CQCPO	0.09	0.04	0.00	0.06	0.07	0.00	0.02	0.02	0.00
10 When do you like visiting the beach?										
Eastern	TVPSS	0.12	0.08	0.04	0.04	0.00	0.00	0.00	0.00	0.00
Sommer	TVPV	0.51	0.69	0.71	0.81	0.54	0.52	0.80	0.62	0.29
End of the year	TVPFA	0.07	0.00	0.00	0.00	0.09	0.09	0.00	0.00	0.21
Other	TVPA	0.24	0.23	0.25	0.15	0.37	0.37	0.17	0.23	0.50
Year-round	TVPO	0.05	0.00	0.00	0.00	0.00	0.02	0.02	0.15	0.00
11 You prefer to come to the beach on										
Weekends	CCPF	0.13	0.04	0.14	0.04	0.15	0.10	0.50	0.10	0.60
Weekdays	CCPES	0.16	0.08	0.09	0.00	0.05	0.08	0.03	0.00	0.00

Attitude							Beaches						
Both	CCPA	0.72	0.88	0.77	0.96	0.80	0.83	0.48	0.90	0.40
12 Compared to the last visit. the beach infrastructure is										
Different	IFRAI	0.32	0.47	0.78	0.82	0.16	0.75	0.80	0.56	0.60
The same	IFRAD	0.68	0.53	0.22	0.18	0.84	0.25	0.20	0.44	0.40
Better	PIFRAM	0.74	0.95	0.60	1.00	1.00	0.78	1.00	1.00	1.00
13 Why?										
Worse Infrastructure	PIFRAP	0.16	0.05	0.40	0.00	0.00	0.22	0.00	0.00	0.00
Don`t know	PIFRANS	0.11	0.00	0.00	0.00	0.00	0.00	0.00	0.00	0.00
The same natural conditions	CNI	0.36	0.66	0.56	0.71	0.32	0.48	0.79	0.61	0.00
Different natural conditions	CND	0.64	0.34	0.44	0.29	0.68	0.53	0.21	0.39	1.00
14 Do you know other beaches?										
No	CNOPN	0.16	0.02	0.09	0.20	0.85	0.75	0.35	0.53	0.80
Yes	CNOPS	0.84	0.98	0.91	0.80	0.15	0.25	0.65	0.48	0.20
15 I know the beaches of:										
center of the coast of Buenos Aires Province	CPC	0.38	0.36	0.28	0.33	0.20	0.17	0.11	0.48	0.44
south of the coast of Buenos Aires Province	CPLS	0.51	0.42	0.50	0.53	0.44	0.44	0.61	0.37	0.44
north of the coast of Buenos Aires Province	CPN	0.11	0.22	0.22	0.14	0.37	0.39	0.29	0.15	0.11
16 The conditions of this beach are the same compare to those ones										
The same	CPROI	0.09	0.08	0.18	0.09	0.50	0.36	0.50	0.33	0.31
Better	CPROM	0.69	0.75	0.75	0.87	0.10	0.32	0.43	0.24	0.19
Worse	CPROP	0.22	0.17	0.07	0.04	0.40	0.32	0.07	0.43	0.50

Table 3b. Questions to measure opinion of the nine studied beaches of two Argentinean municipalities (Necochea and Puerto Madryn)

Opinion	Variable	QU	CE	ES	LI	RA	MI	MU	IN	DO
1 Do you like the sand of this beach?										
Yes	ARES	0.22	0.06	0.00	0.04	0.40	0.53	0.63	0.35	0.95
No	AREN	0.78	0.94	1.00	0.96	0.60	0.48	0.38	0.65	0.05
2 Beach cleanliness										
Dirty	PLS	0.34	0.37	0.17	0.16	0.50	0.68	0.55	0.30	0.45
Clean	PLL	0.66	0.63	0.83	0.84	0.50	0.33	0.43	0.55	0.55
Don't know	PLNS	0.00	0.00	0.00	0.00	0.00	0.00	0.03	0.15	0.00
3 Do you consider this beach safe?										
No	PAPN	0.81	0.67	0.39	0.72	0.80	0.70	0.75	0.82	0.63
Yes	PPS	0.19	0.33	0.61	0.28	0.20	0.30	0.25	0.18	0.38
4 Water cleanliness										
Clean	AGL	0.78	0.82	0.91	0.80	0.60	0.30	0.28	0.68	0.60
Dirty	AGS	0.22	0.18	0.09	0.20	0.40	0.70	0.72	0.32	0.40
5 Water temperature										
Cold	TEMF	0.56	0.71	0.87	0.93	0.63	0.50	0.43	0.38	0.75
Warm	TEMAA	0.44	0.29	0.13	0.07	0.38	0.49	0.50	0.63	0.25
6 Do you smell odors?										
No	OLN	0.91	0.92	0.87	0.88	0.40	0.27	0.40	0.55	0.45
Yes	OLS	0.09	0.08	0.13	0.12	0.60	0.63	0.60	0.45	0.55
7 Do you see animals on the beach?										
Like it	ANAG	0.41	0.35	0.30	0.24	0.10	0.05	0.33	0.05	0.10

Dislike it	ANDG	0.38	0.29	0.35	0.20	0.46	0.51	0.18	0.53	0.35
Do not mind it	ANINID	0.22	0.35	0.35	0.56	0.44	0.43	0.50	0.43	0.55
8 The entries of the beach are:										
Enough	ACCPS	0.69	0.84	0.70	0.68	0.95	0.83	0.43	0.95	0.95
Not enough	ACCPI	0.31	0.16	0.30	0.32	0.03	0.13	0.58	0.05	0.05
Non existent	ACCPNE	0.00	0.00	0.00	0.00	0.03	0.05	0.00	0.00	0.00
9 The public services (restrooms. showers.etc.) of this beach are:										
Adequate	SRSNA	0.34	0.24	0.00	0.00	0.35	0.50	0.08	0.45	0.23
Inadequate	SRSNI	0.03	0.06	0.00	0.00	0.48	0.23	0.40	0.20	0.05
Non existent	SRSNNE	0.63	0.69	1.00	1.00	0.18	0.28	0.53	0.35	0.73
10 Beach resort services										
Good	CONBB	0.47	0.39	0.00	0.08	0.88	1.00	0.11	0.80	0.00
I don't use them	CONBNU	0.34	0.59	0.04	0.00	0.05	0.00	0.12	0.03	0.00
Non existent	CONBNE	0.09	0.00	0.96	0.92	0.00	0.00	0.78	0.17	1.00
11 What do you dislike in this beach?										
Crowding, music, offshore vehicles,dogs, practice of sports,garbage, lack of security and ambulant salesmen presence	DAPS	0.24	0.44	0.37	0.37	0.12	0.31	0.00	0.25	0.04
Wind, stones	DAPN	0.09	0.25	0.15	0.11	0.73	0.55	0.86	0.60	0.64
Nothing	DAPA	0.68	0.31	0.48	0.52	0.15	0.14	0.14	0.16	0.32
12 What do you like in this beach?										
Music, offshore vehicles, dogs, practice of sports	APS	0.49	0.38	0.46	0.55	0.00	0.42	0.04	0.25	0.00
Width, length of the beach, sea, rural aspect	APA	0.39	0.49	0.46	0.42	0.73	0.39	0.56	0.38	0.63
Don't know	APNC	0.12	0.13	0.08	0.03	0.27	0.18	0.40	0.38	0.37

Table 3b. (Continued)

Opinion		Beaches									
13 What would you to like in this beach?											
Tranquility	GEPT	0.00	0.01	0.00	0.00	0.17	0.27	0.25	0.30	0.11
Comfort	GEPC	0.54	0.48	0.59	0.67	0.48	0.52	0.61	0.51	0.71
Infraestructure	GEPCF	0.07	0.04	0.20	0.16	0.16	0.19	0.14	0.02	0.18
Play facilities	GEPJ	0.02	0.26	0.12	0.09	0.18	0.02	0.00	0.17	0.00
14 What did the beach loose?											
Infraestructure	PPI	0.06	0.06	0.17	0.08	0.45	0.14	0.22	0.33	0.57
Dunes, sand, width., fauna	PPN	0.14	0.17	0.14	0.19	0.02	0.28	0.02	0.07	0.00
Don't know	PPNC	0.18	0.37	0.13	0.12	0.48	0.07	0.63	0.52	0.14
15 The dunes are important for:											
Fauna, flora, protection of the coast	IMN	0.65	0.49	0.52	0.74	0.56	0.32	0.18	0.56	0.55
Don't know	IMNS	0.27	0.47	0.28	0.13	0.44	0.61	0.82	0.44	0.45
Sport	IMD	0.08	0.04	0.11	0.13	0.00	0.00	0.00	0.00	0.00
16 Why do beaches loose sand?											
Wind, storm	PPAN	0.54	0.44	0.66	0.52	0.67	0.58	0.85	0.57	0.75
Don't know	PPAI	0.36	0.26	0.22	0.30	0.26	0.25	0.04	0.27	0.08
Infrastructure (buildings. groins. jetties)	PPANS	0.10	0.30	0.13	0.19	0.07	0.17	0.10	0.16	0.21
17 Do people care about the environment?											
Yes	RMNS	0.19	0.28	0.22	0.22	0	0.2	0.13	0.03	0.35
More or less	RMNR	0.16	0.26	0.35	0.43	0.08	0.1	0.13	0.15	0.1
Little	RMNP	0.09	0.24	0.22	0.17	0.13	0.13	0.18	0.28	0.1
No	RMNN	0.56	0.22	0.22	0.17	0.8	0.58	0.58	0.55	0.45

REFERENCES

Aguiló, E., Alegre, J. and Sard, M. (2005). The persistence of the sun and sand tourism model. *Tourism Management 26*: 219–231.

Ariza, E., Jiménez, JA. and Sardá, R. (2008). A critical assessment of beach management on the Catalan coast. *Ocean and Coastal Management 51*(2):141-160.

Ballance, A., Ryan, P.G., andTurpie, JK. (2000). How much is a clean beach worth? The impact of litter on beach users in the Cape Peninsula, South Africa. *South African Journal of Science. 96* (5):210-213.

Botero, C., and Hurtado, Y. (2009). Tourist Beach Sorts as a classification tool for Integrated Beach Management in Latin America *Coastline Reports 13 EUCC – Die Küsten Union Deutschland e.V.: International approaches of coastal research in theory and practice:* 133 – 142.

Cervantes, O., and Espejel, I. (2008). Design of an integrated evaluation index for recreational beaches. *Ocean and Costal Management 51* (5):410-419.

Cervantes, O., Espejel I., Arellano E., and Dellhumeau, S. (2008). Users'perception as a tool to improve urban beach planning and management. *Environmental Management 42*:249-264.

Cochran, W. G., and Cox, G. N. (1965) Diseños experimentales. Trillas. Mexico.

Crow, T., Brown, T., and De Young, R. (2006). The riverside and Berwyn experience: Contrasts in landscape structure, perceptions of the urban landscape, and their effects on people. *Landscape and Urban Planning 75*(3-4):282-299.

Dadon, J. R., (2002). El impacto del turismo sobre los recursos naturales costeros en la costa pampeana. In J. R. Dadon and S. D. Matteucci (Eds.). Zona Costera de la Pampa Argentina, Buenos Aires, Editorial Lugar.

De Ruyck, A.M.C., Soares, A.G. and McLachlan, A. (1995). Factor Influencing Human Beach Choice on Three South African Beaches: A Multivariate Analysis. *GeoJournal 36*(4): 345-352.

Debrot, A.O., and Nagelkerken, I. (2000). User perceptions on coastal resource state and management options in Curaçao. *Revista de Biologia Tropical. 48*: 95-106.

Defeo, O., McLachlan A., Schoeman, D. S., Schlacher, T. A., Dugan, J., Jones, A., Lastrag, M. and Scarpini, F. (2009). Threats to sandy beach ecosystems: A review *Estuarine Coastal and Shelf Science 81*(1):1-12.

Eiser, J.R, Reicher, S.D., and Podpadec, T.J. (1993). What's the beach like? Context effects in judgements of environmental quality. *Journal of Environmental Psychology 13*(4): 343-352.

Enríquez Hernández, G. (2007). Criteria to evaluate the recreational aptitude of beaches in Mexico: a methodological proposal. http://www.ine.gob.mx.

Espejel, I., Espinoza-Tenorio, A.,Cervantes, O., Popoca, I., Mejia, A. and Dellimeau, S. (2008). Proposal for and integrated risk index for the planning of recreational beaches: use at seven mexican arid sites. *Journal of Coastal Research SI 50*:47-51.

James, R.J. (2000). From beaches to beach environments: linking the ecology, human-use and management of beaches in Australia. *Ocean and Coastal Management 43*:495–514.

Kokot, R., and Otero, M. (1999). Factores ambientales y riesgo geológico en el área costera de Puerto Quequén, provincia de Buenos Aires. *Revista de Geología Aplicada a la Ingeniería y al Ambiente 13*:87-100.

Lencek, L. and Bosker, G. (1998). The beach. The history of paradise on earth. London. Secker and Warburg.

Madanes, N., Faggi A., and Espejel, I. (2010). Comparación de valoraciones de playas según la edad de los usuarios. Calidad de Vida UFLO - Universidad de Flores Año I, Número 4, V1, pp.3-24.

Marin,V. , Palmisani, F., Ivaldi, R., Dursi, R. and Fabiano, M. (2009). Users' Perception Analysis for sustainable beach management in Italy. *Ocean and Coastal Management 52*(5):268-277.

Micallef, A., and Williams, A.T. (2003). Application of function analysis to bathing areas in the Maltese islands. *Journal of Coastal Conservation 9*: 147- 158.

Micallef, A., and Williams, A.T. (2002) Theoretical strategy considerations for beach management. *Ocean and Coastal Management 45*:261–275.

Morgan, R., Jones, T.C., and Williams, A.T. (1993) Opinions and perceptions of England and Wales Heritage Coast beach users: some management implications from the Glamorgan Heritage Coast, Wales, *Journal of Coastal Research* 9: 1083–1093.

Morgan, R. (1999). A novel, user-based rating system for tourist beaches. *Tourism Management 20* (4): 393-410.

Municipalidad de Puerto Madryn. Promotores ambientales. http://www.madryn.gov.ar/areas/ecologia/index.php.

Municipalidad de Necochea. Plan Manejo Integral Costero. http://www.necochea.gov.ar/gestion

Nelson, C. and Botterill, D. (2002). Evaluating the contribution of beach quality awards to the local tourism industry in Wales – the Green Coast Award. *Ocean and Coastal Management 45:* 157–170.

Petrosillo, I., Zurlini, G., Corlianò, E., Zaccarelli, N., and Dadamo, M. (2007). Tourist perception of recreational environment and management in a marine protected area. *Landscape and Urban Planning 79, 1*, 15: 29-37.

Piriz, M.L., Eyras, M.C., and Rostagno, C.M. (2003). Changes in biomass and botanical composition of beach-cast seaweeds in a disturbed coastal area from Argentine Patagonia. *Journal of Applied Phycology 15*(1): 67-74.

Priskin, J. (2003). Tourist perceptions of degradation caused by coastal nature-based recreation, *Environmental Management 32* (2003), pp. 189–204.

Ritterfeld , U., and Cupchik, G.C. (1996) Perceptions of interior spaces. *Journal of Environmental Psychology. 16*:349-360.

Roca, E., Riera, C., Villares, M., Fragell, S. and Junyent, R. (2008). A combined assessment of beach occupancy and public perceptions of beach quality: A case study in the Costa Brava, Spain. *Ocean and Coastal Management 51*:839-846.

Roca, E., Villares, M., and Ortego, M.I.. (2009). Assessing public perceptions on beach quality according to beach users' profile: A case study in the Costa Brava (Spain). *Tourism Management 30* (4): 598-607.

Tudor, D.T., and Williams, A.T. (2003). Public perception and opinion of visible beach aesthetic pollution: the utilization of photography, *Journal of Coastal Research* 19:1104-1115.

Uyarra, C. M., Watkinson, A.R. and Côté, I. M. (2009). Managing dive tourism for sustainable use of coral reefs: validating diver perceptions of atractive site features. *Environmental Management* 43(1)1-16.

Van der Weide, J. (1993). A system view of integrated coastal management. *Ocean and Coastal Management 21*:129-148.

Valdemoro, H. I. and Jimenez, J. A. (2006). The influence of shoreline dynamics on the use and exploitation of Mediterranean tourist beaches. *Coastal Management* 34:405-423.

Williams, A. T., Gardner, W., Jones, T. C., Morgan, R., and Ozhan, E. (1993). A psychological approach to attitudes and perceptions of beach users: implications for coastal zone management. In *The first international conference on the Mediterranean coastal environment, MEDCOAST'93, Antalya, Turkey* :217–228.

In: Environmental Management
Editor: Henry C. Dupont

ISBN: 978-1-61324-733-4
© 2012 Nova Science Publishers, Inc.

Chapter 9

SOIL CARBON SEQUESTRATION THROUGH THE USE OF BIOSOLIDS IN SOILS OF THE PAMPAS REGION, ARGENTINA

Silvana Irene Torri[*,1] *and Raúl Silvio Lavado*[1,2]
[1]School of Agriculture, University of Buenos Aires, Argentina
[2]INBA CONICET/FAUBA, Argentina
Av San Martín, Ciudad Autónoma de Buenos Aires, Argentina

ABSTRACT

Carbon sequestration in agricultural soils through the increase of the soil organic carbon (SOC) pool has generated broad interest to mitigate the effects of climate change. Increases in soil carbon storage in agricultural soils may be accomplished by the production of more biomass, originating a net transfer of atmospheric CO_2 into the soil C pool through the humification of crop residues, resulting in carbon sequestration. This Chapter addresses the potential of carbon storage of representative soils of the Pampas region amended with different doses of biosolids, and their soil carbon sequestration potential. Increase in biomass or yields of crops cultivated in sludge amended soils compared to unamended control soils are also discussed. The crops considered in present chapter are a forage, rye.grass (*Lolium perenne*), an annual crop, maize (*Zea mays*) and two trees, pine (*Pinus elliottii*) and eucalyptus (*Eucalyptus dunnii*).

INTRODUCTION

In the last years, concerns about global warming have led to growing interest in developing feasible methods to reduce the atmospheric levels of greenhouse gases (GHGs). The Kyoto Protocol (UN 1998) sets binding targets for industrialized countries for reducing GHGs emissions to mitigate the effects of climate change, allowing sequestration mechanisms (Article 3.4). Among these mechanisms, one of the options that have generated

---
[*] E-mail: torri@agro.uba.ar.

broad interest is carbon sequestration in agricultural soils through the increase of the soil organic carbon (SOC) pool. The global SOC inventory is estimated to be 1200 to 1600 Pg, which is close to the combined amounts stored in terrestrial vegetation (550–700 Pg) and the atmosphere (750 Pg) (Post et al., 1990; Sundquist, 1993). Therefore, even a small percentage change in the SOC pool may easily originate changes in the amounts of atmospheric carbon dioxide. It is widely known that the SOC pool has been considerably depleted by some anthropogenic activities like the conversion of natural to agricultural ecosystems, together with soil degradation processes such as erosion, salinization, and nutrient depletion or imbalances. Furthermore, conventional tillage that employed the mouldboard plow and/or the removal of crop residues aggravated the loss of SOC in agricultural ecosystems (Viglizzo et al, 2011).

Land application of organic amendments like animal manure or sewage sludge is a management practice that enhances the SOC pool in the short term (Torri et al, 2003; Robin et al. 2008; Tian et al. 2009). In the last decades, the production of sewage sludge (SS) has worldwide increased and its accumulation poses a growing environmental problem. Disposal alternatives include soil application, dumping at sea, landfilling and incineration (Sanchez Monedero et al., 2004). Landfilling and particularly land application have become the most widespread method of disposal in most countries because it is generally the most economical outlet (Hong et al, 2009; Ramlal et al. 2009).

Numerous studies have indicated that the use of this waste as a source of organic matter improves the chemical and physical properties of agricultural soils, decreasing bulk density, increasing pore size, soil aeration and root penetrability, water holding capacity and biological, resulting in an increase in crop yields (Khan et al, 2006; Lavado et al, 2007; Gilbert et al, 2011). The soils ability to retain organic carbon is a function of soil particle distribution, cation exchange capacity and the abiotic influence of temperature and precipitation (Grace et al. 2006). Carbon sequestration is defined as any persistent increase in soil organic carbon storage originated from the atmosphere. So, as discussed by Feller and Bernoux's (2008), the high percentage of organic carbon in soils originated from waste application must not be considered as sequestered C, but as "stored C". On the other hand, increases in soil carbon storage in agricultural soils may be accomplished by the production of more biomass. In this way, there is a net transfer of atmospheric CO_2 into the soil C pool through the humification of crop residues, resulting in carbon sequestration (Lal 2004, 2007).

Increase in SOC content by the transfer of C from atmospheric CO_2 to the terrestrial biosphere (soil or vegetation) may contribute to climate change mitigation in the medium-term (Powlson et al, 2011). In contrast to some other options for decreasing net GHG emissions, practices leading to SOC accumulation can start immediately without the need for development of new technologies. This might be achieved by increasing net photosynthesis or by slowing the rate of decomposition of SOC through a change in land management. However, not all practices that increase SOC content represent a transfer of additional C from atmosphere to land. Some practices are simply a movement of C from one pool in the biosphere to another, with no positive or negative implications for climate change. Some authors argue that adding organic materials such as biosolids, crop residues or animal manure to soil does not constitute an additional transfer of C from the atmosphere to land (Lal 2004, 2007; Diacono, Montemurro 2010). However, Powlson et al (2011) postulates that whether SOC increases produced by applying organic residues constitute C sequestration entirely depends on the alternative fate of these materials. For example, if the alternative disposal method is burning, soil application represents additional carbon retention in soil. Similarly, decomposition of organic wastes in landfills leads to the generation of methane under anaerobic conditions (Lou et al, 2011). Unless this is used for energy or heating, application to land will avoid methane emission to the atmosphere, leading to soil C sequestration. On the contrary, if organic wastes are regularly land applied, increases in SOC resulting from their application cannot be regarded as a waste management practice to mitigate climate change. This Chapter addresses the potential of carbon storage of representative soils of the Pampas region amended with different doses of biosolids, and their soil carbon sequestration potential. Increase in biomass or yields of crop plants cultivated in sludge amended soils compared to unamended controls is also discussed.

CHARACTERISTICS OF THE STUDY AREA

The Pampas Region is one of the largest temperate field cropland areas of the Southern Hemisphere and is located between 32° to 39°S and 56 to 67°W. This zone covers more 52 Mha of agriculturally prime quality land. The region also includes marginal or unsuitable areas for cropping, which are devoted to husbandry. These soils developed from loessic material and to a lesser degree from fluvial sediments (Imbellone et al., 2010). The mineralogy of soils varies over the region, but usually illite is the dominant clay mineral (Lavado and Camilion, 1984). Predominant soils are Mollisols formed on loess like materials of eolian origin, and to a lesser degree from fluvial sediments (Imbellone et al., 2010). The humid and semiarid subregions are characterized by Udolls and Ustolls, respectively, with minor occurrence of Aquolls in wet flat areas (Moscatelli and Pazos, 2000). The climate of the region is humid, characterized by long warm summers and mild winters. Mean annual rainfall ranges from 600 mm in the west to 1200 mm in the east, whereas mean annual temperature ranges from 14 °C in the south to 21 °C in the north. These weather conditions allow good development and production of forage and crop species typical of temperate regions. Because of its extension and yield potential, the region is considered as one of the most suitable areas for temperate grain crop production in the world (Satorre and Slafer, 1999). Agriculture is performed on well drained soils, both in the semiarid and humid portions of the region. As in other temperate regions of the world, the replacement of conventional tillage by no-till farming has resulted in better erosion control, water conservation, and nutrient cycling, time savings and reduction in the use of fossil fuels (Díaz Zorita et al., 2002; Thomas et al., 2007).

Soil organic matter has been quoted as one of the most important factors determining crop yield in this region (Barberis et al. 1985; Díaz-Zorita et al. 1999). In the northern portion of the Pampas Region, Michelena et al. (1988) reported an average soil depletion of 35% of the carbon content in the upper 15 cm. This was attributed to a combination of erosion and negative C budgets. Management options to increase soil carbon reserves in soils include reduced and zero tillage, improved rotations, irrigation, perennial and deep rooting crops, and conversion of arable land to grassland or woodland (Smith et al., 2000; Freibauer et al., 2004). These practices will not only remove carbon from the atmosphere, but will also increase quality and fertility of soils and help to reduce erosion and soil compaction as well (Lal et al., 2007).

Buenos Aires City is located in the west of the Pampas region, and is the largest city of the country, annually producing about 1.800.000 metric tons of sewage sludge. At present, sludge is partially aerobically stabilized and mainly discarded in landfarming or, to a minor extent, as a soil amendment on lawns or land filling. Due to legal incompatibilities between administrations and jurisdictions, no massive agricultural uses are current at present.

POTENTIAL OF CARBON STORAGE OF REPRESENTATIVE SOILS OF THE PAMPAS REGION

Although the application of raw organic materials is expected to induce C increases in the short term (Iakimenko et al., 1996; Hemmat et al, 2010), there is considerable controversy about the effects of sewage sludge applications on the SOC pool (Jones et al. 2006). Some authors reported that the strong microbial activity induced by the application of materials with high contents of labile organic substances could mineralize the native SOC (Bernal et al. 1998 a; Freibauer et al. 2004; Soriano Disla 2010). This effect was mainly noticed when SS was applied to soils containing high initial levels of SOC, a phenomenon known as priming effect (Dalenberg and Jager 1989). Conversely, the application of the same amount and type of SS in soils with low levels of initial SOC contributed, in general, to an increase in the short-term SOC pool. An explanation to this was the low density of microbial communities in the latter soils. The objective of this section is to analyse the effects of sewage sludge application on the short-term SOC storage of representative soils of the Pampas region, Argentina. The role of soil properties on carbon storage of recently added carbon is also studied.

1. STUDIES UNDER CONTROLLED CONDITIONS

Soils and Sewage Sludge Characterization

Soil samples of three representative Mollisols (U.S. Soil Taxonomy, USDA, 1999) of the Pampas Region, Argentina were collected. The soils are classified as Typic Hapludoll, Typic Natraquoll and Typic Argiudoll, The soils had different particle size distributions, although the clay-fraction had the same origin and mineralogical composition (Soriano et al., 1991). Composed soil samples (10 sub samples, 0-15 cm depth) were taken from pristine areas and were adequately treated for physical and chemical analysis.

Sewage sludge from the outskirts of Buenos Aires City was provided by the local water operator Aguas Argentinas S.A. The aerobically stabilized sludge used in this experiment was previously dried in holding pools in the sewage sludge treatment plant. The content of potentially trace elements (PTE) in the SS was below the maximum permissible concentration by the Argentinean regulation, whose values are similar to USEPA′s limits. The sludge (SS) was oven-dried at 60°C, ground and sieved (<2 mm). Analytical data (dry mass basis) for soils and SS is presented in Table 1.

Table 1. Selected properties of the soils and sewage sludge (SS)

Soil properties	Typic Hapludoll	Typic Natraquoll	Typic Argiudoll	SS
% clay	19.2	27.6	32.7	
% silt	23.2	43.0	57.5	
pH	5.12	6.21	5.72	5,82
Total C (mg g soil^{-1})	28.6	35.31	24.5	251
Total N (mg g soil^{-1})	2.62	3.81	2.85	19,3
Total P (mg g soil^{-1})	1,07	0.95	1.53	7,2
CE (dS /m)	0.61	1.18	0.70	0,90
CIC (cmol$_{(c)}$. kg^{-1})	20.3	22.3	24.5	11.95
Exchangeable cations				
Ca^{2+} (cmol$_{(c)}$. kg^{-1})	10.2	9.1	12.6	22.5
Mg^{2+} (cmol$_{(c)}$. kg^{-1})	2	5.4	4.3	5.6
Na^{+} (cmol$_{(c)}$. kg^{-1})	0.3	2.1	0.2	
K^{+} (cmol$_{(c)}$. kg^{-1})	2.8	1.6	2.1	1.4

Organic Carbon Evolution in Soils

A pot incubation trial was performed in a greenhouse at air temperature and was arranged in completely randomized blocks with three replications. More details are provided elsewhere (Torri et al, 2003). Total organic C in soil samples was determined by wet oxidation on days 1, 30, 60, 150, 270, and 360 after sludge application (Amato, 1983). At day 0, the application of doses equivalent to 150 t ha^{-1} MS of SS resulted in a significant increase in total carbon levels in the three studied soils compared to controls (Table 2).

During the 30 days following the incorporation of SS, the carbon content decreased rapidly in the three amended soils (Table 2). This decrease was due to degradation mechanisms linked to biological activity. We presume that the more labile components were rapidly mineralized, as reported by Lorenz et al. (2007). After 150 days, the three amended soils appeared to reach a new equilibrium, with carbon contents significantly higher than the respective controls.

Table 2. Total organic carbon in soils amended with SS. For each date, different letters indicate significant differences (Tuckey, p<0.05)

day	Typic Hapludoll control	Typic Hapludoll SS	Typic Natraquoll control	Typic Natraquoll SS	Typic Argiudoll control	Typic Argiudoll SS
0	28,62 c	44,32 a	35,32 c	51,02 a	23,91 c	39,61 a
30	28,52 c	38,06 a	35,14 c	40,89 a	23,02 c	32,19 a
60	28,36 c	36,16 a	35,05 b	40,00 a	22,70 c	32,90 a
150	27,84 c	34,68 a	34,44 b	39,15 a	22,59 c	29,48 a
270	27,33 c	33,47 a	33,61 b	37,68 a	21,14 c	27,96 a
360	27,27 c	33,81 a	33,38 b	37,79 a	20,54 c	27,64 a

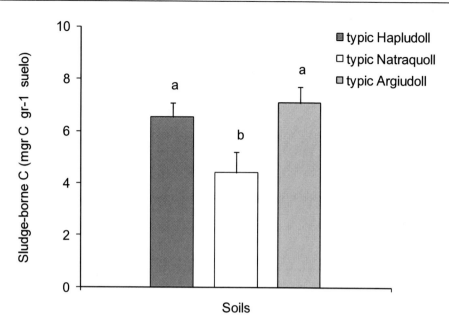

Figure 1. Storage capacity of sludge-borne C of typical soils of the Pampas region 360 days after sludge application.

This total carbon content did not change significantly between days 150 and 360 in the three studied soils (Figure 1), showing the potential of these soils to store sludge-borne C.

As the three soils presented different initial carbon content, Eq. 1 shown below was used to estimate residual sewage sludge carbon in soil at each sampling date.

$$\% \text{ CRSS (t)} = \frac{\text{SOCS (t)} - \text{SOCC (t)}}{\text{SOCS (t=0)} - \text{SOCC (t=0)}} \times 100 \quad \text{[Eq. 1]}$$

where: % CRSS: percent residual organic carbon from sewage sludge in soil; SOCS: organic carbon content in soil from sludge treated soils (mgC. g soil^{-1}); SOCC: organic carbon content from control soil (mgC. g soil^{-1}); t: time after sludge application (days).

The percentage of residual sewage sludge carbon (%CRSS) was analyzed by fitting the experimental data to kinetics models. The first-order exponential kinetic model was originally proposed by Jones (1984) for N mineralization, and was found to be suitable in describing the decomposition process of many types of organic materials due to its versatility (Ajwa y Tabatabai, 1994; Riffaldi et al. 1996). Another single-component model is the hyperbolic model, whereas other authors proposed the use of the double-exponential model to improve the agreement with experimental mineralization data. Nevertheless, the best model is chosen based on statistical analysis.

In this study, the first-order exponential model provided the best fit to carbon mineralization data for the three soils. From this model, soil carbon can be separated into two pools, representing readily mineralizable carbon (CL) which mineralizes at a constant rate (k) and a more resistant carbon (CR) source (Voroney et al., 1989), (Eq. 2).

$$\% \text{ CRSS (t)} = \text{CRS} + \text{CLS e}^{-kt} \quad \text{[Eq. 2]}$$

Table 3. Exponential equation model parameters for carbon mineralization in three representative soils of the pampas region amended with sewage sludge [a]

Soil	CL	CR	k (d^{-1})	r^2 (%)	P
Typic Hapludoll	58.4	41.7	0.035	0.995	<0.001
Typic Natraquoll	71.5	28.5	0.071	0.996	<0.001
Typic Argiudoll	53.0	45.4	0.030	0.913	<0.001

Notes: [a] Values on a dry weight basis.

where: % CRSS: percent residual carbon from sewage sludge in soil; %CLS: initial percent of sewage sludge carbon in the labile pool; %CRS: percent of sewage sludge carbon in the resistant pool; k = first order rate constant (day^{-1}), t: time after sludge application (days).

Table 3 shows the kinetic parameters for sludge-borne C mineralization based on the first order exponential model.

According to the model, organic carbon added through sewage sludge consisted of two fractions of different degree of biodegradability: a labile fraction (53-71%) that mineralized quickly and a resistant fraction (28.5-45.4%), apparently not available to soil microorganisms, that remained in the soils one year after sludge application (Torri et al, 2003). Similar results were observed by other authors in sewage sludge amended soils (Antoniadis 2008; Ojeda et al, 2008). Carbon mineralization from added substrates has been shown to be more rapid in soils with low compared with high clay content (Merckx et al, 1985). Residual substrate and decomposition products may become stabilized by sorption onto mineral particles and by incorporation into soil aggregates, being physically inaccessible to microbial turnover (Christensen, 1996). However, in this study, sewage sludge carbon mineralization did not depend on soil texture in the three studied soils. These results suggest that the recently introduced sludge-organic carbon was located in larger pores and less entangled in aggregates than native soil organic matter. Thomsen et al. (1999) reported that the turnover of organic matter in differently textured soils was better explained by soil moisture parameters than by soil texture. As the water content of the three soils studied was periodically adjusted according to water holding capacity, water availability was high and did not limit microbial activity. Thus, no relationship between soil texture and sewage sludge mineralization was observed during the first year of application. In this way, sludge-borne organic matter characteristics and not soil properties would initially predominate when high doses of sewage sludge are applied to soil, in the zone of sludge incorporation.

Carbon Sequestration by *L. Perenne*

The addition of SS to agricultural land was reported to increase the growth and production of crop plants (Singh, Agrawal, 2008; Torri, Lavado 2009), often exceeding that of well-managed fertilized controls (Dowdy et al., 1978). Increased crop yield is likely to lead to some increase in residue returns to soil and the possibility of a slight increase in SOC content. This small additional accumulation of SOC can be regarded as genuine C sequestration. Similarly, if land application of SS allows yields to be maintained with a smaller input of fertilizer, this represents a saving of GHG emissions. Thus, whilst organic C inputs from manure or residue applications will not be a direct means of sequestering additional C in soil, some indirect climate change benefits are likely to occur. The objective of this section is to analyse the potential of soil carbon sequestration through an increase in biomass and yields of L. perenne plants compared to unamended controls.

A greenhouse experiment was set with three soils (Typic Hapludoll, Typic Natraquoll and Typic Argiudoll, Table 1) amended with sewage sludge (Torri, Lavado 2009). Sewage sludge was applied to each soil at an equivalent field application rate of 150 dry t/ha. Unamended soils were used as control. Pots were sown with L. perenne after a stabilization period of 60 days and harvested 8, 12, 16 and 20 weeks after sowing, by cutting just above the soil surface. Although the growth of L. perenne in the amended soils was initially conditioned by the phytotoxic potential of SS, after 15 days this species grew uniform in sludge amended soils along the growing period, showing no visible symptoms of metal toxicity or nutrient imbalances.

Partial and total dry matter yields of L perenne grown in each treatment in the three soils are shown in Table 4. No significant differences ($p < 0.05$) in terms of aboveground biomass yields were observed between SS treatment and control in the first and second harvest. After that, plants grown in the amended soils exhibited a clear better growth compared to controls. Dry matter yield at the end of the experimental period was, in certain cases, more than 300% as compared to control.

Several factors may have contributed to improve growth in sludge amended soils, especially the increased supply of N and P, in agreement with other authors (Antolin et al 2005; Hseu, Huang 2005), together with an increasing limitation on nutrient supply in control soils with time. In addition, an improvement in the physical and biological properties of amended soils cannot be ruled out (Angers, Carter, 1996).

2. STUDIES UNDER FIELD CONDITIONS

Soil Organic Carbon Evolution

Numerous authors reported significant initial increases in soil organic matter content due to the addition of relatively large amounts of sewage sludge to soil in field conditions, although these values decreased with time due to organic matter mineralization. Sloan et al. (1998) suggested that the rate of organic matter decomposition decreases with time if no additional sludge amendment is applied. Lerch et al. (1992) observed that the highest rate of organic carbon mineralization in sludge-treated soils occurred between 7–11 days after application. Other authors found that 28-31% of the C added through sludge was mineralized within 40-70 days (Barajas-Aceves and Dendooven, 2001; Namkoong et al., 2002).

In a field experiment, sludge-borne C mineralization through the production of carbon dioxide (CO_2) in a desurfaced Vertic Argiudoll of the Pampas region was studied. Treatments included a control, one sludge application (1° year) and two sludge applications (1° and 2° year). Sludge had the same origin as the one used for the pot experiments (Section I), 25 Mg dry matter ha^{-1} was applied in the first year and 10 Mg dry matter ha^{-1} was applied in the second year. In situ soil CO_2-C production was measured during the second year in all plots. Authors estimated a mineralization of 15% of the sludge C applied the first year, and 21% of that applied the second year. Soil organic matter mainly mineralized in spring and summer

(Southern Hemisphere: October to March) ranging the in situ CO_2- C production from 130 to 440 µg C cm^2 day^{-1} (in control soils) while it raised to 208 - 510 µg C cm^2 day^{-1} in sludge amended soils. The second year the CO_2- C production reached 510 µg C cm^2 day^{-1} in that treatment. As only a low sludge proportion was mineralized during the field experiment, authors concluded that most of the sludge components were resistant to microbial degradation in the short term, in good agreement with the results obtained in a greenhouse experiment by Torri et al. (2003) and considered above. Sludge application had no effect on microbial biomass.

Carbon Sequestration by Plants

Maize

The effect of three consecutive annual sludge additions to soil on yield and growth of maize (*Zea mays*) was studied in three farms located in San Antonio de Areco, Buenos Aires Province (Lavado et al, 2008). The soils were typical Argiudolls. In each farm, the experimental design comprised the following treatments: Crop sequence in each farm included wheat - maize - soybean; maize- wheat- maize, and maize - wheat - sunflower. Average yields of maize for each treatment obtained over the three-year period are shown in Figure 2. Analysis of variance confirmed that land application of sewage sludge significantly increased maize yield. Root and stover biomass also significantly increased (data not shown).

The OC levels in control soils ranged from 1.53 % to 2.76, and did not significantly change following the application of sewage sludge along three years. Although these results are in agreement with other published papers (Gaskin et al 2003, Shober et al 2003), other

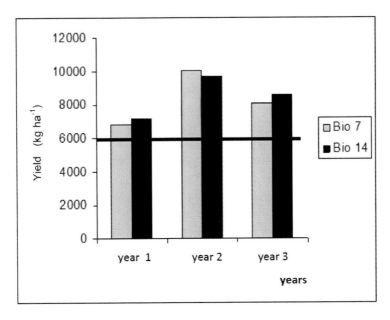

Figure 2. Maize yield obtained over a three-year period with annual application of SS. The horizontal line indicates average yield of controls.

researchers reported significant increases of total SOC after the addition of high rates of SS, i.e. 30 and 50 Mg ha^{-1} (Samaras et al 2008). In this research the lack of SOC variation is possibly due to the low doses applied. These rates were possibly not high enough to increase SOC levels. However, maize responded to SS application, suggesting that 7 Mg ha^{-1} was an adequate dose to significantly increase maize yields compared to unamended control. It is well known that the use of SS waste as a source of organic matter improves the chemical and physical properties of agricultural soils, decreasing bulk density, increasing pore size, soil aeration and root penetrability, water holding capacity and biological characteristics, resulting in an increase in crop yields. In this case, another factor which may have also contributed to increase plant yield might have been the addition of other nutrients commonly found in SS, such as sulphur, magnesium or microelements. In this way, the rate of 14 Mg ha^{-1} may be considered excessively high, for no statistical differences were observed in maize yield between 7 and 14 Mg ha^{-1}. On the other hand, this higher dose may generate a surplus of nitrate that may increase the risk of groundwater contamination by leaching.

Pine and Eucalyptus

Replacing annual crops by tree plantations is considered an important means to mitigate the consequences of global change. Forested ecosystems have a significant potential for sequestering large amounts of C through land management. Kurt et al, (2001) indicated that the amount of carbon sequestered in managed forests will be determined by three factors: (1) the increased amount of C in standing biomass, due to land-use changes and increased productivity; (2) the amount of recalcitrant C remaining below ground at the end of the rotation; and (3) the amount of C sequestered in products created from the harvested wood, including their final disposition. However, this practice is usually conceived as a suitable alternative for low fertile lands, where it is difficult to obtain annual crops acceptable yields if planted under rain-fed conditions. Forest growth may be enhanced through the application of organic wastes with a high content of nutrients, such as sewage sludge. The application of SS on afforested land can increase biological productivity (Singh and Agrawal, 2008; Ramlal et al, 2009), which can be translated into increased fibre yields and, consequently, greater carbon storage. Fast-growing plantations show the most immediate impacts of SS soil amendment (Felix et al., 2008). Another important consideration is cost savings in SS disposal compared to disposal in landfills. Information on sludge amendment, carbon sequestration and storage capabilities of trees in the pampas region is relatively scarce. Taboada et al. (2008) investigated the effect of applying SS on commercial plantations of pine (*Pinus elliottii*) and eucalyptus (*Eucalyptus dunnii*) in moderately degraded soils of the Rolling Pampa. The analysis included the evaluation of the agronomic performance of SS and soil carbon sequestration. Treatments included a control, without application of SS; Bio 7, dose equivalent to 7 t ha^{-1} of SS and Bio 14, dose equivalent to 14 t ha^{-1} of SS, annually applied on surface during three years. Although sludge-born C mineralization was calculated to increase soil organic C by 13 -26 %, analytical measurements indicated that this increase was not significant, from 5 to 10 %. Several reasons were considered for this result, like analytical methods, soil variability and so on (Taboada et al., 2008).

The volume of woody biomass produced progressively increased with time, in both pine and eucalyptus. The magnitude of the increase was five times higher in eucalyptus than in pine. However, wood production did not reach ceiling yields with the addition of SS. Authors concluded that higher doses of SS would originate more positive responses in wood volume. Considering the mineralization rates explained above, the estimated amounts of sludge-bone C that remained in soils after three years were 68.7% and 60.4% for treatments Bio 7 and Bio 14, respectively.

The amounts of carbon sequestered in stems of pine and eucalyptus showed a marked difference between the two species and between the two sludge treatments (Table 4). In Bio 7, eucalyptus accumulated 10 times more C than pine, whereas in Bio 14 treatment, this increment was almost 5 times.

Although the use of SS as a soil amendment produced an increase in biomass production of pine and eucalyptus, plants did not reach their maximum volume of biomass production, possibly limited by the low rates of SS applied or because plantation was young.

Table 3. Response in total aboveground biomass of pine and eucalyptus to the application of Bio 7 and Bio 14 within 3 years

Forest crop	Response in total aboveground biomass		sequestered C	
	Bio 7	Bio 14	Bio 7	Bio 14
	kg ha^{-1}	kg ha^{-1}	kg ha^{-1}	kg ha^{-1}
Pine	498,78	2392,41	249,39	1196,20
eucalyptus	5089,15	13168,15	2544,58	6584,08

Table 4. Partial and total mean values and standard deviation of aerial dry weight (in g) of *L. perenne* grown in control and sludge-treated pots over four harvests (n = 3, ±S.E.). Soils: Typic Hapludoll (H), Natraquoll (N) and Argiudoll (A) ; treatments: C= control, SS= sewage sludge, AS= mixture of sewage sludge and sewage sludge ash. Groups in a column detected as different at the 0.05 probability level (Tukey test) were marked with different letters (a, b, c, etc. for partial cuts; A,B,C, etc. for total yield)

	1° cut			2° cut			3° cut			4° cut			Total aerial biomass	
	aerial dry weight of *L. perenne* (in g) / pot													
H-C	2,43	± 0,055	ab	2,66	± 0,116	ab	3,01	± 0,138	bc	1,73	± 0,103	b	9,83	BC
H-SS	2,45	± 0,078	ab	3,01	± 0,108	a	5,49	± 0,149	a	4,38	± 0,301	a	15,32	A
N-C	1,97	± 0,094	b	1,76	± 0,133	b	1,11	± 0,138	c	0,84	± 0,068	c	5,68	C
N-SS	2,22	± 0,137	ab	2,45	± 0,136	ab	4,99	± 0,229	ab	4,65	± 0,257	a	14,32	A
A-C	2,51	± 0,163	ab	1,96	± 0,162	b	1,58	± 0,161	c	0,78	± 0,024	c	6,82	C
A-SS	3,07	± 0,070	a	3,28	± 0,020	a	5,28	± 0,354	a	4,50	± 0,264	a	16,13	A

Aboveground litter mass is lower in plantations with an age of ten years (Ordóñez et al, 2008), but higher in those with an age of 48 years than that in natural forests (Goma-Tchimbakala, Bernhard-Reversat 2006). The idea that young, aggrading plantation forests may be desirable over old-growth in terms of carbon sequestration has been refuted by Cannell 1999. Nevertheless, even the lowest SS application rate applied in this study resulted in an average net increase in biomass C sequestration (Table 3). The results of this study suggest that tree species selection, age and SS application rate will influence the amounts of carbon sequestration.

Conclusion

The application of sewage sludge supposed increments in the levels of SOC in the studied soils of the Pampas region, showing the potential of these soils to store sludge borne carbon. Carbon storage capacity was not related to initial SOC contents or texture and was apparently of low magnitude according to the biosolids doses used. The improved fertility originated in sludge amended soils enhanced plant growth, resulting in a genuine transfer of C from the atmosphere to vegetable biomass. Under the temperate climate conditions of the Pampas region, about one-third of plant material added to soil is retained at the end of one year. In this way, C sequestration through biomass increment in sludge amended soils may be a positive result for CO_2 atmospheric reduction, but with few effects on C sequestration on annual crops. However, the application of sewage sludge in forestry plantations, particularly eucalyptus, is a possibility that deserves being studied, since it is an alternative with great potential for combining biomass carbon sequestration and a safe sludge disposal with no major health risks.

References

Ajwa H. A. y M. A. Tabatabai. 1994. Decomposition of different organic materials in soils. *Biol.Fertil. Soils*, 18, 175-182.

Amato M., 1983. Determination of 12C and 14C in plant and soil, *Soil Biology and Biochemistry 15* (1983), pp. 611–612.

Angers, D.A. and Carter, M.R. 1996. Aggregation and organic matter storage in cool, humid agricultural soils. In: *Structure and Organic Matter Storage in Agricultural Soils* (eds M.R.Carter and B.A.Stewart), pp. 193–211. CRC Press, Boca Raton, FL.

Antolin M.C, I. Pascaul, C. Garcia, A. Polo and M. Sanchez-Diaz, 2005. Growth, yield and solute content of barley in soils treated with sewage sludge under semiarid Mediterranean conditions, *Field Crops Res*. 94: 224–237.

Antoniadis, V. 2008. Sewage sludge application and soil properties effects on short-term zinc leaching in soil columns. *Water, Air, and Soil Pollution* 190: 35-43.

Barajas-Aceves M, Dendooven L. 2001. Nitrogen, carbon and phosphorus mineralization in soils from semiarid highlands of central Mexico amended with tannery biosolids. *Bioresour. Technol.* 77: 121–130.

Barberis LA Chamorro E Baumann Fonay C Zourarakis D Canova D, Urricariet S 1985. Respuesta del cultivo de maíz a la fertilización nitrogenada en la Pampa Ondulada, campañas 1980/81 ± 1983/84. II Modelos predictivos y explicativos. *Revista de la Facultad de Agronomía* 6, 65±84.

Bernal MP, Sánchez-Monedero MA, Paredes C, Roig A 1998 Carbon mineralization from organic wastes at different composting stages during their incubation with soil. *Agric Ecosyst Environ.* 69:175–189.

Cannell, M.G.R. 1999. Environmental impacts of forest monocultures: water use, acidification, wildlife conservation, and carbon storage. *New Forests* 17: 239–262.

Christensen, B.T. Carbon in primary and secondary organomineral complexes. In *Advances in Soil Science Structure and Organic Matter Storage in Agricultural Soils;* Carter, M.R., Stewart, B.A., Eds.; CRC Lewis Publishers: Boca Raton, FL, 1996; 97–165.

Dalenberg JW, Jager G .1989. Priming effect of some organic additions to 14C-labeled soil. *Soil Biol. Biochem.* 21:443–448.

Diacono, M.; Montemurro, F. 2010. Long-term effects of organic amendments on soil fertility: a review. *Agronomy for Sustainable Development* 30:401-422.

Díaz Zorita, M., G.A. Duarte, and J.H. Grove. 2002. A review of no-till systems and soil management for sustainable crop production in the subhumid and semiarid Pampas of Argentina. *Soil Tillage Res.* 65:1–18.

Diaz-Zorita M Buschiazzo BEand Peinemann N 1999. Soil organic matter and wheat productivity in the Semiarid Argentine Pampas. *Agronomy Journal* 91, 276±279.

Dowdy R.H., Larson W.E., Titrud J.M. and Latterell J.J. 1978. Growth and metal uptake of snap beans growth on sewage previous termsludgenext term amended soil. A four year study, *J. Environ. Qual.* 7: 252–257.

Felix E., D.R. Tilley, G. Felton and E. Flamino. 2008. Biomass production of hybrid poplar (Populus sp.) grown on deep-trenched municipal biosolids, *Ecol. Eng.* 33: 8–14.

Feller C., Bernoux M. (2008) Historical advances in the study of global terrestrial soil organic carbon sequestration, *Waste Manage.* 28, 734–740.

Freibauer, A., Rounsevell, M.D.A., Smith, P., Verhagen, J., 2004. Carbon sequestration in European agricultural soils. *Geoderma* 122, 1-23.

Gaskin, J.W., Brobst, R.B., Miller, W.P., Tollner, E.W. 2003 Long-term biosolids application effects on metal concentrations in soil and bermudagrass forage. *Journal of Environmental Quality* 32: 146-152.

Gilbert, P., Thornley, P., Riche, A.B. 2011. The influence of organic and inorganic fertiliser application rates on UK biomass crop sustainability. *Biomass and Bioenergy* 35: 1170-1181.

Goma-Tchimbakala J, Bernhard-Reversat F (2006) Comparison of litter dynamics in three plantations of an indigenous timber-tree species (Terminalia superba) and a natural tropical forest in Mayombe, Congo. *Forest Ecology and Management* 229: 304–313.

Grace PR, Ladd JN, Robertson GP, Gage S (2006) SOCRATES – a simple model for predicting long-term changes in soil organic carbon in terrestrial ecosystems. *Soil Biology and Biochemistry* 38, 1172–1176.

Hemmat A., N. Aghilinategh, Y. Rezainejad, M. Sadeghi. 2010. Long-term impacts of municipal solid waste compost, sewage sludge and farmyard manure application on organic carbon, bulk density and consistency limits of a calcareous soil in central Iran *Soil and Tillage Research*, 108: 43-50.

Hong, J., Otaki, M., Jolliet, O. 2009. Environmental and economic life cycle assessment for sewage sludge treatment processes in Japan. *Waste Management* 29: 696-703.

Hseu] Z.Y. and C.C. Huang. 2005. Nitrogen mineralization potentials in three tropical soils treated with biosolids, *Chemosphere* 59: 447–454.

Iakimenko, O., Otabbong, E., Sadovnikova, L., Persson, J., Nilsson, I., Orlov, D., Ammosova, Y. 1996. Dynamic transformation of sewage sludge and farmyard manure components. 1. Content of humic substances and mineralisation of organic carbon and nitrogen in incubated soils. *Agriculture, Ecosystems and Environment* 58 (2-3), pp. 121-126.

Imbellone, P A, J E Giménez, JL Panigatti. 2010. Suelos de la Región Pampeana. *Procesos de formación*. INTA. 320 p.

Jones, C. 1984. Estimation of an active fraction of soil nitrogen. Commun. *Soil Sci. Plant Anal.* 1984, 15, 23–32.

Khan, J., Qasim, M., Umar, M., 2006. Utilization of sewage sludge as organic fertiliser in sustainable agriculture. *Journal of Applied Science* 6, 531–535.

Kurt H. Johnsen, D. Wear, R.Oren, R.O. Teskey, F. Sanchez, R. Will, J. Butnor, D. Markewitz, D. Richter, T. Rials, H.L. Allen,J. Seiler, D. Ellsworth, C. Maier, G. Katul, and P.M. Dougherty. 2001. Carbon sequestration and southern pine forests. *Journal of Forestry.* 99(4): 14-21.

Lal R. 2004. Soil carbon sequestration to mitigate climate change, *Geoderma* 123, 1–22.

Lal R. 2007. Carbon management in agricultural soils, *Mitigation and Adaptation Strategies for Global Change* 12, 303–322.

Lal, R. 2001. World cropland soils as a source or sink for atmospheric carbon. *Adv. Agron.* 71, 145–191.

Lavado, R S, M F González Sanjuán, M B Rodríguez and M A Tabeada. 2008. Respuesta de secuencias de cultivos a la aplicación continua de biosolidos. XXI congreso Argentino de la Ciencia del Suelo. *Proceedings in CD*. ISBN: 978-987-21419-9-8.

Lavado, R.S., Camilion, M.C., 1984. Clay minerals in salt affected soils of Argentina. *Clay Res.* 3, 68–74.

Lerch R.N., K.A. Barbarick, L.E. Sommers and D.G. Westfall, 1992. Sewage sludge proteins as labile carbon and nitrogen source. *Soil Sci. Soc. Am. J.* 56: 1470–1476.

Lorenz, K.; R. Lal, C.M. Preston and K.G.J. Nierop, 2007. Strengthening the soil organic carbon pool by increasing contributions from recalcitrant aliphatic bio(macro)molecules, *Geoderma* 142: 1–10.

Lou Z, Wang L, Zhao Y. 2011. Consuming un-captured methane from landfill using aged refuse bio-cover. *Bioresource Technology*, 102: 2328-2332.

Merckx, R.; den Hartog, A.; van Veen, J.A. 1985. Turnover of root derived material and related microbial biomass formation in soils of different texture. *Soil Biol. Biochem.*, 17, 565–569.

Michelena RO Irurtia CB Pittaluga A Vavruska F and Sardi MEB 1988. Degradación de los suelos en el sector norte de la Pampa Ondulada. *Ciencia del Suelo* 6, 60-66.

Moscatelli, G. and Pazos, M.S., 2000. Soils of Argentina: Nature and Use. *Proceedings of International Symposium on Soil Science: Accomplishments and changing paradigm towards the 21th. Century*. April 2000. Tailandia. Pp 81.

Namkoong W., E.-Y. Hwang, J.-S. Park and J.-Y. Choi. 2002. Bioremediation of diesel-contaminated soil with composting. *Environ. Pollut.* 119: 23–31.

Ojeda, G., Alcañiz, J.M., Le Bissonnais, Y. 2008 Differences in aggregate stability due to various sewage sludge treatments on a Mediterranean calcareous soil. *Agriculture, Ecosystems and Environment* 125:. 48-56.

Ordóñez JAB, de Jong BHJ, garcía-Oliva F, Aviña FL, Pérez JV, et al. (2008) Carbon content in vegetation, litter, and soil under 10 different land-use and land-cover classes in the Central Highlands of Michoacan, Mexico. *Forest Ecology and Management* 255: 2074–2084.

Post, W.M., T.H. Peng, W.R. Emanuel, A.W. King, V.H. Dale y D.L. DeAngelis. 1990. The global carbon cycle. *Am. Scientist* 78: 310-326.

Powlson, D.S., Whitmore, A.P. , Goulding, K.W.T. 2011. Soil carbon sequestration to mitigate climate change–a critical re-examination to identify the true and the false. *European Journal of Soil Science*, 62, 42–55.

Ramlal, E., Yemshanov, D., Fox, G., McKenney, D. 2009 A bioeconomic model of afforestation in Southern Ontario: Integration of fiber, carbon and municipal biosolids values. *Journal of Environmental Management* 90 (5), pp. 1833-1843.

Riffaldi R., A. Saviozzi y R. Levi-Minzi. 1996. Carbon mineralization kinetics as influenced by soil properties. *Soil Biol. Biochem.* 22, 293–298.

Robin P, Ablain F, Yulipriyanto H, Pourcher AM, Morvan T, Cluzeau D, Morand P. 2008. Evolution of non-dissolved particulate organic matter during composting of sludge with straw. *Bioresour Technol* 99:7636–7643.

Samaras, V., Tsadilas, C.D., Stamatiadis, S. 2008. Effects of repeated application of municipal sewage sludge on soil fertility, cotton yield, and nitrate leaching. *Agronomy Journal* 100: 477-483.

Sánchez-Monedero M.A.; C. Mondini, M. de Nobili, L. Leita and A. Roig. 2004. Land application of biosolids. Soil response to different stabilization degree of the treated organic matter, *Waste Manage.* 24: 325–332.

Satorre, E.H. and Slafer, G.A. 1999. Wheat production systems of the Pampas. En: *Wheat: Ecology and Physiology of Yield Determination*. Eds.: E.H. Satorre and G.A. Slafer, The Haworth Press, New York, pp. 333-348.

Shober, A L, R C Stehouwer and K E Macneal. 2003. On-farm assessment of biosolids on soil and crop tissue quality. *J. Environ. Qual.*: 1873-1880.

Singh R.P., Agrawal M. 2008. Potential benefits and risks of land application of sewage sludge. *Waste Management*, 28: 347-358.

Sloan J.J., R.H. Dowdy and M.S. Dolan. 1998. Recovery of biosolids-applied heavy metals sixteen years after application. *J. Environ. Qual.* 27 (1998), pp. 1312–1317.

Smith P., D.S. Powlson, J.U. Smith, P.D. Falloon and K. Coleman, 2000. Meeting Europe's climate change commitments: quantitative estimates of the potential for carbon mitigation by agriculture. *Global Change Biol.* 6: 525–539.

Soriano-Disla, J.M., Navarro-Pedreno, J., Gomez, I. 2010. Contribution of a sewage sludge application to the short-term carbon sequestration across a wide range of agricultural soils. *Environmental earth sciences*, 61: 1613-1619.

Sundquist, E. T. 1993. The global carbon dioxide budget. *Science* 259: 934-94 1.

Taboada, M. A. E. Borodowski and R.S. Lavado 2008. Secuestro de carbono utilizando biosólidos en cultivos de maíz y plantaciones forestales. XXI congreso Argentino de la Ciencia del Suelo. *Proceedings in CD*. ISBN: 978-987-21419-9-8.

Lavado, R.S., Rodríguez, M., Alvaro, R., Taboada, M.A., 2007. Transfer of potentially toxic elements from biosolid-treated soils to maize and wheat crops. *Agriculture, Ecosystems And Environment.* 118: 312-318.

Lavado, R S, M F González Sanjuán, M B Rodríguez and M A Taboada. 2008. Respuesta de secuencias de cultivos a la aplicación continua de biosolidos. XXI congreso Argentino de la Ciencia del Suelo. *Proceedings in CD.* ISBN: 978-987-21419-9-8.

Tejada, M., Gonzalez, J.L. 2007. Application of different organic wastes on soil properties and wheat yield. *Agronomy Journal* 99: 1597-1606.

Thomas, G.A., G.W. Titmarsh, D.M. Freebairn and B.J. Radford. 2007. No-tillage and conservation farming practices in grain growing areas of Queensland: A review of 40 years of development. *Aust. J. Exp. Agric.* 47:897–898.

Thomsen, I.; Schjonning, P.; Jensen, B.; Kristensen, K.; Christensen, B.T. 1999. Turnover of organic matter in differently textured soils. II. Microbial activity as influenced by soil water regimes. *Geoderma,* 89, 199–218.

Tian G, Granato TC, Cox AE, Pietz RI, Carlson CR Jr, Abedin Z (2009) Soil carbon sequestration resulting from long-term application of biosolids for land reclamation. *J. Environ. Qual* 38:61–74.

Torri S, Alvarez R, Lavado R. 2003. Mineralization of Carbon from Sewage sludge in three soils of the Argentine pampas. Commun. Soil Sci. and Plant Anal. 34: 2035-2043.

Torri S, Lavado R. 2009. Plant absorption of trace elements in sludge amended soils and correlation with soil chemical speciation. *Journal of Hazardous Materials,* 166: 1459–1465.

Viglizzo, E.F., Frank, F.C., Carreño, L.V., Jobbágy, E.G., Pereyra, H., Clatt, J., Pincén, D., Ricard, M.F. 2011. Ecological and environmental footprint of 50 years of agricultural expansion in Argentina. *Global Change Biology* 17: 959-973.

Voroney R.P., E.A. Paul and D.W. Anderson, 1989. Decomposition of wheat straw and stabilization of microbial products, *Canadian Journal of Soil Science* 69: 63–77.

In: Environmental Management
Editor: Henry C. Dupont

ISBN: 978-1-61324-733-4
© 2012 Nova Science Publishers, Inc.

Chapter 10

DEVELOPING AN ECOSYSTEM-BASED HABITAT CONSERVATION PLANNING PROTOCOL IN SALINE WETLAND

Jennifer Hitchcock and Zhenghong Tang[*]
College of Architecture, University of Nebraska-Lincoln, Lincoln, NE, US

1. INTRODUCTION

The ecosystem approach to management is in its integrative form a management method that considers the system as a whole instead of as individual components. This holistic approach focuses on habitats and system integrity, and on an objective aimed at the health and integrity of the ecosystem (Currie 2010). Ecosystem-Based Management is a unique management style that is intended to overcome the shortfalls of single-sector management and contains the following characteristics (NatureServe 2010):

- Integrates ecological, social, and economic goals,
- Recognizes humans as key components of the ecosystem,
- Considers ecological instead of political boundaries,
- Addresses the complexity of natural processes and social systems,
- Uses adaptive management approach in the face of resulting uncertainties,
- Engages multiple stakeholders in a collaborative processes to define problems and find solutions,
- Incorporates understating of ecosystem processes, and
- Is concerned with the ecological integrity and the sustainability of both human and ecological systems.

---

[*] Corresponding Author: Phone: E-mail: ztang2@unl.edu , (402) 472-9281; Fax: (402) 472-3806.

Originally, ecosystem-based management was designed to manage people's impacts on the oceans. But, it can and should be extended and adapted to any ecosystem where existing policies and management practices have proved insufficient to sustain nature's services.

Nebraska's eastern saline wetlands are one of the most critically endangered ecosystems within the state. In addition to being endangered themselves, the saline wetlands are home to several at-risk species including the federally endangered Salt Creek tiger beetle (*Cicindela nevadica lincolniana*), the federally endangered Interior Least Tern (*Sterna antillarum athalassos*), the federally threatened Piping Plover (*Charadris melodus*), and the state endangered Saltwort (*Salicornia rubra A. Nels.*). Of the more than 20,000 acres that once extended Lancaster and Saunders' Counties, less than 4,000 acres remain (LaGrange *et al.* 2003). Without active conservation and planning efforts, this landscape and the species that rely upon it will continue to diminish.

Conservation of these landscapes is eminent. In 2003, the Saline Wetlands Conservation Partnership was created for their protection. Since that time, the US Fish and Wildlife Service has released the Draft Recovery Plan for the Salt Creek tiger beetle, and designated much of saline wetlands its critical habitat. The next step in the conservation process for these wetlands is the creation of a Habitat Conservation Plan. To launch this conservation plan's development, this project creates a Habitat Conservation Planning Guide.

Three phases were used to develop the Habitat Conservation Planning Guide for Nebraska's Eastern Saline Wetlands. Phase I created individual Habitat Conservation Planning Guides from the scientific literature, the US Fish and Wildlife Service's Endangered Species Habitat Conservation Planning Handbook, and Benton County, Oregon's Benton County Prairie Species Habitat Conservation Plan Revised Draft. Phase II combined these individual guides into a single Habitat Conservation Planning Guide. Finally, Phase III localized the single guide for Nebraska's Eastern Saline Wetlands using four local plans: the Lincoln/Lancaster County Comprehensive Plan, Environmental Resources Section, the Salt Creek Tiger Beetle (*Cicindela nevadica Lincolniana*) Draft Recovery Plan, the Implementation Plan for the Conservation of Nebraska's Eastern Saline Wetlands, and the Little Salt Creek Watershed Master Plan.

The outcome of the analysis was a Habitat Conservation Planning Guide Checklist for Nebraska's Eastern Saline Wetlands for use in the development or review of a Habitat Conservation Plan for Nebraska's Eastern Saline Wetlands. This guide ensures all components required by the US Fish and Wildlife Service, as well as those necessary for the plan's quality are included. Additionally, a Components Glossary has been created to further define each of the guide's components.

2. STUDY AREA

Nebraska's Eastern Saline Wetlands is a truly unique and invaluable landscape. These wetlands provide critical habitat for several species of wildlife. Their conditions uniquely allow salt-tolerant vegetation to thrive hundreds of miles away from the sea. And, were instrumental in the location and boom of the City of Lincoln. The saline wetlands are a piece of Lancaster County's history, and like a rare family heirloom, residents of Lincoln and Lancaster County need to protect this endangered landscape. Without the work of advocates and private landowners to protect these landscapes; Nebraska's eastern saline wetlands will disappear forever. The salt found in Nebraska's eastern saline wetlands originated from the Western Interior Seaway that was present during the late Cretaceous period. Trapped saline groundwater flows up through porous Dakota Sandstone, from deeper shale rock formations created during the Permian Period (Harvey *et al.* 2007 and Lincoln Parks and Recreation 2010). Saline groundwater has migrated thousands of years along these flow paths into the Salt Creek, Little Salt Creek, and Rock Creek watersheds. Salts from the groundwater accumulate in the surrounding soils, thus creating the saline wetlands.

Creating a Multi-Species Habitat Conservation Plan (MSHCP) for the Eastern Saline Wetlands will help protect and increase these landscapes, while Lincoln and Lancaster County continue to grow and develop. As landscapes continue to change, a MSHCP will be an invaluable tool for the people of Lancaster County. Currently, private landowners that have saline wetland habitat used by any of the threatened or endangered species that use it must obtain authorization from the US Fish and Wildlife Service (USFWS) on a case by case basis for any land-disturbing activities. Creation of a MSHCP will allow the public to obtain authorization locally through Lancaster County without going to the USFWS. This simplifies the permit process, saving landowners time and money. Additionally, creation of the MSHCP for the Eastern Saline Wetlands fits directly with the three guiding principles listed in the Environmental Resources section of the Lincoln/Lancaster County Comprehensive Plan (2006). These principles are: 1) Maintain the richness and diversity of the county's urban and rural environments; 2) Be broadly inclusive; and 3) Focus attention on unique landscapes. The Comprehensive Plan takes into consideration the effects of natural phenomena not only upon localized development, but also upon the community as a whole, upon private ownership issues, and upon recreational opportunities. The Plan thus commits Lincoln and Lancaster County to preserve unique and sensitive habitats and endorses creative integration of natural systems into developments. (Lincoln/Lancaster County Comprehensive Plan 2006).

According to the Nebraska Game and Parks Commission (2005), wetlands are "those areas that are inundated or saturated by surface or groundwater at a frequency and duration sufficient to support, and that under normal circumstances do support, a prevalence of vegetation typically adapted for life in saturated soil conditions." Wetlands are characterized by both wet and dry conditions that vary from season to season, as well as year to year. The dynamic of wet and dry interacting with natural process, such as flooding, fires, and grazing, allow wetlands to be productive (Nebraska Game and Parks Commission 2005). Additionally, Nebraska's eastern saline wetlands are further characterized by saline soils, salt-tolerant vegetation, and unique wildlife.

Soils: There are generally two zones of soil salinity found in and around saline wetlands. High salinity zones appear as salt encrusted mudflats that are devoid of vegetation. Zones of low salinity are fully vegetated with salt-tolerant plants (Nebraska Game and Parks Commission 2005).

Vegetation: According to Gilbert *et al.* (1994), salt tolerant plant species found in Nebraska's eastern saline wetlands include:

- Spearscale (Atriplex subspicata)
- Inland Salt Grass (Distinchlis spicata var. stricta)
- Sea Blite (Suaeda dpressa)

- Prairie Bulrush (Scirpus maritimus var. paludosus)
- Saltwort (Salicornia rubra)
- Narrow Leaved Cattail (Typha angustifolia)
- Saltmarsh Aster (Aster subulatus var. ligulatus)
- Seaside Heliotrope (Heliotropium curassavicum)
- Texas Dropseed (Sporobolus texanus)

Saline wetlands also include some invasive species. Invasive species to Nebraska's eastern saline wetlands include (Schneider *et al.* 2005):

- Reed Canary Grass (*Phalaris arundinacea*)
- Narrow Leaved Cattail (*Typha angustifolia*)
- Phragmites (European Variety)
- Salt Cedar (*Tamarix spp.*)

Wildlife: Saline wetlands provide migration, breeding, and feeding habitat for more than 230 species of birds (Lincoln Parks and Recreation 2010). In May 2008, Nebraska Game and Parks recorded over 60 species of birds at the Frank Shoemaker Marsh during their annual Birding Day (Lincoln Parks and Recreation 2010). Mammal, fish, and amphibian species found within the surrounding tall grass prairie ecosystem, may be found within the saline wetlands. Additionally there is an entire array of insects that use the saline wetlands (Schneider et al. 2005). In addition, the Salt Creek Tiger Beetle (Cicindela nevadica var. lincolniana), Encoptolophus subgracilis, Salt Creek Grasshopper (Trimerotropis salina), and the Desert Forktail Damselfly(Ischnura barberi) are three very rare and at-risk species of insects found in the saline wetlands. Threatened or Endangered species that use the eastern saline wetlands for habitat include:

- Federal Endangered Salt Creek Tiger Beetle (*Cicindela nevadica var. lincolniana*)
- State Endangered Saltwort (*Salicornia rubra*)
- Federal *Endangered Inte*rior Least Tern (*Sterna antillarum athalassos*)
- Federal Threatened Piping Plover (*Charadrius melodus*)

3. CORE FUNCTION AND VALUE OF SALINE WETLAND

There are six primary functions listed by the Nebraska Game and Parks Commission (2005) that contribute to their overall value of saline wetlands for residents of the City of Lincoln, Lancaster County, the State of Nebraska, and the World.

a) *Improving Water Quality:* Saline wetlands improve stream water quality by collecting and filtering rain and runoff waters. The collection/filtering process slows water down, allowing chemical processes to settle pollutants out. Because of these cleansing functions; wetlands are increasingly being used for water pollution control and wastewater treatment (Nebraska Game and Parks Commission 2005).

b) *Providing Habitat for Wildlife, Fish, and Plants:* Nebraska's eastern saline wetlands provide habitat for federal and state listed threatened and endangered species, migratory birds, as well as hundreds of familiar mammal, fish, and reptile species (Lincoln Parks and Recreation 2010).

c) *Reducing Flooding and Soil Erosion:* The water collection process that helps improve water quality also reduces flooding and soil erosion by allowing stored water to seep deep into the ground, slowing its release into the stream. This slow release of water reduces peak flows which greatly reduces soil erosion (Lincoln parks and Recreation 2010). Studies show that saline wetland locations in conjunction with soil types provide indicators that flood control functions are being provided by these wetlands (Nebraska Game and Parks Commission 2005).

d) *Supplying Water:* Some rain and runoff waters seep deep enough into the ground to recharge groundwater stores. The slow release of collected waters into the creeks help maintain stream flows that benefit municipal and agricultural water users and provide water for livestock (Nebraska Game and Parks Commission 2005).

e) *Providing Recreational and Educational Opportunities:* Recreational and educational opportunities provided by saline wetlands include, but are not limited to: Bird watching; Nature study; Waterfowl and Pheasant hunting; Photography; Hiking, etc.

f) *Producing Food and Fiber:* Many former saline wetlands have been drained for use as agricultural crop land. Some existing saline wetlands are used as livestock grazing sites in conjunction with the Nebraska Game and Parks Commission's Grazing and Wildlife initiative.

The eastern saline wetlands of Lancaster and Saunders Counties are the most limited and endangered wetland type in Nebraska (LaGrange *et al.* 2003). According to the Nebraska Game and Parks Commission (2005), 168 of 188 uncultivated wetland sites were considered to have a high or moderate vulnerability to future wetland degradation or loss. Categories of threats include: drainage or filling, stream-bed degradation, agricultural conversion or use, residential or commercial development, transportation, and water pollution. Serious long-term threats include stream channel deepening that result in erosive gullies that eventually drain wetlands and lead to low water table levels (Nebraska Game and Parks Commission 2005). Gilbert *et al.* (1994) identified four categories of saline wetlands as described below.

Category 1: Site currently provides saline wetland functions of high value or has the potential to provide high values following restoration or enhancement measures, and meets one or more of the following criteria: a) Salt creek tiger beetle present; b) The presence of one or more rare or restricted halophytes; c) Identified as having historical significance by the Nebraska state historical society; d) Contains at least one saline wetland association as part of the site's flora and is not highly degraded or has potential to maintain or improve saline wetland; e) Contain no saline wetland plant associations but high potential exists for restoring the historical salt source.

Category 2: Given current land use or degree of degradation, the site currently provides limited saline wetland functions and low values. Restoration potential is low. Category 2 criteria includes: a) Currently contaminated by hazardous or toxic waste or is or has been used for municipal or industrial solid waste disposal and has limited potential for providing high functions and values through restoration; b) Contains at least one saline wetland plant association as part of the site's flora and is highly degraded and has limited potential for the long-term maintenance or improvement of saline vegetative characteristics through enhancement or restoration; c) Contains no saline wetland plant associations and provides low functions and values due to degradation and has low potential for restoration of the historic salt source.

Category 3: Site is functioning as a freshwater wetland having freshwater plant communities on a saline soil. Currently provides freshwater wetland values and no feasible restoration measures exist to reestablish the historic salt source and saline plant associations.

Category 4: Site is functioning as a freshwater wetland having freshwater plant communities on non-saline hydric soil.

Not Categorized (NC): Insufficient data for categorization or interagency work teams were unable to gain access.

Saline wetlands provide unique core function and values for the at-risk species. At-risk species found in Nebraska's eastern saline wetlands include several state and federally listed threatened or endangered species, as well as several rare or imperiled species. Schneider *et al.* (2005) have categorized these species into Tier I and Tier II species.

Tier I: Tier I species include those species found in Nebraska that are globally or nationally most at-risk of extinction. Schneider *et al.* (2005) used the following selection criteria for determining a species Tier I status.

- State and Federally listed species.
- Species ranked by NatureServe and the Natural Heritage Network as either globally critically imperiled, imperiled, or vulnerable.
- Declining Species - Species whose abundance and/or distribution has been declining across much of their entire range.
- Endemic Species - Species whose entire range of distribution occurs within or primarily within Nebraska.
- Disjunct Species – Species whose populations in Nebraska are widely disjunct (200 miles or more) from the species' main range of distribution.

Tier II: According to Schneider et al. (2005) Tier II species include those that did not make the Tier I criteria, but were ranked by the Nebraska Natural Heritage Program as either State Critically Imperiled, State Imperiled, or State Vulnerable. Tier II species are typically not at risk from the global or national perspective, but are rare or imperiled within Nebraska.

4. HABITAT CONSERVATION PLANNING

The Endangered Species Act began as the Endangered Species Preservation Act in 1966. In 1973, Congress passed the Endangered Species Act of 1973. The creation of a Habitat Conservation Plan (HCP) is the first step in the Incidental Take Permit process. Habitat conservation planning integrates development and land-use activities with ecological conservation (U.S. Fish and Wildlife Service 2005). The purpose of the HCP process is to provide a framework for people to complete projects while conserving at-risk species; encourage the private sector to develop long-term conservation plans; reduce conflicts between Federally listed species and non-Federal land development; stimulate "creative partnerships" between the public and private sector, and protect non-listed species in conjunction with mitigation measures for listed species (U.S. Fish and Wildlife Service 1996 and 2005). There are significant benefits for habitat conservation planning (U.S. Fish and Wildlife Service 1996 and 2005):

- Protect habitats and prevent the decline of sensitive species.
- Help to maintain healthy ecosystems and valuable green space.
- Create a regional approach that streamlines the permit process; saving time and money.
- Protects property owners from legal liability.
- Allows a broad range of activities under the umbrella of the permit's legal protection.
- Maximizes flexibility needed to develop innovative mitigation programs
- Minimizes Endangered Species Act compliance by replacing individual project reviews with comprehensive, area-wide review.
- Allows Development/Redevelopment within the endangered species' habitat.
- Allows private property owners to be included with the county's incidental take permit; saving them time and money.
- Protects non-Federal landowners through the "no surprises assurances."

However, there are also big challenges to develop a successful habitat conservation plan. The challenges may come from the follow aspects: a) Difficult to meet the needs of economic development alongside endangered species conservation; b) Difficult to build consensus; c) Problematic to integrate interests of all parties; d) Hard to clarify biological issues; e) Can easily become excessively complex; trying to satisfy too many land-use or endangered species issues in one effort; f) Funding must be acquired for habitat conservation and/or mitigation projects.

HCP is a collaborative planning process which include multiple permittees and stakeholders. It is critical to implement ecosystem-based management principles in HCP process (in Table 1).

Table 1. Examples of Permittees and HCP Size

Permittees	HCP Size
Individual landowner	Straightforward HCP
Natural resource owner	Straightforward HCP
Local or Regional Authorities	Regional HCP
One or two private landowners	Large-Scale HCP
City or county governments	Large-Scale HCP
Local agencies working jointly	Large-Scale HCP
Private groups	Large-Scale HCP

5. METHOD TO DEVELOP HABITAT CONSERVATION PLAN PROTOCOL

The goal of this project is to create a protocol for use by the Saline Wetland Partnership in the creation of a Habitat Conservation Plan for Nebraska's eastern saline wetlands. The research methods include the following three phases: *Phase I:* Review scientific, legal, and practical elements that affect the creation of a HCP, and create individual plan protocols for each element. *Phase II:* Combine individual protocols to create a single Habitat Conservation Plan Protocol. *Phase III:* Localize the HCP Protocol for Lincoln/Lancaster County's eastern saline wetlands. The following sections will provide detail statements for the research methods to develop HCP protocol and indicators.

Phase I : A) Scientific Element: A scientific study was chosen as an element in which to create the Habitat Conservation Planning Protocol in order to provide scientific support for the components chosen for the protocol. Science is most often on the cutting edge of development. As with other fields, this holds true for planning practices. By including a scientific element, the protocol ensures the inclusion of the most up-to-date planning methods and practices. Based on the literature review (from Brody, 2003), this study created a conceptual model for what makes a high quality ecosystem plan, and then looked at the ability of local comprehensive plans in Florida to incorporate principles of ecosystem management. The results of the study measured the relative strengths and weaknesses of local plans, to achieve the objectives of ecosystem management, and provide direction on how communities can improve their environmental framework (Brody, 2003). The Ecosystem Plan Coding Protocol was used as the individual plan protocol for the scientific element. Brody's study was chosen as the scientific element because he builds upon the standard conceptions of plan quality (factual basis, goals, and policies), by adding inter-organizational coordination and capabilities, and implementation. These additions to plan quality tailor Ecosystem Plan Coding Protocol to ecosystem management, by extending plan quality to more effectively capture the principles of ecosystem management. Examples of these additional components include collaboration and conflict, ecological monitoring, enforcement, and commitment to putting the plan into place. This phase's study ensures that plans created using the HCP Protocol incorporate the principles of ecosystem-based management and that the plan will be high quality.

Phase I: B) Legal Element: The Endangered Species Habitat Conservation Planning Handbook by the U.S. Fish and Wildlife Service was reviewed as the legal element of phase I. This handbook outlines the legal requirements for creating a Habitat Conservation Plan and applying for an Incidental Take Permit as set forth by the Endangered Species Act of 1973 and the U.S. Fish and Wildlife Service. These legal requirements, along with other suggested items to include, form the basis of the individual plan protocol for the legal element. Using components from the U.S. Fish and Wildlife handbook ensures the Lancaster County HCP will meet all legal requirements.

Phase I: C) Practical Element: A "real" world Habitat Conservation Plan was chosen for a practical element in creating the planning protocol. Both science and the law can often be ambiguous. A practical element provides a model for how one entity interpreted the legal requirements along with their results to problems that arose during the plan development process. The Benton County Prairie Species Habitat Conservation Plan, Revised Draft from

Benton County, Oregon was used for the practical element of phase I. This plan was chosen, because when critiqued using protocol from the previous studies, the plan rated as a high quality plan. The plan was also chosen because of several similarities between Benton County and Lancaster County, and the HCP itself and the plan that will be created for Lancaster County. Both Benton County and Lancaster County were historically prairie ecosystems. What remains of the ecosystems in each of these counties face similar threats such as: invasive species, stream channel alterations, operation management activities (e.g. roadside spraying, mowing, and recreation), and general habitat fragmentation. Additionally Benton County's HCP meets the three guiding principles of Lincoln/Lancaster County's Environmental Resources Section of the Comprehensive Plan: a) Maintain the richness and diversity of the county's urban and rural environments; b) Be broadly inclusive; c) Focus attention on unique landscapes. There are also similarities between the Benton County plan and the plan that will be created for Lancaster County. Both plans cover multiple species that include species of invertebrates, birds, and plants. Both plans will be about the same size, and deal with similar covered activities including: transportation, land use planning, and parks and natural areas. The practical element individual plan protocol included themes, concepts, as well as specific items taken from the Benton County plan. Including these components ensured the feasibility of components taken from the scientific and legal elements. Some components were added which were not addressed by the other elements, but were necessary to provide additional clarity, as well as add a measure of public education to the plan.

Phase II: The individual protocols from the scientific, legal, and practical elements were combined to create a single HCP Protocol. Overlapping themes and concepts from the three elements were combined when appropriate, and components from the legal and practical elements were categorized into one of the five conceptions of plan quality (factual basis, goals and objectives, inter-organizational coordination and capabilities of ecosystem management, policies, tools and strategies, and implementation).

Phase III: In order to localize the HCP Protocol to Lincoln/Lancaster County and Nebraska's eastern saline wetlands, four local area plans were analyzed with the HCP Protocol. These four plans included: a) Environmental Resources section of the Lincoln/Lancaster County Comprehensive Plan; b) Salt Creek Tiger Beetle (*Cicindela nevadica lincolniana*) Draft Recovery Plan; c) Implementation Plan for the Conservation of Nebraska's Eastern Saline Wetlands; d) Little Salt Creek Watershed Master Plan. In this study, components of the HCP Protocol that were not applicable to any of these four plans were removed from the protocol.

6. RESULTS

This study develops the HCP protocol with measurable indicators. This protocol can be practical guidance for local planners to integrate ecosystem-based conservation concepts into their daily decision-making. The planning guide should be used as a checklist, ensuring that each component is addressed within the Habitat Conservation Plan. Components have been ranked with those required by the US Fish and Wildlife Service (USFWS) at the top. Components that are not required, but are necessary for plan quality may be found further down the checklist. Finally, the planning guide has been divided into six sections corresponding to the five conceptualizations, plus the — Additional Items section (Table 2).

Table 2. Habitat Conservation Plan Protocol

1. Factual Basis		
1.1. Resource Inventory		
1.1.1 Location, HCP Coverage/ Plan Area (required by USFWS)	1.1.2 Species Occurrence (required by USFWS)	1.1.3 Ecological Functions (required by USFWS)
1.1.4 Covered Species/ Biological Data (required by USFWS)	1.1.5 Species Range and Distribution (required by USFWS)	1.1.6 Unlisted Species (required by USFWS)
1.1.7 Ecological Zones/ Habitat Types and History	1.1.8 Areas of High Biodiversity/Species Richness	1.1.9 Habitat Corridors
1.1.10 Vegetation Classified	1.1.11 Wildlife Classified	1.1.12 Vegetation Cover Mapped
1.1.13 Invasive/Exotic Species	1.1.14 Soils Classified	1.1.15 Wetlands Mapped
1.1.16 Climate Described	1.1.17 Indicator/Keystone species	1.1.18 Ground Water Resources
1.1.19 Surface Hydrology	1.1.20 Geomorphic Investigation	1.1.21 Listed Species Not Covered in this HCP
1.1.22 Graphic Representation of Transboundary Resources		
1.2. Ownership Patterns		
1.2.1 Conservation Lands Mapped	1.2.2 Entities	1.2.3 Management Status Identified for Conservation Lands
1.2.4 Existing and Proposed Lands for Des-ignation as Conservation Areas	1.2.5 Distribution of Species within Conser-vation Network	
1.3. Human Impacts		
1.3.1 Covered Activities (required by USFWS)	1.3.2 Indirect Effects (required by USFWS)	1.3.3 Effects on Critical Habitat (required by USFWS)
1.3.4 Alternatives (required by USFWS)	1.3.5 Existing Applicable Environmental Regulations Described (required by USFWS)	1.3.6 Population Growth
1.3.7 Road Density	1.3.8 Fragmentation of Habitat	1.3.9 Development of Wetlands
1.3.10 Nutrient Loading	1.3.11 Water Pollution	1.3.12 Alteration of Waterways
1.3.13 Impacts	1.3.14 Carrying Capacities Measured	
2. Goals and Objectives		
2.1 Purpose of HCP	2.2 Vision and Goals (present and clearly specified)	2.3 Biological Goal
2.4 Conservation Measures	2.5 Presence of Measureable Objectives	
3. Inter-Organizational Coordination and Capabilities		
3.1 Public Meetings and Outreach (required by USFWS)	3.2 Agreements (required by USFWS)	3.3 Integration with Other Plans/Policies (required by USFWS)
3.4 Overview of Conservation Planning Process	3.5 Covered Entities and Lands within the HCP	3.6 Organizations and Stakeholders Identified
3.7 Organization and Stakeholder	3.8 Information Sharing and Data	3.9 Links between Science

Coordination	Management	and Policy
3.10 Conflict Management Process		
4. Policies, Tools, and Strategies		
4.1 Regulatory Standards and Relationship to Recovery (required by USFWS)	4.2 Relocating Project Facilities within Project Area (required by USFWS)	4.3 Phasing (required by USFWS)
4.4 Adaptive Management (required by USFWS)	4.5 Buffer Requirements (required by USFWS)	4.6 Mitigation Policies (required by USFWS)
4.7 Restrictions and Best Management Practices	4.8 The Greenprint Challenge: Imple-mentation Strategies	4.9 Public Education Programs
5. Implementation		
5.1 Costs and Funding (required by USFWS)	5.2 Schedule (required by USFWS)	5.3 Regular Plan Updates and Assessments (required by USFWS)
5.4 Monitoring Measures (required by USFWS)	5.5 Analysis of Existing Data (required by USFWS)	5.6 Quantifying Impacts (required by USFWS)
5.7 Consistency in Mitigation Standards (required by USFWS)	5.8 Unforeseen Circum-stances/Extraordinary Circumstances (required by USFWS)	5.9 Designation of Responsibility
6. Additional Items		
6.1 Sources (required by USFWS)	6.2 Titled — Draft and dated (required by USFWS)	6.3 Acronyms
6.4 Glossary	6.5 Term of Incidental Take Permit	6.6 HCP and Incidental Take Permit Renewal
6.7 Suspension/Revocation		

The results of the analysis of scientific literature, with the US Fish and Wildlife Services, Endangered Species Habitat Conservation Planning Handbook, and Benton County Oregon's, Benton County Prairie Species Habitat Conservation Plan, Revised Draft was a Habitat Conservation Planning Protocol that included every item covered by each of the three element sources. This study suggests five conceptions including: *1) Factual basis:* Used to asses existing and projected conditions, identify problems, and provide an informational base upon which goals and objectives rely. Includes an inventory of existing resources, issues, policies, and stakeholder's interests within the ecosystem. Factual basis is further categorized into resource inventories to understand what is there, ownership patterns to characterize existing habitat management, and human impacts that identify resource problems from human development associated with the ecosystem. *2) Goals and Objectives:* Are catalysts for action. Goals and objectives help prioritize issues and problems facing a community. They should be spatially specific and very detailed in their aims to protect the functionality of the ecosystem and its unique landscapes and species. *3) Inter-Organizational Coordination and Capabilities:* Recognize that planning problems extend beyond political boundaries and jurisdictions. "Coordination and capabilities are essential to defining plan quality because they measure to what degree a local community is able to recognize the transboundary nature of natural systems, and coordinate with other parties both within and outside of its jurisdictional lines" (Brody 2003). *4) Policies, Tools, and Strategies:* Are used to set forth specific principles of land use design or development management. Policies should be derived from goals and objectives, but focus directly on government action. Policies, tools, and strategies set forth the actions to protect the ecosystem. *5) Implementation:* Explains how an adopted plan can endure through regulations and collective action. Implementation conceptualizes commitment to implementing the final plan in the future. It does not tell how well the plan will actually be implemented. In addition, the Handbook was analyzed and all required and recommended elements listed were added to the protocol. Lastly, the Benton County Oregon HCP was analyzed and all themes, concepts, as well as specific items covered and included in the plan were added to the protocol. Many of these items corresponded to the legal requirements set forth by US Fish and Wildlife Service, but many non-required elements were used for additional information, clarification, or support purposes.

Components taken from the three element sources were combined and categorized under the five conceptualizations. Many of these items had overlapping themes or concepts and therefore were combined as the individual protocols were combined.

The single protocol developed in phase II was localized using four local area plans for Lincoln/Lancaster County Nebraska. These plans included: Environmental Resources section of the Lincoln/Lancaster County Comprehensive Plan; Salt Creek Tiger Beetle (*Cicindela nevadica lincolniana*) Draft Recovery Plan; Implementation Plan for the Conservation of Nebraska's Eastern Saline Wetlands; and Little Salt Creek Watershed Master Plan.

During the localization process several components did not relate to conditions or the environment of Nebraska's eastern saline wetlands. These components were removed from the protocol. An additional category was added for Additional Items that did not fit into the five conceptualizations, but were either required by the US Fish and Wildlife Service or were deemed important to the HCP for other reasons. Finally, many of the components were identified as examples of types of Goals (for example), were generalized within the protocol and individual items removed from the protocol. These examples are listed in the Recommendations Section below. The resulting Habitat Conservation Planning Protocol includes the remaining components, justifications for why each was deemed necessary to HCP development, and a list of sources that further clarify, justify, or give examples.

7. POLICY IMPLICATIONS AND RECOMMENDATIONS

During the analysis and development of the Habitat Conservation Planning Protocol, several general recommendations for HCP development and environmental planning were discovered. Some of these recommendations provide additional justification for included protocol elements, others relate to ecosystem management in general, and still others provide examples for use when developing a HCP. The following is a list of recommendations for preparing a high quality Habitat Conservation Plan.

1. Prepare a strong existing resource inventory section to increase overall plan quality and give the community a sense of responsibility for existing resources.
2. Incorporate Geographic Information Systems (GIS) technology into plans for use as an analytical tool and aid in educating people about complex problems and issues.
3. Increase monitoring activities, such as adaptive management, to ensure the plan will change with the environmental conditions, and to help identify and mitigate adverse impacts.

4. Ensure goals and policies are clear and specific in order to guide the implementation process.
5. Choose incentive-based policies to allow parties to meet their objectives while protecting natural resources. Examples include: density bonuses; clustering away from habitats; transfer of development tights; preferential tax treatments; and mitigation/habitat banking.
6. Use education-based policies to help change behavior, build public awareness, and generate proactive ecosystem management practices.
7. Plan at the ecosystem level to consider the full range of interactions among habitat components.
8. Assess the cumulative effects of impacts over time and/or space.
9. Ensure chosen mitigation measures are truly effective.
10. Ensure the preservation of the ecological integrity of habitats. Ecological integrity includes: protecting communities over single-species, protecting biological diversity, and protecting ecosystem functions and services.

CONCLUSION

The results of this study provide a Habitat Conservation Planning Protocol for Nebraska's Eastern Saline Wetlands for the Saline Wetland Partnership to use for the creation of a Multi-Species Habitat Conservation Plan for the saline wetlands of Lancaster County, Nebraska. This protocol along with its list of recommendations will serve as a tool during the HCP process, and ensure the resulting HCP will be a high quality plan.

By incorporating practical elements from the Benton County Prairie Species Habitat Conservation Plan Revised Draft, required elements from the USFWS's Endangered Species Habitat Conservation Planning Handbook, with required quality assurance components from previous studies, Implementing the Principles of Ecosystem Management through Local Land Use Planning, using this protocol will assure the Saline Wetland Partnership that all legal and practical requirements will be met, as well as the guaranteed quality of the resulting MSHCP. The protocol was further analyzed for its use specifically for Nebraska's eastern saline wetlands through the localization process using the Environmental Resources section of the Lincoln/Lancaster County Comprehensive Plan, the Salt Creek Tiger Beetle Draft Recovery Plan, the Implementation Plan for the Conservation of Nebraska's Eastern Saline Wetlands, and the Little Salt Creek Watershed Master Plan.

Each component listed in the protocol should be addressed in the Eastern Saline Wetlands Habitat Conservation Plan. Additionally, the protocol lists the general reason why each component should be included and corresponding resources that address that particular component. Additionally, this planning protocol is a flexible tool, that with little to no modification may be applied as a HCP planning protocol for other the creation of HCP's for other landscapes throughout Nebraska. A good amount of conservation work has already been accomplished for the endangered saline wetland landscape, but there is still much to do. It is hoped that the creation of this HCP protocol will become a catalyst in the creation of an eastern saline wetland habitat conservation plan for Lancaster County.

The protocol gives guidance for considering environmental resources, as policy and developmental decisions are made. The HCP will allow for the creation of public-private partnerships as well as an intra-county partnerships between Lancaster and Saunders Counties. Also, a HCP for the saline wetlands will allow for careful management, ensuring their long-term viability.

REFERENCES

Associated General Contractors of America (2007). *Nebraska threatened and endangered species identification guide*. http://www.nlc.state.ne.us/epubs/R6000/H053-2007.pdf

Beckmeyer, R.J. and Huggins, D.G. (1998). *Checklist of Kansas damselflies*. http://www.emporia.edu/ksn/v44n1-march1998/slideshow/pages/ v44n1p12n32_gif.htm

Biggs, K. (2010). California Damselflies aka California Zygoptera. Retrieved June 7, 2010, http://southwestdragonflies.net/damsels/2_Zygoptera.html

Brody, S.D. (2003). Implementing the principles of ecosystem management through local land use planning. *Population and Environment*, Vol. 24, No 6, July 2003 Human Sciences Press, Inc.

Brust, M.L., Hoback, W.W., Wright, R.J. (2008). *The grasshoppers of Nebraska*. University of Nebraska Extension.

Currie, D. (2010). Ecosystem-based management in multilateral environmental agreements: progress towards adopting the exosystem approach in the international management of living marine resources. Retrieved July 5, 2010 from. http://assets.panda.org/downloads/ wwf_ecosystem_paper_ final_wlogo.pdf

Department of the Interior (2010). *Approach to adaptive management*. Retrieved July 2, 2010 from http://www.doi.gov

Environmental Protection Agency (1993). Habitat evaluation guidance for the review of environmental impact assessment documents. http://www.epa.gov/Compliance/resources/ policies/nepa/habitat-evaluation-pg.pdf

Farrar, J. and Gersib, R. (1991). Nebraska salt marshes: the last of the least. http://www.lincoln.ne.gov/city/parks/parksfacilities/wetlands/links/saltmarsh.pdf

Gilbert, M., Stutheit, R.,Hickman, T., Wilson, E., Taylor, T. (1994). Resource categorization of Nebraska's eastern saline wetlands.

Haddock, M. (2010). Kansas wildflowers and grasses: saltmarsh aster. http://www. kswildflower.org/flower_details.php?flowerID=280

Harvey, F.E., Ayers, J.F., Gosselin, D.C. (2007). Ground water dependence of endangered ecosystems: Nebraska's eastern saline wetlands. *Ground Water* 45:6. DOI: 10.1111/j.17456584.2007.00371.x.

LaGrange, T., Grenich, T. Johnson, G. Schulz, D., Lathrop, B.(2003). Implementation plan for the conservation of Nebraska's eastern saline wetlands.

Lincoln Convention and Visitors' Bureau (2010). *History*. Retrieved June 20, 2010 from http://www.lincoln.org/visiting/whyvisit/history.

Lincoln Parks and Recreation (2010). What are saline wetlands? Retrieved June 5, 2010, from http://www.lincoln.ne.gov/city/parks/parksfacilities/wetlands/ wetlandsinfo.htm

Lincoln Parks and Recreation (2010). Saline wetlands conservation partnership. Retrieved June 20,2010, from http://www.lincoln.ne.gov/city/parks/ Mparksfacilities/wetlands/wetlandspartnership.htm

NatureServe (2010). Ecosystem-based management tools network. Retrieved July 2, 2010 from http://www.ebmtools.org

Nebraska Game and Parks Commission (2005). Guide to Nebraska's wetlands and their conservation needs. http://outdoornebraska.ne.gov/wildlife/ programs/wetlands/pdf/wetlandsguide.pdf

Schneider, R., Humpert, M., Stoner, K., Steinauer, G. (2005). The Nebraska legacy project a comprehensive wildlife conservation strategy. http://outdoornebraska.ne.gov/wildlife/programs/legacy/review.asp

U.S. Department of Agriculture (2009). Plant fact sheet Texas dropseed. http://www.plant-materials.nrcs.usda.gov/pubs/txpmcfs9183.pdf

U.S. Department of the Interior. U.S. Fish and Wildlife Service (2003). *Endangered species act of 1973: as amended through the 108th congress*. Washington: GPO.

US Fish and Wildlife Service 1996. *Endangered species habitat conservation planning handbook*.

U.S. Fish and Wildlife Service (2005). *Habitat conservation plans: section 10 of the endangered species act.* http://www.fws.gov/Endangered/pdfs/HCP/HCP_Incidental_Take.pdf

U.S. Fish and Wildlife Service (2008). *A history of the endangered species act of 1973*. http://www.fws.gov/endangered/factsheets/history_ESA.pdf

U.S. Fish and Wildlife Service (2009). *Draft recovery plan for the Salt Creek tiger beetle*. U.S. Fish and Wildlife Service, Lakewood, CO.

In: Environmental Management
Editor: Henry C. Dupont

ISBN: 978-1-61324-733-4
© 2012 Nova Science Publishers, Inc.

Chapter 11

A PROCESS TO DETERMINE ONE ORGANIZATION'S ENVIRONMENTAL MANAGEMENT SYSTEM

Lindsay Thompson and Shirley Thompson[*]
Natural Resources Institute, University of Manitoba,
Winnipeg, Manitoba, Canada

ABSTRACT

Implementing an Environmental Management System (EMS) can reduce operational costs for an organization as well as increase employee morale. An EMS is a component of an organization's overall management system that commonly includes an environmental policy, environmental aspects, objectives, targets, actions, significant aspects, training and auditing. Reasons for different approaches to environmental management include managerial interpretations of environmental issues as threats or opportunities, top management attitudes toward the environment, organizational champions pushing for a more proactive environmental approach, stakeholder pressures and regulations. An EMS provides a means to benefit business and environmental goals simultaneously, if it can integrate the two into day-to-day decisions. But how is this done? This paper provides a case study of integrating sustainability in one medium sized organization through both quantitative and qualitative methods to develop an EMS.

INTRODUCTION

Demands that organizations have better environmental performance continue to increase over time (Hart, 2005; Savitz and Weber, 2006). Organizations are challenged to integrate environmental demands with business needs (Rueda-Manzanares, Aragon-Correa and Sharma, 2008). An environmental management system (EMS) provides a means to benefit business and environmental goals simultaneously, if it can integrate the two into day-to-day decisions.

---

[*] Corresponding Author: phone: (204) 474-7170 fax: 204-261-0038. e-mail: s_thompson@umanitoba.ca.

The purpose of an EMS is to reduce an organization's environmental impact and increase its sustainability. An EMS is a component of an organization's overall management system that commonly includes an environmental policy, environmental aspects, objectives, targets, actions, significant aspects, environmental programmes, legal requirements, training and awareness, control, plans for addressing non-conformities, monitoring, management reviews and auditing (Morrow and Rondinelli, 2002; Rondinelli and Vastag, 2000). Environmental management systems are designed to improve environmental performance, which can reduce costs by lowering compliance costs, reducing waste, and improving efficiency and productivity (Ambec and Lanoie, 2008; Hart, 1995; Hart and Ahuja, 1996).

Reasons for different approaches to environmental management include managerial interpretations of environmental issues as threats or opportunities, top management attitudes toward the environment, organizational champions pushing for a more proactive environmental approach, stakeholder pressures and regulations (Bansal and Roth, 2000). Research demonstrates that environmental performance can lead to a competitive advantage through improvements in legitimacy (Bansal and Clelland, 2004), strengthening firm reputation (Hart, 1995; Miles and Covin, 2000), product differentiation (Ambec and Lanoie, 2008; Porter and van der Linde, 1995), international competitive advantages (Hart, 1995; Miles and Covin, 2000), greater appeal to consumers (Miles and Covin, 2000), the development of new market opportunities and better access to markets (Ambec and Lanoie, 2008), selling of pollution control technology (Ambec and Lanoie, 2008) and the creation of entry barriers (Dean and Brown, 1995; Hart, 1995; Russo and Fouts, 1997).

Environmental performance also offers regulatory advantages by leading to greater flexibility to adapt to legislative changes (Bansal and Bogner, 2002), by reducing or avoiding legal liabilities (Hart, 1995; Rooney, 1993) and through the ability to influence environmental laws and regulations (Faucheux et al., 1998; Hart, 1995; Hillman and Hitt, 1999; Miles and Covin, 2000). Conversely, negative implications are connected to poor environmental performance.

Events, such as an oil spill, can have a large effect on firm profitability with potential liabilities, fines, penalties, and clean-up costs that investors react to (Bansal and Clelland, 2004). As well, Bansal and Clelland (2004) found that the firms considered to be environmentally illegitimate experienced higher unsystematic risk. Konar and Cohen (2001) determined that legal chemical releases reported to the TRI had a significant negative effect but a 10 percent reduction in emissions resulted in a $34 million increase in market value. The question remains how to integrate business and environmental concerns. This paper provides a case study of integrating sustainability in one medium sized organization through both quantitative and qualitative methods to develop an EMS.

The quantitative approach rates the priorities based on the definition of sustainable development: "Sustainable development is development that meets the needs of the present without compromising the ability of future generations to meet their own needs (Brundtland, 1987, p. 43)". The qualitative methods include surveys of employees and managers and focus groups. The integration of these qualitative and quantitative methods into an EMS is discussed below.

METHODS

Five environmental areas were considered in this qualitative and quantitative analysis namely: energy use, transportation, waste production, water usage, and green purchasing.

The method is provided in two sections: 1) quantitative approach; and 2) qualitative approach.

1) Quantitative Analysis to Determine Sustainability

To help determine the economic rating and other impacts, baseline data was collected including a benchmark energy audit and waste audit to determine the feasibility of various actions that the organization will use the data to monitor the success of the EMS. The three pillars of sustainability (Brundtland, 1987) provided the framework for rating impacts over three areas, namely: *environment, social and economic*. Each impact was rated on a three point scale from zero to two and the three scores were added, with six being highest priority and zero being lowest priority.

Environmental impact ratings were assigned. A rating of two signified environmental harm with impacts causing irreversible damage, releasing a persistent toxin or carcinogen, utilizing non-renewable energy, or virgin materials. A moderate environmental impact activity received a rating of one for releasing a toxic chemical that is not persistent or a carcinogen, using renewable energy sources, or over consuming recycled materials. A minor or sustainable environmental impact is reversible, non-toxic, or uses limited recycled materials and would receive a rating of zero.

Social impact on culture and health ratings considered environmental justice. Environmental justice literature discusses how dominant cultures often affect marginalized cultures by destroying their livelihoods and cultural integrity through different means. An action that causes long-term negative impacts on culture or health received a rating of two. An action that caused moderately negative impacts on culture or health received a rating of one. A minor social impact that caused minimal, short-term negative impact on culture or health received a rating of zero.

Economic impact rating was based on payback period. An action requiring short payback period of less than two years received a rating of two. An action requiring two to five year payback period received a rating of one. A payback period over 5 years would receive a rating of zero.

2) Qualitative Approach to Determine Priorities

The qualitative approach involved a survey, focus group and many joint management meetings.

a. Survey

To determine the areas and level of interest in the environment, an electronic survey was emailed out to every employee, which consisted of 33 questions. The survey questions focused on environment, health and quality of life to determine significant environmental aspects. Sixty-one employees of 150 employees responded to the environmental survey; the response rate was 40.3%. The average survey completion time was 10 minutes.

b. Focus Group

After receiving the survey results, a focus group session was held with a number of employees and managers to determine the environmental priorities of the employees and managers. At this meeting the findings of the survey were reviewed.

c. Meetings with Management

A number of meetings with management provided feedback as to what was the priority of upper management and the financial limits to environmental action. Ultimately the management determined the EMS targets and priorities with the knowledge of the survey, focus group and sustainability rating.

Findings

The five environmental aspects that make up the EMS are discussed in the five sections below. The findings for the quantitative sustainable rating are summarized in table 1 and then compared to qualitative findings in table 2 as to whether it is rated as a high, moderate or low priority.

Table 1. Quantitative Rating of Sustainability by Organization

Area	Aspect	Environmental	Social	Economic	Sustainability Rating Score	Overall Rating
Energy	Heating	2	2	1	5	Moderate
	Lighting	2	2	1	5	Moderate
	Windows	2	2	0	4	Low
Transportation	Bicycle	2	2	2	6	High
	Carpool	2	2	2	6	High
	Transit	2	2	2	6	High
Waste	Styrofoam	2	0	0	2	Low
	Paper usage	2	2	2	6	High
	Composting	1	0	0	1	Low
Green Purchasing	Green purchasing	2	2	2	6	High
Water	Water usage	1	2	2	5	High

Table 2. Comparison of Sustainability Rating, Employee Survey Results, and EMS Action Plan for the Five Environmental areas

Area	Category	Sustainability Rating	Employee Survey & Focus groups	Action Plan
Energy Use	Heating	High	High	- Future initiative
	Lighting	High	High	- Apply for Manitoba Hydro's Commercial Lighting Program to retrofit existing light fixtures - Involve students in designing a new lighting scheme for the Home Office - Retrofit existing light fixtures - Reduce the number of light fixtures
Transportation	Carpool	High	Moderate	- Participate in the Commuter Challenge - Research viability of carpooling and bicycle storage - Explore charging nominally for parking
	Bicycle	High	Moderate	- Participate in the Commuter Challenge - Research viability of bicycle storage - Explore charging nominally for parking
Transportation	Public transportation	High	Moderate	- Participate in the Commuter Challenge - Research viability of subsidized transit passes - Explore charging nominally for parking
Waste	Styrofoam	Low	Moderate	- Reduce use of disposable cups and plates - Educate about methods for reducing paper usage - Conduct a waste audit - Reduce paper usage
	Paper usage	High	Low	- Increase green purchasing - Increase the variety of products that are recycled
	Composting	Low	Low	- Implement a composting program
Green Purchasing	Green purchasing	High	Low	- Provide training for the EMS Coordinator about green purchasing
Water	Water usage	High	Low	- Install sink aerators - Replace urinals with waterless urinals - Replace toilets with low flush

Table 3. Environmental Management System applied to different aspects

Aspect	Objective	Target	Activities	Responsibility
Energy	Reduce energy consumption	Yearly 5% reduction in energy usage	- Apply for Commercial Lighting Program to retrofit existing light fixtures - Involve students in designing a new lighting scheme for the Home Office - Conduct an energy audit - Reduce the number of light fixtures	- EMS Coordinator - Maintenance
Transportation	Encourage employees to take alternative transportation	Yearly 10% increase in alternative transportation	- Participate in the Commuter Challenge - Research viability of subsidized transit passes, carpooling and bicycle storage - Explore charging nominally for parking	- EMS Coordinator
Waste Generation	Reduce, divert and prevent waste and pollution	Yearly 10% reduction in overall waste	- Implement a composting program - Reduce use of disposable cups and plates - Educate about methods for reducing paper usage - Conduct a waste audit - Increase the variety of products that are recycled - Reduce paper usage	- EMS Coordinator - Elsie Bear's Kitchen
Purchasing	Purchase more green products	Yearly 20% increase in green purchasing	- Increase green purchasing - Provide training for the EMS Coordinator about green purchasing	- EMS Coordinator
Water Usage	Reduce water usage	Yearly 10% reduction in water usage	- Install sink aerators - Replace urinals with waterless urinals - Replace toilets with low flush	- EMS Coordinator - Maintenance

Table 4. The Company's Energy Usage and Estimated Savings from EMS Activities

Energy	Indicator	Energy usage*	Percent of bill	Cost per annum	Target	Reduction*	Consumption*	Saving
Electricity	Lights, hot water and miscellaneous equipment	1,197,504	38%	$55,376	13%	157,911	1,039,593	$7,302
	Heating equipment	71,280	2%	$3,296	0%	0	71,280	$0
	Cooling equipment	303,534	10%	$14,036	0%	0	303,534	$0
	Total	1,572,318	50%	$72,708	13%	157,911	1,493,702	$7,302
Natural gas	Lights, hot water and miscellaneous equipment	33,203	1%	$1,328	0%	0	33,203	$0
	Heating equipment	1,800,993	49%	$72,009	0%	0	1,800,993	$0
	Cooling equipment	0	0%	$0	0%	0	0	$0
	Total	1,834,196	50%	$73,337	0%	0	0	$0
	Grand Total	3,406,514	100%	$146,045	13%	170,326.00	3,236,188	$7,302

Notes: *The energy was measured using kilowatt-hour equivalent (kWhe). Electricity is measured in kilowatt hour (kWh) and natural gas is converted to the equivalent energy in kilowatt hour (kWhe).

The rationale behind each aspect's rating is provided in each section below. The qualitative survey and focus group findings are summarized in table 2 in a very succinct way with a more full description for each aspect in each of the five sections below. Finally the targets that were determined for each aspect with the objectives, activities, etc are provided in Table 3. In each aspect below, how the target and activities were determined by either quantitative or qualitative means are discussed.

Generally, the survey response rate demonstrated the high level of interest in the environment and support for addressing the organization's environmental impacts. In total, thirty-five of the surveyed employees (57%) rated their level of interest in environmental issues as high, very high, or extremely high, while twenty-four had moderate interest (39%) and two employees had slight interest (3%). None of the employees surveyed stated that they have no interest in environmental issues.

Environmental initiatives were strongly supported as sixty (98%) of employees surveyed stated that it is important or very important for the organization to address its impact on the environment while one (2%) stated that it is less important.

Energy Use

The energy index for the organization's building, where 150 people work, is 32.34-kWh/sq. ft, which is slightly higher than the average of 29.80-kWh/sq. ft for similar office buildings. Heating accounts for 98.6% of the natural gas consumption and lights, hot water, and miscellaneous equipment account for the majority of electricity (75.8%) (Table 3). Lighting accounts for most of the electrical use at 38% of total energy use. Hawken, Lovins, and Lovins (1999), recommend beginning retrofitting initiatives with lighting as it can alter the building's need for heating and cooling.

Natural gas is a non-renewable energy source (environment rating of two) that is toxic and its production, and transportation causes significant negative health and social impacts (social rating of two). In terms of economics, the payback period for a new energy efficient boiler is two years to five years (rating of one for payback), as it would reduce heating costs by 30% and has a rating for heating of five. New windows would reduce natural gas use and therefore have the same significant negative social and environmental impact ratings as natural gas. The payback period for new windows would be more than five years (economic rating of zero) resulting in an overall sustainability rating of four (Table 1).

Unlike natural gas' use in heating, the electricity used to provide lighting is considered to be renewable (environment rating of one). The majority of electrical energy in Manitoba is produced at fourteen generating stations located on the Nelson River, the Winnipeg River, the Saskatchewan River, and the Laurie River (Manitoba Hydro, nd). Hydroelectricity generation in Canada causes significant social and health impacts (social rating of two) as hydro generation has resulted in major negative consequences for northern people which are mainly Aboriginal peoples (First Nation and Métis) by flooding large areas of land permanently, causing mercury contamination, impacting hunting and trapping, and disrupting communities (Kulchyski and Neckoway, 2006; Boyd, 2003). According to Manitoba Hydro, replacing existing lighting with energy efficient lighting could have a payback period of less than two years (economic rating of two) resulting in an overall sustainability rating of five (Table 1).

The heating system, windows, and lighting are a great concern for employees because they are not at comfortable levels rather than because they are energy inefficient. The environmental survey identified that employees are most concerned with the thermal comfort due to poor regulation of temperature and drafts from windows and secondly the lighting. In total, 11 of the employees surveyed were uncomfortable (18%) and 23 were less comfortable (38%) with the temperature and air exchange in their workspace. Twenty-two employees were comfortable (37%) and four were very comfortable (7%). Clearly, the temperature was not well regulated with many employees complained that: office space was too cold (22 employees), office space was too hot (18 employees), cannot control heating and cooling (6 employees), too stuffy because there is not enough ventilation; (6 employees), and too much dust (4 employees). Focus group participants recommended new windows that open for comfort and safety, improved thermal climate and cleaner and increased ventilation.

To reduce the environmental, social, and economic impact, the organization decided to reduce energy consumption by 5% annually by retrofitting and replacing existing light fixtures and conducting an energy audit (Table 2). As lighting accounts for 38% of total energy usage the savings from achieving the 5% target in reducing energy consumption are estimated to be $7,302/year. Heating and windows were not selected as initial actions, as they require major renovations to the building. However, a low cost solution of changing the set points for temperature may reduce energy while improving thermal comfort.

Transportation

Commuting to work by car creates noise and air pollution with negative social impacts on health, stress, families and communities. Benefits of more active transportation include improved fitness and health for employees. As well, alternative transportation provides an opportunity to use the organization's parking spaces for other uses such as green spaces or increase revenue by renting more parking spaces to neighborhood businesses. By people shifting from personal vehicles, the extra parking spaces (currently given to employees at no charge) would be available to generate revenue for daily or monthly parking leases. Market research is also required to determine the feasibility and possible monthly cost for leasing additional monthly parking in the Home Office parking lots.

Single vehicle transportation has a significant environmental impact including the release of toxic air pollution, being energy intensive, and mostly relying on non-renewable energy (rating of two for environmental impact). The greatest contributor to air pollution in Canada is the transportation sector (Boyd, 2003). The production and transportation of gasoline causes significant negative health and social impacts (social rating of two). The methods for encouraging alternative transportation, as further discussed, all have a payback period that would be less than one year (rating of two) resulting in an overall sustainability rating of six (Table 1).

According to the environmental survey, most employees drive to work: transportation baseline data collected from the environmental survey demonstrates that 48 employees (80%) travelled to work by personal vehicle; six employees used public transportation (10%); two employees carpooled (3%); one employee walked to the Home Office (2%); one employee carpooled and used public transportation (2%); one employee used their own vehicle and public transportation (2%); and one employee used a personal vehicle in the winter and walked in the summer (2%).

To help shift employees from driving their personal vehicles to bicycling, taking public transportation, or carpooling, the transportation target is a yearly 10% increase in alternative transportation. This shift will be encouraged by: participating in a yearly alternative transportation challenge, researching the viability of subsidized transit passes, carpooling and bicycle storage; and exploring charging nominally for parking (Table 2). The organization's target should be easily attainable as 24 of the surveyed employees (40%) would be willing to take transit to work if their monthly passes were subsidized. If the organization were to organize carpooling, 30 surveyed employees (49%) would be willing to carpool to work. In addition, if the organization were to offer bicycle storage facilities, 23 surveyed employees (38%) would be willing to cycle to work. Alternative transportation was both considered highly sustainable in the quantitative rating and was of interest to employees. Nevertheless, many employees may have trouble modifying their behavior.

Waste Production

Waste production indicates over consumption of resources and impacts the environment negatively through leachate, landfill gas, toxins and gases emitted as a result of incineration and product degradation, etc. As recycled materials can act as substitutes for raw materials in many manufacturing processes, recycling helps to reduce the amount of raw materials extracted and processed, as well as their disposal in landfills. Organizations can reduce waste by decreasing the use of disposal products and reducing paper use, as well as promoting recycling and composting materials over land filling these materials. Reducing the amount of paper this organization uses would mitigate negative environmental impacts, as the forest industry is a contributor to air and water pollution, waste production, and non-renewable energy consumption (environment rating of two). Conventional forestry results in deforestation, relies on non-renewable energy consumption to power machinery and for transportation, and contributes to greenhouse gas production (Kissinger and Rees, 2006). Conventional forestry significantly impacts hunting and trapping, which are traditional and important cultural activities of Aboriginal peoples (social rating of two). Reducing paper usage would have a payback period of less than one year (economic rating of two) although it would require training to promote behavioral or technological changes to print double sided, review papers on-line, reuse paper for scrap and send online memos. The overall rating for reducing paper usage is high (sustainability rating of six) (Table 1).

According to the results from the organization's waste audit the organization uses approximately 1.5 lbs of styrofoam per day. Styrofoam is produced from petrochemicals and is not recyclable (environment rating of two). Styrofoam usage has minimal social impacts (rating of 0 for social impacts). Replacing disposables with biodegradables has a long payback period that is greater than five years (economic rating of zero). The overall sustainability rating for reducing and replacing styrofoam is low (sustainability rating of two) (Table 1).

Composted organic waste displaces potentially toxic chemical fertilizers to enrich soil (environment rating of one) although the social impact of not composting is low (social rating of zero). The payback period for composting would take more than five years for payback (economic rating of zero). Implementing a composting program requires employee education and training for the employees of the cafeteria. The compost pickup organization provides clear directions on what materials they can compost. The overall sustainability rating for composting is low (sustainability rating of one) (Table 1). As funding was available for this item by partnering with the Natural Resource Institute at the University of Manitoba and the Province of Manitoba's Waste Reduction and Pollution Prevention Fund, a composting program was initiated. Thus, although composting was not seen as a priority in the qualitative research or had a high sustainability rating it was implemented.

Employees discussed the need to reduce paper usage in the environmental survey and the focus group session. "I feel the organization could improve on cutting back on the amount of paper we use. We could do more correspondence on-line." The focus group identified waste reduction and increased recycling as priorities but did not mention composting or reducing the usage of styrofoam cups. We did not specifically ask about styrofoam in the organization environmental survey although one employee requested, "Can we please get rid of the styrofoam cups!" Employees did not mention composting as a means to reduce waste in the environmental survey.

Composting was included in the companies EMS to reduce the amount of waste being land filled: the target for the organization was set at a 10% yearly reduction in overall waste by implementing a composting program; reducing the use of disposable cups and plates; educating organization employees about methods for reducing paper usage; conducting a waste audit; increasing the variety of products that are recycled; and reducing paper usage (Table 2). The organization now has weekly commercial composting pickup by a local organization. The private organization that runs the organization's kitchen could choose to include composting as a requirement under their contract.

Water Usage

Unnecessary water consumption is a significant contributor to pollution, waste, and energy usage. Canadians use the second largest amount of water per capita (Boyd, 2003). The average Canadian uses 343 litres per day, which equals 500,000 litres annually per household (Body, 2003). As part of Canadian society, the organization does have a number of wasteful water practices that can be remediated with effort (environment rating of one). Water wastage has a negative social impact for people who still live off the land and clean water is essential to Aboriginal and all cultures (social rating of two). Affordable and accessible strategies, such as installing sink aerators and fixing leaks, are available to significantly reduce water consumption with less than a year payback (economic rating of two). The overall sustainability rating for reducing water wastage is high (sustainability rating of five) (Table 1).

Employees discussed methods for reducing water consumption during the organization's environmental survey. However, this issue was not identified as a priority issue in the survey or focus groups. The most common solution identified for eliminating water wastage was to install low flow toilets. Focus group participation also discussed methods for reducing water usage during the focus group and management discussed methods during meetings. The target for the organization was set at 10% yearly reduction in water usage by installing sink aerators; replacing urinals with waterless urinals; and replacing toilets with low flush models (Table 2).

Green Purchasing

Green purchasing reduces over consumption of resources and increases sustainability. Source reduction of waste and toxic chemical use is the best way to reduce environmental impacts such as greenhouse gas emissions, waste generation and toxic chemical exposures (Mohareb, Warith, and Narbaitz, 2004; Min and Galle, 1997). Cleaning, office, and other products, often include carcinogens, neurotoxins, and heavy metals that negatively impact health (social impact of two) (Government of Ontario, 2007), or cause undue environmental impacts (environment rating of two). Paper that is typically used by the organization is from virgin forests. There is a short monetary payback for green purchasing as the organization can purchase recycled paper and green cleaning supplies for the same cost as non-environmentally friendly paper and cleaners (economic rating of two). Green cleaning products and recycled paper are both great options for reducing an organization's environmental impacts as they replace products that can cause large environmental and health impacts. The overall sustainability rating for green purchasing is high (sustainability rating of six) (Table 1).

The survey asked employees whether the organization's kitchen should offer more organic and fair trade options, which was selected by 19 (33%) of the responding employees. Fifteen employees indicated that the organization should use more green cleaning products (26%). Changing from the current situation of various departments purchasing their supplies in small units to green purchasing for the entire office could easily reduce the organization's environmental impact. To achieve their target, the organization will have to provide training for the accounts departments to increase their knowledge about green purchasing such as what is considered a green product and the benefits of purchasing green (Chen, 2005). The organization plans to eventually incorporate bulk purchasing into their purchasing plan. Based on participatory methods the target for the organization was to increase green purchasing by 20% annually.

CONCLUSION

The sustainability rating system and the data gathered during the participatory process were both valuable for the process of identifying aspects and determining actions and targets. Combining these methods effectively integrated business and environmental issues for the organization. The sustainability rating system was valuable for ensuring that the organization considered environment, social/culture, and economic concerns in their EMS. The participatory process was important to build support for the EMS and education on environmental health issues for employees and management.

This analysis comparing quantitative and qualitative analysis showed that different priorities can result from different methods. For example, only two of the activities, heating and lighting, were viewed as both having high employee and management interest during the participatory research as well as being sustainable. However, many highly sustainable activities had moderate employee interest (Table 2). Some activities were considered unsustainable due to cost, which was a valuable input of the sustainability calculator.

Many of the activities that were calculated to be highly sustainable included bicycling, carpooling, transit, lighting and reduced paper and water usage require that both the employees modify their behavior and the organization provide supports such as showers for bike users (Table 4). The activities that were most requested by the employees including heating, and windows are not lifestyle issues but rather require major renovations for the building. Heating and windows were identified as future actions for the organization but are currently cost restrictive. Composting and reducing styrofoam received a low sustainability rating, due largely to the economic aspect, and had the lowest demand and interest by employees but composting was addressed due to governmental project funding being provided in that area. Thus, government incentives and regulation can skew priorities.

It should be noted that environmental health aspects were not fully taken into consideration by the sustainability calculator that was applied. If it had, there may be more impetus to make different changes, as air quality is a significant factor for environmental health. Air quality issues were raised by employees as their major concern. These issues have important economic and social impacts as poor air quality results in more employee sick days and less efficiency while at work. Although the sustainability calculator is useful as is, it would be improved by taking into account occupational and environmental health.

REFERENCES

Ambec, S., and Lanoie, P. 2008. Does it pay to be green? A systematic overview. *The Academy of Management Perspectives,* 22(4): 45-62.

Anton, W.R., Deltas, G., and Khanna, M. (2003). Incentives for Environmental Self-regulation and Implications for Environmental Performance. *Journal of Environmental Economics and Management*, 48, 632-654.

Banerjee, S. B. (2001).Managerial Perceptions of Corporate Environmentalism: Interpretations From Industry and Strategic Implications for Organizations. *Journal of Management Studies,* 38 (4), 489-513.

Bansal, P., and Bogner, W. C. 2002. Deciding on ISO 14001: Economics, institutions, and context. *Long Range Planning*, 35(3): 269.

Bansal, P., and Clelland, I. 2004. Talking trash: Legitimacy, impression management, and unsystematic risk in the context of the natural environment. *Academy of Management Journal,* 47(1): 93-103.

Bansal, P., and Roth, K. 2000. Why companies go green: A model of ecological responsiveness. *Academy of Management Journal*, 43(4): 717-736.

Bartlett, J.G., Iwasaki, Y., Gottlieb, B., Hall, D., and Mannell, R. (2007). Framework for Aboriginal-guided Decolonizing Research Involving Métis and First Nations Persons with Diabetes. *Social Science and Medicine*, 65, 2371-2382.

Boyd, D. (2003). *Unnatural law: rethinking Canadian environmental law and policy.* Vancouver: UBC Press.

Brundtland, G. (1987): *Our common future. Report to the World Commission on Environment and Development*. Oxford: Oxford University Press.

Buysse, K., and Verbeke, A. (2009). Proactive Environmental Strategies: A Stakeholder Management Perspective. *Strategic Management Journal*, 24 (5), 453-470.

Chen, C-C. (2005). Incorporating Green Purchasing Into the Frame of ISO 14000. *Journal of Cleaner Production*, 13, 927-933.

Child, J., and Tsai, T. (2005). The Dynamic Between Firms' Environmental Strategies and Institutional Constraints in Emerging Economies: Evidence from China and Taiwan. *Journal of Management Studies*, 42 (1), 95-125.

Crawford, J. (1985). What is Michif?: Language in the metis tradition. J. Peterson, and J. Brown (Eds.), *The New Peoples: Being and becoming Metis in North America* (pp. 231-241). Winnipeg: University of Manitoba Press.

Cuddihy, J., Kennedy, C., and Byer, P. (2005). Energy Use in Canada: Environmental Impacts and Opportunities in Relationship to Infrastructure Systems. *Canadian Journal of Civil Engineering*, 32, 1-15.

Dean, T. J., and Brown. R. L. 1995. Pollution regulation as a barrier to new firm entry: Initial evidence and implications for future research. *Academy of Management Journal*, 38: 288-303.

Dias-Sardinha, I., and Reijnders, L. (2001). Environmental Perfomance Evaluation and Sustainability Performance Evaluation or Organizations: An Evolutionary Framework. *Eco-Management and Auditing*, 8, 71-79. doi:10.1002/ema.152

Gonzalez-Bonito, J. O. (2006). A Review of Determinant Factors of Environmental Proactivity. Business Strategy and the Environment. *Business Strategy and the Environment*, 15 (2), 87-102.

Government of Ontario. (2007). Cleaning products. Retrieved October 21, 2009 from: http://www.ene.gov.on.ca/en/myenvironment/home/cleaningproducts.php

Hart, S. L. 1995. A natural-resource-based view of the firm. *Academy of Management Review*, 20(4): 986-1014.

Hart, S. L., and Ahuja, G. 1996. Does it pay to be green? An empirical examination of the relationship between emission reduction and firm performance. *Business Strategy and the Environment*, 5: 30-37.

Hawken, P., Lovins, A., and Lovins, L. (1999). *Natural capitalism: Creating the next industrial revolution*. Boston: Little, Brown and Organization.

Henriques, I., and Sadorsky, P. 1999. The relationship between environmental commitment and managerial perceptions of stakeholder importance. *Academy of Management Journal*, 42(1): 87-99.

Hillman, A. J., and Hitt, M. A. 1999. Corporate political strategy formulation: A model of approach, participation, and strategy decisions. *Academy of Management Review*, 24(4): 825–842.

Kissinger, M.F., and Rees, W. (2006). Wood and Non-wood Pulp Production: Comparative Ecological Footprinting on the Canadian Prairies. *Ecological Economics*, 62, 552-558. doi: doi:10.1016/j.ecolecon.2006.07.019.

Konar, S., and Cohen, M. A. 1997. Information as regulation: The effect of community right to know laws on toxic emissions. *Journal of Environmental Economics and Management*, 32: 109–124.

Kulchyski, P., and Neckoway, R. (2006). The town that lost its name: the impact of hydroelectric development on Grand Rapids, Manitoba. Retrieved from: http://policyalternatives.ca/documents/Manitoba_Pubs/2006/Grand_Rapids.pdf

Lee, S. Y., and Rhee, S. (2007). The Change in Corporate Environmental Strategies: A Longitudinal Empirical Study. *Management Decision*, 45 (2), 196-216.

Manitoba Healthy Living. (2004). Injuries in Manitoba: A 10-Year Review. Retrieved from: http://www.gov.mb.ca/healthyliving/injury/review.html

Manitoba Hydro. Energy efficient boilers. Retrieved October 21, 2009 from: http://www.hydro.mb.ca/your_business/hvac/hvac_boilers.pdf

Manitoba Hydro. History and timeline. Retrieved October 21, 2009 from: http://www.hydro.mb.ca/corporate/history/hep_1970.html

Miles, M. P., and Covin, J. G. 2000. Environmental marketing: A source of reputational, competitive, and financial advantage. *Journal of Business Ethics*, 23(3): 299-311.

Morrow, D., and Rondinelli, D. (2002). Adopting Corporate Environmental Management Systems: Motivations and Results of ISO 14001 and EMAS Certification. European Management Journal, 20, 159-171. doi: 10.1016/S0263-2373(02)00026-9.

Porter, M. E., and Van der Linde, C. 1995. Toward a new conception of the environment-competitiveness relationship. *Journal of Economic Perspectives*, 9(4): 97-118.

Province of Manitoba. (2007). Province funds development of Manitoba Métis Federation waste reduction programs. Retrieved October 21, 2009 from: http://news.gov.mb.ca/news/index.html?archive=2007-02-01anditem=1201

Ramos, T., and Joanaz de Melo, J. (2006). Developing and Implementing an Environmental Performance Index for the Portuguese Military. *Business Strategy and the Environment*, 15, 71-86. doi: 10.1002/bse.440.

Rondinelli, D., and Vastag, G. (2000). Panacea, Common sense, or Just a Label? The Value of ISO 14001 Environmental Management Systems. *European Management Journal*, 18, 499-510. doi: 10.1016/S0263-2373(00)00039-6.

Rooney, C. 1993. Economics of pollution prevention: How waste reduction pays. *Pollution Prevention Review*, 3(Summer): 261-276.

Rueda-Manzanares, A., Aragon-Correa, J. A., and Sharma, S. 2008. The influence of stakeholders on the environmental strategy of service firms: The moderating effects of complexity, uncertainty and munificence. *British Journal of Management*, 19(2): 185-203.

Russo, M. V., and Fouts, P. A. 1997. A resource-based perspective on corporate environmental performance and profitability. *Academy of Management Journal*, 40(3): 534-559.

Savitz, A.W., and Weber, K. 2006. *The Triple Bottom Line: How Today's Best-Run Companies are achieving Economic, Social, and Environmental Success-and how you can too*. San Francisco: Jossey-Bass.

Sharam, S. and Vrendenburg, K. (1998). Proactive Corporate Environmental Strategy and the Development of Competitively Valuable Organizational Capabilities. *Strategic Management Journal*, 19 (8), 729-753.

Sharma, S. (2000). Managerial Interpretations and Organizational Context as Predictors of Corporate Choice of Environmental Strategy. *Academy of Management Journal*, 43, 681-697.

Sharfman, M.P., Shaft, T. M. and Tihanyi, L. (2004). A Model of the Global and Institutional Antecedents of High-level Corporate Environmental Performance. *Business and Society,* 43 (1), 6-36.

Thompson, L. (2009). Creating a culturally relevant environmental management system for a Métis workplace (Master's Thesis, University of Manitoba, 2009.

U.S. Green Building Council. (2005). *U.S. Green Building Council.* Retreived from: http://www.usgbc.org

United Nations. (1987). *Our common future* (Brundtland report), Oxford University Press, New York, NY.

In: Environmental Management
Editor: Henry C. Dupont

ISBN: 978-1-61324-733-4
© 2012 Nova Science Publishers, Inc.

Chapter 12

ARE ECOSYSTEM MODELS AN IMPROVEMENT ON SINGLE-SPECIES MODELS FOR FISHERIES MANAGEMENT? THE CASE OF UPPER GULF OF CALIFORNIA, MEXICO

Alejandro Espinoza-Tenorio[1], Matthias Wolff[1] and Ileana Espejel[2]

[1]Leibniz-Zentrum für Marine Tropenökologie GmbH, Fahrenheitstr, Bremen, Germany
[2]Facultad de Ciencias, Universidad Autónoma de Baja California, Ensenada, Mexico

ABSTRACT

We review the recent applications of ecosystem models (EMs) as tools for fisheries management in the Upper Gulf of California (UGC), Mexico. EMs are compared with single-species model applications in the UGC, as a basis for assessing the benefits of each ecosystem model as a tool for evaluating management alternatives capable of diminishing impacts on marine ecosystems. The strengths and weaknesses of different types of EMs and their ability to evaluate the systemic mechanisms underlying observed shifts in resource production are also examined with respect to Ecosystem-Based Fisheries Management (EBFM) general goals. Findings showed that ecosystem modeling has increasingly resulted in support for EBFM in the UGC. However, outputs also proved evidence on that EMs are facing a most complicated situation than single-species models regarding the lack of data. Thus, the step from single-species models to EMs is a stage of the management of the area that does not require the elimination of the first approach, but rather the use of both approaches in a complimentary manner. The challenge is the integration of current ecosystem information to detect the gaps in the collective knowledge on the UGC. Insights from this study are valuable in defining a planning model scheme that supports ecosystem-based management policies in local fisheries.

[*] Corresponding Author: Tel.: +52 646 174 5925; fax: +52 646 174 4560. E-mail: ileana.espejel@uabc.edu.mx (I. Espejel).

Keywords: Ecosystem model, Single-species model, Upper Gulf of California.

INTRODUCTION

Ecosystem-Based Fisheries Management (EBFM) is a relatively recent concept that has been promoted by international resource management agreements to sustain healthy marine ecosystems and the fisheries that they support (e.g., Law of the Sea Convention, Code of Conduct for Responsible Fisheries). By reversing the order of management priorities (i.e., starting with the ecosystem rather than the target species), EBFM represents a holistic approach and emphasizes an understanding of the reciprocal and complex interactions between humans and marine resources [1]. Ecosystem models (EMs) are tools used within this overall framework to evaluate ecosystem properties and provide practical information on the potential effects of changes in EBFM practices on the ecosystems [2]. A large number of ecosystem models have been designed in recent decades with the goal of best representing the basic features of an ecosystem, given a limited number of data sources [3,4]. Currently, a standard procedure in fisheries management includes fitting an ecosystem model to data and then describing the impacts of fishing pressure and environmental variability on populations and ecosystems [5].

In general, updating fisheries management in developing countries like Mexico from single-species population models to ecosystem models has been gradual. The main modeling efforts continue to be localized to specific areas, such as the Upper Gulf of California (UGC), which is one of the most diverse marine ecosystems on Earth [9] and one of the best studied in Mexico [7]. Extreme climatic conditions, along with a lack of a connection to the open ocean, have led to particular physical characteristics in the UGC, such as ample tide intervals (10 m), shallow areas and extreme temperature ranges (8 - 30 °C), as well as elevated turbidity, evaporation and salinity indices [8]. These attributes are favorable for several species, which promotes a highly diverse ecosystem. Unfortunately, the biodiversity of the UGC has deteriorated due to human activities related to the diversion of water from the Colorado River for irrigation and municipal uses and due to the increase of artisanal and industrial fishing activities [9]. With the aim of preserving the biodiversity of the area while also planning for the development of fisheries, management strategies have commonly focused on single-species regulations, such as seasonal and spatial closures, minimum size limits, fishing effort controls, and quotas. Currently, conservation efforts have being implemented to repair the damage done to the marine ecosystems. The most significant effort occurred when the Upper Gulf of California and the Colorado River Delta Biosphere Reserve (UGCandCRDBR) was established in 1993, and the most recent effort occurred when the refuge area of Mexico's only endemic marine mammal, the vaquita (*Phocoena sinus*), was established in 2005 [10].

However, the application of fisheries policies in the UGC has not been effective in the recovery of fish or other protected species [11], and even in recent years, the pressure on fisheries resources, such as shrimp, fish and elasmobranchs, has increased dramatically [12]. For several reasons (e.g., insufficient involvement of local communities in management, conservation, and enforcement measures), the success of both single-species and ecosystem-based regulations have been limited, but, according to Morales-Zárate et al. [13], one of the most crucial reasons is a lack of an understanding of ecosystem processes.

Because adaptive planning should proactively anticipate the need to change management practices, learn from experience, and adopt strategies accordingly [14], we carried out a literature review on the current state-of-the-art assessments of this type in the UGC. Ecosystem models were compared with single-species model applications in the UGC, as a basis for assessing the benefits of each ecosystem model as a tool for evaluating management alternatives that are capable of diminishing fishing impacts on marine ecosystems. Although EBFM continues to evolve with a broad discussion of definitions and interpretations of concepts, methods, and scopes [15], there are basic principles that are commonly expressed in the literature such as society participation, the balance of conservation and use, integrated management, spatio-temporal scales, and ecosystem dynamics [16]. Thus, this study discusses results with respect to general goals of EBFM that are directly involved with the main contributions of the ecosystem models, such as ecosystem dynamics and spatio-temporal scales, as well as goals addressing the search for a balance between conservation and use. We think that insights from this study are valuable in defining a planning model scheme that supports ecosystem-based management policies in local fisheries.

ECOSYSTEM MODELING

Starfiel et al. [17] described a model as a purposeful representation of a system that consists of a reduced number of system elements, the internal relationships between these elements, and the relationships between system elements and the surrounding environment of the system. In fisheries science, the first models were single-species population models. These were required to make predictions about the direct responses of target populations and the incidental mortality of other biota for evaluating alternative management choices [18]. However, these models did not provide measurements of indirect effects of the use of a target resource on other biota and the whole ecosystem. Ecosystem models were thus designed to provide information on how ecosystems are likely to respond to changes in management practices involving ecological processes, environmental variables, and social-ecological relationships [2].

However, there is not yet a global consensus on how ecosystem models may be directly used within a standard framework for decision-making in EBFM [19]. EBFM is still being developed in a resource management framework [15] and not only requires an understanding of the system, but must also deal with high levels of uncertainty, divergent interests of stakeholders, and, often, a great urgency for decision-making [20]. Furthermore, the difficulties of constructing, parameterizing, calibrating, and validating ecosystem models complicate their approbation [21]. Contrary to stock assessment models, which have provided robust biological reference points for setting control mechanisms (e.g., spawning-stock biomass, maximum sustainable yield) [18], ecosystem modeling has no methodological benchmarks against which one may evaluate these models [2]. As a result of these unresolved challenges, the structural assumptions of these models are usually not considered, and their reliability is poorly understood [22]. Consequently, the value of ecosystem models may generally be over-estimated in the decision-making process, with the danger of giving more credibility to the outputs of these models than they may warrant [23].

The formal separation of EMs is not always easy, and there are many ways to categorize these models [2,4]. Plagányi [24] designed four categories according to criteria from EBFM, especially in regards to the level of complexity: Extensions of Single-Species Assessment Models (e.g., [25,26]), Dynamic Multi-Species Models (e.g., MSVPA [27]; IBM [3]; MULTSPEC [28]), Dynamic System Models (e.g., OSMOSE [29]; ATLANTIS [30,31]), and Whole Ecosystems Models (e.g., Ecopath with Ecosim [2,32]; Loop Analysis [33,34]).

METHODS

To describe the current state-of-the-art approaches in the UGC, single-species and ecosystem models were identified through an electronic bibliographic search within international web searches, Mexican websites, and libraries. The ecosystems models were evaluated in this study with regard to their compliance with the three EBFM goals: (1) understanding the key processes of ecosystem dynamics, (2) using the spatio-temporal scales that are adequate for management, and (3) allowing for a balance between resource conservation and use (Table 1).

Table 1 General goals of ecosystem-based fisheries management (EBFM) directly involved with ecosystem modeling, according to Espinoza-Tenorio et al. [16]

Category	Description
Ecosystem dynamics	EBFM requires the understanding of key processes and relationships controlling the ecosystem. The consideration of ecological and environmental variables and their interdependencies allows us to describe and model ecosystem responses to internal or external perturbations. This understanding enables us to conserve ecosystem structure and function in order to maintain the evolutionary potential of species and ecosystems. EBFM should also examine how perturbations affect distant ecosystem interconnections
Spatio-temporal scales	EBFM should be undertaken at appropriate resolutions to suitably represent ecosystem processes that operate on different temporal scales and affect different spatial extents. For those processes with unclear boundaries (e.g., migration and water currents), it is critical to consider the drivers of change operating both between geographic scales, as well as those occurring over long term periods
The balance of conservation and use	The ecosystem approach should seek an appropriate balance between conservation and use, where the priority is to obtain and maintain long-term socioeconomic benefits without compromising the ecosystem. To achieve this balance, EBFM recognizes that change is inevitable, and therefore, it must be adaptive in its development over time as circumstances change or as new information becomes available. The ecosystems may be managed within the limits of their functioning and some level of precaution is inevitably required, especially for impacts that are potentially irreversible over long time periods

RESULTS

To represent the fishing dynamics of the UGC, a range of single-species models, from rather simple models limited to commercial fisheries resources such as shrimp [35] and elasmobranchs [36,37], to those which add different species, such as the endemic totoaba (*Totoaba macdonaldi* [38,39]), were used. These single-species models have focused on providing advice on the robustness of management procedures in the UGC. For instance, García-Juárez et al. [35] developed a dynamic model of potential catch quotas for blue shrimp (*Litopenaeus stylirostris*) in the buffer zone of the Biosphere Reserve to estimate the expected shrimp yield and future biomass, including fishing season as a factor and CPUE as an abundance index. This assessment projected levels of biomass under three catch quota management scenarios, recommending the adoption of quotas of 2200 t, which could be increased in some exceptions to 2400 t. A constant harvest rate of 59% of the available annual biomass was also recommended for the management of this resource in the protected area.

In the UGC, most of the single-species models come from official research, which provided technical and scientific advice to the fisheries sector [40]. In this sense, the national institute of fisheries research (INAPESCA, by its Spanish acronym) has based its management decisions on commercial fisheries through indirect assessments, which estimate the parameters of pre-determined selectivity curves simultaneously across mesh size and size class from catch data [41]Kirkwood, G.P. and Walker, T.I., 1986. , Gill net selectivities for gummy shark, Mustelus antarcticus Günther, taken in south-eastern Australian waters. Aust. J. Mar. Freshw. Res. 37, pp. 689–697 Full Text via CrossRef or represent the rate of harvesting through the catch-per-unit effort (CPUE, [42]).

Other contributions of single-species models have been oriented toward conducting historical reconstructions of the stocks to establish crucial parameters and relationships and to describe the current stock status (assessment). Smith et al. [37], for example, determined that, given decades of largely unrestricted exploitation of elasmobranchs in east coast of Baja California, population declines and shifts in size structure are likely to have occurred among those species with the lowest fecundities and latest ages at maturity.

Table 2. Ecosystems models in the Upper Gulf of California according to the arrangement suggested by Plagányi (2007)

Category	Description	Model and local references
Extensions of Single-Species Assessment Models	Models that expand on current single-species assessments, including only a few additional aspects such as the consideration of species related to target resources or fishing effort	[43-47]
Whole Ecosystems Models	This category includes approaches that attempt to take into account all trophic levels in the ecosystem to examine the energy flow among components. Models often contain up to 30 species/groups and may include additional socioeconomic variables	- Ecopath with Ecosim [13,48-50] - Loop Analysis [51]

Ecosystem Modeling in the UGC

In order to represent the ecosystems in the UGC, two types of ecosystem modeling exercises were performed, as shown in Table 2. These ecosystem models are discussed (below) in regards to their main local contributions to ecosystem dynamics, spatio-temporal scales, and the search for a balance between conservation and use.

Ecosystem Dynamics

With respect to the understanding of the key process controlling the ecosystems in the UGC, only in recent decades has the number of models that combine environmental and ecological factors in the UGC increased. The majority of environmental disturbance studies have evaluated the consequences of the drastic reduction in fresh water flow by US dams from the Colorado River. In this sense, the use of extended Single-Species Models has provided evidence for a negative correlation between shrimp catch and discharge from the Colorado River [43]. To investigate the causes of the totoaba fishery collapse, previous studies have analyzed trends in catch, abundance, and fishing mortality in relation to the flow of the Colorado River and a diversity of climatic indexes (e.g. Pacific Decadal Oscillation Index–PDOI) and have reconstructed fishing effort time series [46]. Their results have confirmed the importance of the cessation of the flow from the Colorado River on the catch decrease and have revealed a new strong correlation between catch / abundance and large temporal and spatial scale processes, as shown by the PDOI.

The whole ecosystem dynamic has been addressed by the application of Whole Ecosystem models. For example, Morales-Zárate et al. [13] showed that the UGC is highly dynamic, more complex, and probably a more mature ecosystem than other similar marine systems around Mexico. Using 50 functional groups to represent the UGC, Lozano [49] used mass-balanced models (EwE) to show that this ecosystem is highly dependent on lower trophic levels, mainly detritus-benthic components, which appear to largely control biomasses. Other subsequent whole ecosystem approaches have identified key species when analyzing the function of the system. For instance, Espinoza-Tenorio et al. [51] suggested that fishes, such as milk fish (*Micropogonias megalops*), bass (*Cynoscion othonopterus*), and mackerel (*Scomberomorus concolor*) should be considered as key species when designing management strategies because these species function as buffer species that reduce the direct effects of disturbance on higher trophic levels.

Spatio-Temporal Scales

In general, spatial and temporal variations are usually considered separately in ecosystem models. Temporal dynamics are robustly incorporated into all categories of ecosystem models, only distinguishing among those that consider time in a continuous manner from those with discrete time steps that define abstract or physical periods (mostly years, but also hours or decades). For instance, to evaluate the probability of recovery of the endemic vaquita, the implementation of a net phase-out in 5, 10, and 15 years was considered in an Extension of a Single-Species Model [44].

Although limited by the lack of temporal series of information, Whole Ecosystem models also used (temporal periods) time series to simulate fisheries management scenarios. Lercari [50] for example, used ECOSIM (EwE) simulations to suggest that a reduction of 65% in the industrial shrimp fleet, a value similar to the governmental recommendation (50%, [52]), is required within the next 20 years to achieve sustainability of the fishery.

In contrast, the objective of other EwE exercises was to reconstruct the past UGC ecosystem conditions for two periods of time, 1900 and 1990-2002, and thus evaluate the environmental consequences associated with drastic and historical disturbances [49].

While Extensions of Single-Species models do not consider space, the use of EwE and its spatial module, Ecospace, allowed studying the distribution of biological groups and the ecosystem effects of the marine protected areas implemented in the region [48]. This ability of EwE to represent spatial variability is shared with methodological arrangements, such as Loop Analysis-Geographic Information Systems [51], which indentified how fisheries resources (crustaceans, mollusks, fishes, and protected species), as well as the complexity of the trophic nets in which they are embedded, vary across four types of fishing seascapes (Figure 1).

THE BALANCE OF CONSERVATION AND USE

In the UGC, there have not been any practical instances where an ecosystem model was applied at the level of a whole fisheries management plan.

Nevertheless, Extensions of Single-Species models seem to facilitate the search for a balance between conservation and use. For instance, the fishery tendency analysis developed by Rodríguez-Quiroz et al. [47] has successfully shown that the Biosphere Reserve and the recently declared Vaquita Refuge area are important grounds for artisanal fishing. Consequently, they also explained that shrimp capture in both protected areas has maintained a continuous level of production with economic incentives, making it attractive to fishermen, despite recent restrictions on their activities.

Another example oriented toward understanding the conservation and use relationship in the UGC is the shrimp fishing fleet behavior analysis developed by Cabrera and González [45], which has been presented to the fishing co-operatives as an approach to address social concerns with respect to the technical efficiency of the vessels and co-operatives.

The most realistic representations of Whole Ecosystem models have been generally restricted to specific management strategies.

In these cases, particular interest has focused on the ecosystem's response to specific impacts, such as discards and by-catch. For instance, Morales-Zárate et al. [13] found that most functional groups (29) are impacted more by predation and competition than by fishing pressure in the UGC. Using a pyramid apex angle as an index of ecosystem structure [32], they also showed that use of the ecosystem is balanced. Subsequent Whole Ecosystem models have also been focused on resource exploitation and biodiversity conservation [50,51].

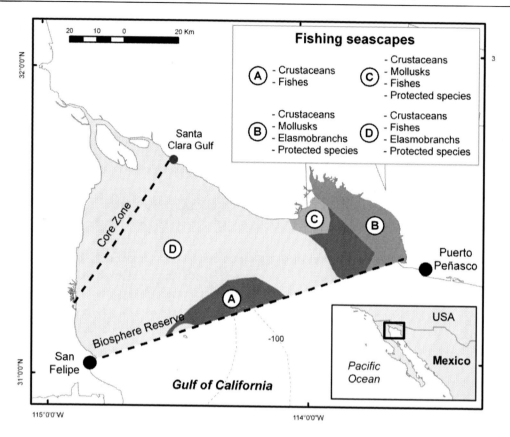

Figure 1. Upper Gulf of California regionalization according to the methodological modeling arrangement Loop Analysis-Geographic Information Systems (modified from [51]).

DISCUSSION AND RESEARCH DIRECTIONS

In the absence of long-term monitoring and complex biological data for fishery stock assessment, modeling of the UGC has been oriented toward the establishment of the maximum and minimum catch rates and the understanding of selectivity properties of fishing gear, providing a basis for the implementation of standard mesh-sizes as a regulatory tool for fishery management. Thus, most of these single-species models have proposed specific actions (e.g., total allowable catch) that steer the stock toward a desired status (short-term forecasting); however, few studies have make long-term predictions of the likely future status of stocks under various management scenarios to reach desirable levels (long-term forecasting) [40].

In this context, ecological models are geared primarily toward the three major ecological processes that underlie the forces governing populations: competition, predation, and environmental disturbance. In this sense, Extensions of Single-Species models, which have mainly been restricted to catch data and specific ecological relationships such as predation, have been more recently oriented to specific environmental disturbances, such as the cessation of flow from the Colorado River [43]. The outputs from these models have been used in subsequent, more complex, models to improve, for example, the quality of biomass and productivity information. Whole Ecosystem models have thus addressed competition, predation, and environmental disturbance processes in a more general way, mainly through quantitative and qualitative trophic structure approaches. An approach that has not yet been explored in the UGC is to investigate the hypotheses that have emerged from previous Whole Ecosystem models. In this sense, Extensions of Single-Species models can be used to refine estimates from more complex models on key ecosystem aspects, such as if the biological community is impacted more by depredation and competition than by fishing pressure [13].

Spatial and temporal variations have different grades of advancement in the UGC. The lack of large time series of fisheries information has limited the development of single-species models, now restricted to the ecosystem assessments. However, temporal simulations from Extensions of Single-Species Models and Whole Ecosystem models have evaluated management scenarios, which typically range between a half a year to 20 years in length. Although better computing capacity has recently allowed for spatial modeling, the use of spatial models continues to be limited because of the necessity to represent spatial variability, increased uncertainty, and large data demands, which few approaches are capable of handling. The ability to represent spatial variability is present in Whole Ecosystems models, which face local restrictions because of the lack of georeferenced databases. To overcome such limitations, Extensions of Single-Species Models might explore the spatial strategies that are currently being proposed for the UGC by conservation strategies, such as the refuge area of the vaquita [10], or by Whole Ecosystems models, such as the ecological function of particular fishing seascapes of high biodiversity, like Bahía Aldair [51]. Single-species approaches, such as the dynamic model, which was already proved by García-Juárez et al. [35] for the Biosphere Reserve, can also be useful for analyzing spatial processes.

Due to the presence of both commercially important species and highly endangered species, a balance between conservation and use is a priority for the UGC. However, there have not been any practical instances in which an EM has been applied at the level of a whole-fisheries management plan. This is understandable because finding a balance between conservation and use creates further uncertainty [20] whereby researchers frequently underestimate precautionary limits and managers often do not consider uncertainty [24]. In contrast to the conceptual and abstract index used in Whole Ecosystem models, Extensions of Single-Species Models in the UGC seem to facilitate such a balance due to their capacity to provide estimates of recognized target reference points, such as fishing mortality rates or the risk of extinction [44]. Therefore, both types of models can be used complementarily to facilitate a balance between conservation and use. For instance, Lercari's and Arreguín-Sánchez's [50] EwE model resulted in effort allocation consistent with those proposed by single-species population models [35] and a decrease in the industrial shrimp and gillnet fishing fleets and an increase in the artisanal shrimp fishery.

Like the search for a balance between conservation and use, society participation and integrated management are two EBFM goals that are not directly involved with the understanding of the system, but lead to the incorporation of social aspects into EMs. Worldwide, these three goals are the least attended by ecosystem modeling [16] because these goals are part of a societal process involving high levels of uncertainty, which is generally avoided by ecological modelers. Although the UGC is no exception to this tendency, some Whole Ecosystem models have been used (e.g., [49,51]), facilitating the involvement of the users through the use of their particular ecological knowledge about fisheries dynamics in the process of building these models. In addition, simple models, such as Extensions of Single-

Species Approaches (e.g., [45,47]), seem to be the most promising approaches because they facilitate the participation of the fishery sector due to their simplicity and lower data requirements.

In the UGC, neither Dynamic Multi-Species Models nor Dynamic System Models have been applied. For more detailed studies that combine environmental and ecological factors (e.g., environmental effects of the Colorado River flow cessation), dynamic system models can be useful because they involve both induced changes by environmental disturbances [5] and the effects of ecological driving forces, such as competition and predation [29-31]. Moreover, although Dynamic Multi-Species Models often do not incorporate environmental forcing [4] and usually ignore any potential effects of changing prey populations on the predators themselves [24]. They do provide complete ecological representations of species or groups at higher trophic levels such as vaquita and totoaba in the UGC. However, to implement Dynamic Multi-Species Models or Dynamic System Models, more ecosystem information that is not currently available to the UGC is necessary.

Conclusion

Ecosystem modeling has increasingly resulted in support for EBFM in the UGC. Thus, the current state-of-the-art of ecosystem modeling approaches in the UGC is beginning to form the whole picture of the ecosystem. However, if single-species models are limited by lack of data used to assess performance of the UGC, the situation with ecosystem models seems to be more complicated because it deals with a high level of uncertainty. The step from single-species models to EMs is a stage of the management of the area that does not require the elimination of the first approach, but rather the use of both approaches in a complimentary manner. If the next stage in the UGC is to explore hypotheses that have emerged from the previous models, it will be necessary to promote the integration of current ecosystem information to detect the gaps in the collective knowledge. Because the dynamics of most of the Mexican marine ecosystems have, in the best cases, only been modeled once, the UGC is an exceptional study area. Lessons learned in UGC should be considered when designing model-structured strategies in other marine ecosystems.

Acknowledgments

This study was funded by Deutscher Akademischer Austausch Dienst (DAAD) under a PhD's scholarship to A.E-T. We thank Jorge Paramo for all suggestions and helpful comments. The language corrections of AJE are also appreciated.

References

[1] Pikitch, E.K.; Santora, C.; Babcock, E.A. et al. Ecosystem-based fishery management. *Science* 2004, 305, 346-347.

[2] Hollowed, A.B.; Bax, N.; Beamish, R. et al. Are multispecies models an improvement on single-species models for measuring fishing impacts on marine ecosystems? *ICES J. Mar Sci.* 2000, 57, 707-719.

[3] Grimm, V. Ten years of individual-based modeling in ecology: what have we learned and what could we learn in the future? *Ecol. Modell.* 1999, 115, 129-148.

[4] Keyl, F.; Wolff, M. Environmental variability and fisheries: what can models do? *Rev Fish Biol. Fish.* 2008, 18, 273-299.

[5] Hilborn, R. The state of the art in stock assessment: where we are and where we are going. *Sci Mar.* 2003, 67, 15-20.

[6] International Union for Conservation of Nature and Natural Resources (IUCN). *Islands and protected areas of the Gulf of California (Mexico)*; IUCN: Gland, Switzerland, 2005; pp 14.

[7] Espinoza-Tenorio, A.; Espejel, I.; Wolff, M. *Capacity Building to Achieve Sustainable Fisheries in Mexico*. Ocean Coast Manage. submitted.

[8] Brusca, C.R. The Gulf of California—an overview. In *A seashore guide to the northern Gulf of California*; Brusca, C.R.; Kimrey, E.; Moore, W.; Arizona-Sonora Desert Museum: Tucson, 2004; pp 203.

[9] Sala, E.; Aburto-Oropeza, O.; Reza, M.; Paredes, G.; López-Lemus, L.G. Fishing down coastal food webs in the Gulf of California. *Fisheries* 2004, 29, 19-25.

[10] DOF–Federal Gazette. Acuerdo mediante el cual se establece el área de refugio para la protección de la vaquita (Phocoena sinus). Adopted on 8 September 2005.

[11] Cisneros-Mata, M.A. Sustainability and complexity: from fisheries management to conservation of species, communities and spaces in the Sea of Cortez. In *A seashore guide to the northern Gulf of California*; Brusca, C.R.; Kimrey, E.; Moore, W.; Arizona-Sonora Desert Museum: Tucson, 2004; pp 203.

[12] Cudney, R., Turk, P.J. Pescando entre mareas del Alto Golfo de California. Una guía sobre la pesca artesanal, su gente y sus propuestas de manejo; Centro Intercultural de Estudios de Desiertos y Océanos: Sonora, 1998; pp 166.

[13] Morales-Zárate, M.V.; Arreguín-Sánchez, F.; López-Martínez, S.E.; Lluch-Cota, E. Ecosystem trophic structure and energy flux in the Northern Gulf of California, México. *Ecol. Modell.* 2004, 174, 331-345.

[14] Kaufman, L.; Karrer, B.L.; Peterson, H.C. Monitoring and evaluation. In *Ecosystem-based management for the oceans*; McLeod, L.K.; Leslie, H.M.; Island press: Washington, DC., 2009; pp 115-128.

[15] Caddy, J.F. Fisheries management in the twenty-first century: will new paradigms apply? *Rev. Fish Biol. Fish.* 1999, 9, 1-43.

[16] Espinoza-Tenorio, A.; Wolf, M.; Taylor, M.H.; Espejel, I. What model suits Ecosystem - based Fisheries Management? – A plea for a structured planning process. *Rev. Fish Biol. Fish.* submitted.

[17] Starfield, A.M.; Smith, K.A.; Bleloch, A.L. *How to model it: problem solving for the computer age*; McGraw-Hill: New York, 1990; pp 206.

[18] Hilborn, R.; Walters, C.J. *Quantitative fisheries stock assessment: Choice, dynamics and uncertainty*; Thomson Science: New York, 1992; pp 592.

[19] Breckling, B.; Müller, F. Current trends in ecological modeling and the 8th ISEM conference on the state-of-the-art. *Ecol. Modell.* 1994, 75/76, 667-675.

[20] Taylor, L.B.; Wade, R.P.; De Master, P.D.; Barlow, J. Incorporating uncertainty into management models for marine mammals. *Conserv. Biol.* 2000, 14, 1243-1252.

[21] Pauly, D.; Christensen, V.; Walters, C. Ecopath, Ecosim, and Ecospace as tools for evaluating ecosystem impact of fisheries. *ICES J. Mar. Sci.* 2000, 57, 697-706.

[22] Wätzold, F.; Drechsler, M.; Armstrong, W.C. et al. Ecological-Economic Modeling for Biodiversity Management: Potential, Pitfalls, and Prospects. *Conserv. Biol.* 2006, 20, 1034-1041.

[23] Bart, J. Acceptance criteria for using individual-based models to make management decisions. *Ecol. Appl.* 1995, 5, 411-420.

[24] Plagányi, E.E. *Models for an ecosystem approach to fisheries*; Food and Agriculture Organization of the United Nations: Rome, 2007; pp 108.

[25] Routledge, R. Mixed-stock vs. terminal fisheries: a bioeconomic model. *Nat. Res. Model.* 2001, 14, 523-539.

[26] Tjelmeland, S.; Lindstrom, U. An ecosystem element added to the assessment of Norwegian spring spawning herring: implementing predation by minke whales. *ICES J. Mar Sci.* 2005, 62, 285-294.

[27] Sparholt, H. Using the MSVPA/MSFOR model to estimate the right-hand side of the Ricker curve for Baltic cod. *ICES J. Mar. Sci.* 1995, 52, 819-826.

[28] Tjelmeland, S.; Bogstad, B. MULTSPEC ± a review of a multispecies modeling project for the Barents Sea. *Fish Res.* 1998, 37, 127-142.

[29] Shin, J.Y.; Cury, P. Exploring fish community dynamics through size-dependent trophic interactions using a spatialized individual-based model. Aquat Living Resour. 2001, 14, 65-80.

[30] Fulton, E.A.; Smith, A.D.M.; Johnson, C.R. Biogeochemical marine ecosystem models I: IGBEM—a model of marine bay ecosystems. *Ecol. Modell.* 2004, 174, 267-307.

[31] Fulton, E.A.; Parslow, J.S.; Smith, A.D.M.; Johnson, C.R. Biogeochemical marine ecosystem models II: the effect of physiological detail on model performance. *Ecol. Modell.* 2004, 173, 371-406.

[32] Pauly, D.; Christensen, V. Stratified models of large marine ecosystems: a general approach and an application to the South China Sea. In *Large Marine Ecosystems: Stress, Mitigation and Sustainability*; Sherman, K.; Alexander, L.M.; Gold, B.D.; AAS Press: Washington, DC., 1993; pp 376.

[33] Ortiz, M.; Wolff, M. Application of loop analysis to benthic systems in northern Chile for the elaboration of sustainable management strategies. *Mar. Eco Prog. Ser.* 2002, 242, 15-27.

[34] Montaño-Moctezuma, G.; Li, H.W.; Rossignol, P.A. Alternative community structures in a kelp-urchin community: A qualitative modeling approach. *Ecol. Modell.* 2007, 205, 343-354.

[35] García-Juárez, A.R.; Rodríguez-Dominguez, G.; Lluch-Cota, D.B. Blue shrimp (Litopenaeus stylirostris) catch quotas as a management tool in the Upper Gulf of California. *Ciencias Marinas* 2009, 35, 297-306.

[36] Márquez-Farias, J.F. Gillnet mesh selectivity for the shovelnose guitarfish (Rhinobatos productus) from fishery-dependent data in the artisanal ray fishery of the Gulf of California, Mexico. *J. Northw Atl. Fish Sci.* 2005, 35, 443-452.

[37] Smith, W.D.; Bizzaro, J.J.; Cailliet, G.M. The artisanal elasmobranch fishery on the east coast of Baja California, Mexico: Characteristics and management considerations. *Ciencias Marinas* 2009, 35, 209-236.

[38] Cisneros-Mata, M.A.; Montemayor-López, G.; Román-Rodríguez, M.J. Life history and conservation of Totoaba macdonaldi. *Conservation Biology* 1995, 9, 806-814.

[39] Pedrín-Osuna, O.A.; Córdova-Muruete, J.H.; Delgado-Marchena, M. Crecimiento y mortalidad de la totoaba, Totoaba macdonaldi, del alto golfo de California. *Ciencia Pesquera* 2001, 14, 131-140.

[40] Instituto Nacional de Pesca (INP). Sustentabilidad y pesca responsable en México, evaluación y manejo; SAGARPA: Mexico, 2006; pp 560.

[41] Kirkwood, G.P.; Walker, T.I. Gill net selectivities for gummy shark, Mustelus antarcticus Günther, taken in south-eastern Australian waters. *Aust J. Mar. Freshw Res.* 37, 1986, 689-697.

[42] Schaefer, M.B. Some considerations of population dynamics and economics in relation to the management of marine fishes. *J. Fish Res. Board Can.* 1957, 14, 669-681.

[43] Galindo-Bect, M.S.; Glenn, E.P.; Page, H.M.; et al. Penaeid shrimp landings in the upper Gulf of California in relation to Colorado River freshwater discharge. *Fish Bull.* 2000, 98, 222-225.

[44] Ortiz, I. Impacts of fishing and habitat alteration on the population dynamics of the vaquita, (Phocoena sinus); Master thesis; University of Washington: DC., 2002; pp 72.

[45] Cabrera, H.R.M.; González, J.R.C. Manejo y eficiencia de la pesquería del camarón del Alto Golfo de California. Estudios Sociales, Revista de Investigación Científica 2006, XIV, 123-138.

[46] Lercari, D.; Chávez, E.A. Possible causes related to historic stock depletion of the totoaba, Totoaba macdonaldi (Perciformes: Sciaenidae), endemic to the Gulf of California. *Fish Res*. 2007, 86, 136-142.

[47] Rodríguez-Quiroz, G.; Aragón-Noriega, E.A.; Ortega-Rubio, A. Artisanal shrimp fishing in the biosphere reserve of the Upper Gulf of California. *Crustaceana* 2009, 82, 1481-1493.

[48] Lercari, D. Manejo de los recursos del ecosistema del norte del Golfo de California: integrando explotación y conservación; PhD Thesis; CICIMAR-IPN: Mexico, 2006; pp 173.

[49] Lozano, H. Historical ecosystem modeling of the upper Gulf of California (Mexico): following 50 years of change; PhD Thesis; The University of British Columbia: Vancouver, 2006; pp 266.

[50] Lercari, D.; Arreguín-Sánchez, F. An ecosistema modeling approach to deriving viable harvest strategies for multispecies management of the Northern Gulf of California. *Aq Cons. Mar. Fresh Ecos.* 2009, 19, 384-397.

[51] Espinoza-Tenorio, A.; Montaño-Moctezuma, G.; Espejel, I. Ecosystem-based analysis in a marine protected area where fisheries and protected species coexist. *Environ. Manag*. 2010, 45, 739-750.

[52] CONAPESCA. Pesca sustentable del camarón en el Golfo de California. Propuesta de reestructuración de la flota camaronera de altamar; SAGARPA: Mexico, 2004, Manuscript.

In: Environmental Management
Editor: Henry C. Dupont

ISBN: 978-1-61324-733-4
© 2012 Nova Science Publishers, Inc.

Chapter 13

A STUDY ABOUT THE ADOPTION OF THE PRACTICE OF CLEANER PRODUCTION IN INDUSTRIAL ENTERPRISES CERTIFIED ISO 14001 IN BRAZIL

José Augusto de Oliveira, Otávio José de Oliveira and Sílvia Renata de Oliveira Santos
UNESP – São Paulo State University, Brazil
CEUCLAR – Centro Universitário Claretiano de Batatais, Brazil

ABSTRACT

The growth of developing countries has significantly driven the industrial activity, which has generated employment and development. However, this scenario also has its downside, in other words, due to lax laws and consumers still little conscious of the importance of effective environmental protection. The industry has been the main degrading environment in these nations. However, some technical programs and management tools have been used to minimize this serious problem, especially if the environmental management system ISO 14001:2004 and Cleaner Production (CP). For these reasons, this book chapter has as main objective to present good practices of cleaner production adopted by four industrial companies operating in Brazil with high profile and certified according to ISO 14001:2004. The companies act in the area of cellulose and paper and chemical. Their management systems ISO 14001:2004 will be characterized and the major practices of Cleaner Production will be reported. In general terms, it was observed that the EMS has greater coverage of these companies that ISO 14001:2004 due to demands from the processes related to cleaner production. They use specific methodologies and specific classifications to the elements of CP and do not follow exactly the elements proposed by UNIDO (United Nations Industrial Development Organization).

It was observed that the companies have achieved significant results by adopting the cleaner production achieved environmental and financial gains from the reuse of production inputs, internal recycling in the production process, cost reduction with treatment and disposal of waste, reducing spending with environmental liabilities promoted by prevention of the pollution, among others, which are the main benefits expected by the practice of CP.

Keywords: Environmental Management, ISO 14001:2004, Cleaner Production, Industrial Companies, Brazil.

INTRODUCTION

In a scenario that increasingly needs measures and actions of environmental nature, companies seek to adhere to systems and tools that assist them in that regard. The industrial bough is the sector with the highest incidence of environmental impacts and therefore need to take significant actions for environmental responsibility. The environmental management system (EMS) based on ISO 14001:2004 can reap environmental benefits, economic and market organizations, which can also be observed by establishing procedures for cleaner production (CP) by industries.

In this context, this chapter presents and analyzes the activities of EMS and CP of four industrial companies located in the Brazilian state of Sao Paulo - Brazil, certified according to ISO 14001:2004, aiming to highlight good practices and to characterize their environmental actions . Companies are considered success stories in CP in the country.

According to the National Institute of Geography and Statistics (IBGE) (2008), the Gross Domestic Product (GDP) of São Paulo, represented 33% of Brazilian GDP, and 30% of that amount belongs to the industrial activity, which explains the choice of firms in the survey, which is composed of two national industries of pulp and two multinational chemical companies.

1. ENVIRONMENTAL MANAGEMENT SYSTEMS

According to Seiffert (2005), Environmental Management Company can be considered as an adaptation of organizations to a new way of planning, considering in his list of goals and objectives, strategies that include prevention and minimization of environmental impacts of direct and indirect result of their productive activities. Corporate Environmental Management is being increasingly discussed in the business world. Going from just a strategic externality, the environmental issue is gaining a vital position, with emphasis on current business sector (ROSEN, 2001). This practice is done through the systematization of human, technological, financial and intellectual organizations, generating so-called Environmental Management Systems (EMS), which are the main elements of Corporate Environmental Management.

The Responsible Care (RC) program, established in 1980 by the Canadian Chemical Producers Association, through a private initiative as a result of discrediting the image of the industry to consumers at the time, is not known strictly as an EMS in the literature, it is considered a policy Corporate Social Responsibility (CSR), but covers and objective improvements in economic performance, social and environmental (BERLAND; LOISON, 2008) and is widely used in industrial field as an EMS.

The RC was developed and is regarded as an environmental policy business, according to three important aspects of the program: the growth of companies' goal of a fitness to environmental issues, providing tools to measure the company's environmental performance, high incidence of focus on environmental issues in general and the possibility of verification of corporate environmental performance at both the domestic and foreign industry (MELNYK, *et al*., 2002). The name given to the certification granted to the RC in the industries is Verification.

2. ENVIRONMENTAL MANAGEMENT SYSTEM BASED ON ISO 14001 :2004

One of the best known Environmental Management Systems in the world, is the EMS based on ISO 14001:2004. This rule proposes requirements to be met by companies to deploy the EMS and so, after a specific audit, obtaining ISO 14001:2004 certification. The structure of ISO 14001: 2004 is presented in Table 1.

Table 1. Structure of ISO 14001:2004

0 Introduction
1 Purpose and scope
2 Normative References
3 Terms and definitions
4 Requirements for environmental management system
4.1 General Requirements
4.2 Environmental Policy
4.3 Planning
4.4 Implementation and operation
4.5 Checking
4.6 Management review
Annex A - Guidance for use of this standard
Annex B - Correspondence between ISO 14001:2004 and ISO 9001:2000
Annex C - Bibliography

Source: ISO 14001 (2004).

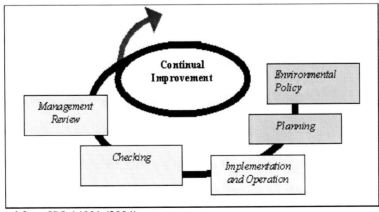

Source: Adapted from ISO 14001 (2004).

Figure 1. Key elements of an EMS based on ISO 14001:2004.

Companies deploy EMS according to the appropriateness of their procedures to the requirements of ISO 14001:2004, committing themselves to continuous improvement over its stated goals in its environmental policy, as can be seen in Figure 1.

According to Matthews (2003), the main elements mentioned in Figure 1, will be joined as follows:

- Planning: policies, impacts and environmental targets;
- Execute: environmental activities and environmental documentation;
- Check: environmental audits and environmental performance evaluation, and
- Act: environmental training and environmental communication.

3. CLEANER PRODUCTION

Another important instrument of environmental nature widely used in industrial field, and this will be covered in this chapter, in addition to the research topic, though it wouldn't hardly be an EMS, but efficiently meets the expectations of business adaptation to an environmentally responsible behavior is CP.

The CP was developed after the establishment of UNEP (United Nations Environment Programme) in 1972, which is the leading authority on global environment and is part of the UN (United Nations). The national UNIDO (United Nations Industrial Development Organization) is also a UN body responsible for controlling global manufacturing. The CP is also encouraged and monitored the state of São Paulo CETESB (Environmental Company of São Paulo).

The CP is the set of measures applied continuously within a corporate strategy geared towards the economic, environmental and social, through the focus of processes, products and services in preventing pollution, minimizing the use of production inputs and from waste disposal and recycling internal production processes (UNEP, 2007).

The CP is a very effective environmental strategy on environmental and economic aspects, providing significant benefits to organizations that use it efficiently.

The great advantage of the CP to other environmental management tools is the controversy with the old system end-of-pipe, in other words, the inclusion of environmental measures at the end of the production process. The CP's main objective is the prevention of pollution, increasing efficiency and minimizing the use of raw materials, recycling of the inputs used in the production process and managing waste disposal (MARTIN, RIGOLA, 2001), as illustrated Figure 2.

4. CASE STUDIES

This session will present the case studies that are composed of two Brazilian pulp industries and two chemical companies. There will be a general characterization of the organizations, the presentation of EMS's and the activities of CP of each case.

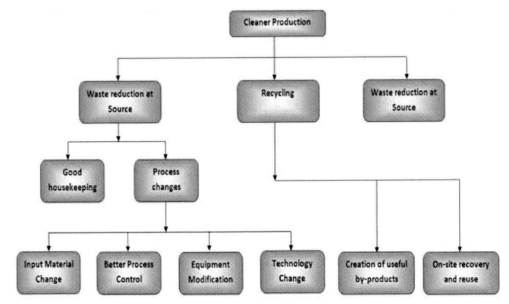

Source: Adapted from UNEP (2007).

Figure 2. Scope of work of the methodology of Cleaner Production (CP).

All the companies studied have some characteristics that are appropriate for a cross-analysis, making it suitable for a comparison study of information relevant to the topic. Among the key features common to the companies analyzed, it is worth mentioning:

- These companies are from the Brazilian industrial sector in state of Sao Paulo;
- They are both large item sales, for the number of employees;
- The companies had a business conduct adopting environmental principles before certification of EMS by ISO 14001:2004;
- The companies have ISO 14001:2004 certification;
- The companies have Integrated Management System (IMS);
- They adopt procedures for CP, even if not using the same nomenclature and methodology of UNIDO; and
- They have success stories of CP in Brazil, according CETESB.
- It is important to mention that not all companies have authorized the disclosure of names or other information enabling their identification.

4.1. Case 1

The company 1 was founded in January 1865, it is the world's chemical market with a portfolio of 8000 products covering the agricultural, functional, construction, paints, plastics and performance and has a turnover of 1.9 million Euros in Brazil.

The organization 1 has a worldwide recognition for their environmental practices, which can be evidenced by the inclusion of its shares in the Dow Jones of New York Stock Exchange. Table 2 illustrates the main certifications of the company in Brazil.

Table 2. Certification given to the company

Certificate	Object/ certification area	Year
ISO 9001: 2008	Quality Management System	2009
ISO TS 16949:2002	Quality Management System	2009
ISO 14001: 2004	Environmental Management System	2009
Q1 Ford Award	Quality Standard Favorite	2004
Verification	Responsible Care Program	2004

Note itself that the company does not have an OHSAS 18001 certification, but it is still followed an IMS, the RC, which incorporates other management systems. The company is concerned with certifiable management systems, which also demonstrates its good position in the domestic market, since the aforementioned certifications are extremely important for the relationship with suppliers, customers and society in general.

Deserve to be highlighted in Table 2 because they meet with the objectives of this chapter, the ISO 14001:2004 and the RC. Observe that the RC was obtained 5 years before the ISO 14001:2004 certification. This information confirms that the environmental manager of the company did and what the literature on EMS mentions that the RC program is a more comprehensive environmental policy that the EMS based on ISO 14001:2004.

The company did not reveal indicators arising from the EMS based on ISO 14001:2004, but cited some of the main benefits that this system after the company earned ISO 14001:2004 certification:

- Improved corporate image;
- Improvement in compliance of the company;
- Managing and reducing significant environmental impacts;
- Optimizing the use of production inputs, thereby reducing costs;
- Reuse of waste, adding value to production;
- Incorporation of environmental issues, to conduct business as a whole;
- Encourage environmental awareness of employees;
- Minimizing the amount of fines and environmental liabilities of the company;
- Reduction of risks during production;
- Improved communication between the company and surrounding society.

According to the environmental manager of a company, the program covers all RC policies and all requirements of ISO 14001:2004, also acting on other aspects as transport and logistics, emergency preparedness and response to broader, more emphasis on environmental protection and also environmental protection with regard to worker health. So, it took only a few adaptations of the old EMS to meet the requirements of the standard and thus obtaining ISO 14001:2004 certification.

This organization has procedures for CP that were implemented after the EMS based on ISO 14001:2004. Using the nomenclature and methodology particular, CP 1 functions in the organization through a spreadsheet that stores and manages information about the company's overall production system.

Table 3. The main projects of CP from company 1

Local Application	Project Description
Production Line	Reprocessing of a chemical product used in the production of the company avoid costly transport and disposal of this substance
Production Line	Training of operators to the manufacturing process, streamlining and encouraging waste minimization
A factory	Installation of a centrifuge that separates solid waste from wastewater, thus enabling the reuse of material in the manufacturing process itself, avoiding its disposal
A factory	Specific reduction of specific waste generation by 25% on a production unit

This tool, like CP, has goals of economic and environmental improvements to the processes of a company, also following the objectives of RC. With the spread of these objectives at all levels of the organization, each employee is responsible for identifying improvements in their area of the production process, thus passing this information to the environmental department and supplying the spreadsheet.

The company uses 13 of the 20 steps proposed by UNIDO. In possession of this information and using the tool Brainstorming and Six Sigma program, the department assesses the environmental or ecotime opportunities and their economic viability, technical and environmental improvement projects proposed to be assessed with the consent of senior management. The environmental department is responsible for planning the implementation of the opportunities and finally approved by the management and support activities. The main projects of CP adopted by the company 1 are illustrated in Table 3.

By observing Table 3, note the special attention of the company in training its employees, thus spreading an organizational culture that takes the environment as a strategic objective. Moreover, the focus of the CP in the company is to reuse waste, allowing reduced use of raw materials and effluent disposal, causing environmental and economic benefits to the organization.

According to the environmental manager of the company, it has already been observed significant quantitative and qualitative benefits. But were only made available for research, qualitative, among them one can mention:

- New design in the company's production system, reducing costs with the rationalization of raw material and adding value by reducing waste;
- Generation of environmental and economic values to the company
- Improved management and control of production as a whole, with greater emphasis on the issue of wastes; and
- Reduction of negative environmental impacts and financial resources that the company suffered in its production activities.

As there is full support from top management to the promotion of these procedures in the company and that each project implemented generates a financial return that exceeds the amount invested for its implementation, the manager of the company's environmental claims that there are significant difficulties for the deployment of CP procedures in a company 1.

Table 4. Approvals granted to the company

Certificate	Object/ certification area	Year
ISO 9001: 2008	Quality Management System	2000
ISO 14001: 2004	Environmental Management System	2004
OHSAS 18001	Health and Safety Management System	2006
FSC	Forest Stewardship Council	2008

4.2. Case 2

The company 2 was born in February 2009 from the merger of two large Brazilian organizations and now leads the world market for pulp and renewable forestry business. Employing approximately 14,600 employees, the organization has a second average revenue of 3.04 billion Euros and a production capacity of 5.4 million tons per year of pulp and 313,000 tons per year of paper.

This enterprise also had their shares incorporated in the Dow Jones of New York Stock Exchange and the Corporate Sustainability Index, which shows a national and international recognition of sustainable practices of the company.

The company has an IMS certificate, as shown in Table 4.

According to Table 4 shows the main certifications IMS company, these are the latest versions of each standard, which are still in their validity. For these certificates, there is concern the company's Quality, Environment and Health and Safety and also with forest management, which is the main activity of the second branch of the organization 2.

Observe the company's concern with the second certification of their management systems, showing their position in national and international markets, while those standards are required by some suppliers and customers.

The organization 2 reported that only two qualitative indicators which the EMS based on ISO 14001:2004 provided. For the company, these are the most important benefits for the company:

- Improved corporate image, but only the year of certification;
- Improving the culture of the company to environmental issues;
- Re-use of chemicals, thus avoiding its disposal, and pollution caused by it;
- Reduction of emission of effluents
- Improved control and management of gaseous emissions, liquid and solid;
- Assistance in certification of forest management, and
- Improved management of environmental aspects and impacts.

According to the environmental manager of company 2, ISO 9001:2008 helped a lot the process of certification under ISO 14001:2004, each time that the standardization and adaptation of processes to the requirements of ISO were already established. Also according to the manager, as had already implemented environmental initiatives and governed by an EMS, will not face significant difficulties in obtaining ISO 14001:2004 certification, requiring only a few adjustments to the standard of documentation required.

Table 5. Major projects of CP in the company 2

Local application	Project description
Boiler	Increasing the efficiency burning biomass with reduced consumption of auxiliary fuel NG / Oil in order to reuse / recycle 30% of the waste ash
Fiber Line	Fiber Line Loss Reduction of C fibers
Waste	Reduction of 20% in waste generation in 2007 of causticizing
STP - sewage treatment station	Reduction of 10% in the generation of biological sludge in effluent treatment plant
Production	Reuse of chemicals for extraction of pulp wood preparation

This is the only company in the sample who took this survey to be officially signed the CP, formalizing their commitment to the six principles in the UNEP in 2004. The organization disseminates the practice of CP in all its sectors, establishing March-objectives that are managed by rounds of projects. Each round, which lasts about a year, has a particular focus, for example, focus on the reduction of solid waste disposal. The company uses 17 of the 20 steps proposed by UNIDO. According to environmental manager, the three steps is not formally used are intrinsic to all proposed methodology and are therefore considered unnecessary. There is also a ecotime the environment department and responsible for the EMS in the company. This team makes use of an interactive Excel spreadsheet to supply, manage and stock up information on the production system. After this step, a brainstorming is done on the opportunities, thus making your selection.

The Table 5 shows the main procedures of CP adopted by the organization 2.

It can be observed in Table 5 the adoption of CP projects focused on reducing waste and also the reuse of production inputs. According to the environmental manager of the company in February, this is due to the fact that most of these projects were implemented in year round or focusing on waste, to pursue the policy of the company CP.

According to the environmental manager of the company, can be found relevant and important benefits to the organization after the implementation of CP. Quantitative indicators cannot be provided as a matter of trade secrets, however, among the qualitative benefits, include:

- Reuse of chemicals used in the extraction of cellulose;
- Increased efficiency of biomass burning, thereby reducing fuel consumption;
- Reduced loss of wood fiber;
- Optimization of production, reducing time and increasing productive capacity
- Production environment cleaner;
- Significant reduction in odor emissions and pollution;
- Reducing emissions gaseous, liquid and solid;
- Improved quality of life for employees, and
- Encouraging the production.

Table 6. Certificates awarded to the company

Certificate	Object/ certification area	Year
ISO 9001: 2000	Quality Management System	2010
ISO 9001: 2008	Quality Management System	2009
ISO 9001: 2008	Quality Management System	2010
ISO 14001: 2004	Environmental Management System	2009
ISO/IEC 17025: 2005	General requirements for competence of calibration and testing laboratories	2010
OHSAS 18001	Health and Safety Management System	2009

The organization noted some difficulties in the implementation of CP, such as the high costs of implementing technologies, manpower and costly production losses resulting from the break periods during the implementation of technologies in the production line.

4.3. Case 3

The organization 3 is a unit of a U.S. based group that has existed for over one hundred years. This unit exists in Brazil since 1946 and are a major group subsidiaries worldwide. The company employs approximately 3,444 employees in Brazil and has an average sales 837 million Euros, with a range of about 1,000 products.

The certificates awarded to the company's management systems 3 are shown in Table 6.

The company has a three IMS certified in all its provisions, Table 6 presents the latest version and show well, the continuing responsibility of the organization to produce quality, and environmental and social responsibilities, in addition to the certification granted to the laboratory, which is very important branch of the company's activity, since its consequences involve chemicals relevant to the target and its use.

This feature shows an advantageous position in the market, since the certification of the IMS is considered, and responsibility to quality, environment and health and safety, a search for better strategic position against competitors.

As a matter of private and confidential, the company chose not to mention numerical indicators of the environmental benefits received by the EMS based on ISO 14001:2004, but in relation to quality, are:

- Improved corporate image;
- Reduction of emission of volatile organic compounds;
- Reduction of energy consumption;
- Reduced consumption of water, and
- Increased reuse and recycling of production inputs

According to environmental manager of the company in March there were no difficulty striking also to obtain ISO 14001:2004 certification, each time the company already had a policy of environmental conduct underpinned by important principles and actions that promoted environmental responsibility into corporate culture and production dynamics.

Table 7. The main projects of CP in Company 3

Local application	Project description
Process of digital manufacturing	Printing plates and other through digital, eliminating four of the eight stages of manufacture, and two of these four steps emitting harmful gases
Transport boxes	Replacing wooden shipping crates for non-returnable returnable cans
Reduction of internal transport	Reduction of logistics activities unnecessary and aggregates
Cardboard packaging for polymer	Replacement of cardboard packaging for polymers for packaging made from the same polymer, which is incorporated into the product
Sponge	Creating cleaning sponge made with natural materials and recyclable polymer

The organization is not recognized as a signatory to the CP, but as CETESB, the company is considered a successful case of CP in Brazil. This can be evidenced by the adoption of procedures that promote environmental and economic improvements in the productive process. It is worth noting that the company's main focus for the adoption of these procedures is the economic question, ie the financial return, but with this practice, the company also obtains environmental gains.

The senior management created an internal regulation in the company, stipulating the rules for the participation of CP projects, stimulating the implementation of a campaign by awarding the best projects. Thus, each employee or team for each sector, seeks to identify improvements in the productive process.

This information is transmitted to the environmental department, which evaluates and relays to top management. In a few cases, the initiative of the projects part of ecotime himself or environmental department. To assist in the identification and selection of opportunities, the company uses the Six Sigma program.

The organization C uses 11 of the 20 steps proposed by UNIDO for the implementation of projects in CP. Note that the analysis of economic viability is the focus for the implementation of any project in the company.

The main CP practices performed by the organization 3 are illustrated in Table 7.

In Table 7, we can see the representation of some of the main projects of CP by the company according to the environmental manager. But only in 2009 the company has deployed more than 40 projects of CP and it has already checked in all its history of conducting such actions significant economic and environmental benefits. The company did not publish their numeric indices, but among the key benefits realized by the CP of a qualitative order, are:

- Reduction of greenhouse gas emissions that cause global warming;
- Energy saving;
- Reduced consumption of wood for making pallets;
- The reduction of fuel and therefore reducing the emission of greenhouse gases;
- Reuse cardboard and consequently reducing the deposition of such material;
- Reduced use of raw materials;

Table 8. Certificates awarded to the company

Certificate	Object/ certification area	Year
ISO 14001: 2004	Environmental Management System	2009
FSC	Forest Stewardship Council	2009

- Reduced water use, and
- Increased reuse and recycling of relevance in general inside the company

According to environmental manager of the company, as the feasibility analysis for implementation of every project is focused on the economy, that is, each project must have a financial return to pay the investment, are not faced considerable difficulties in the implementation of CP in the company.

4.4. Case 4

The fourth case study is one of the pulp company founded by an industry group located in the state of Sao Paulo in 1986. With a contingent of 450 employees, this organization has a production capacity of 240,000 tons per year and did not inform their average income. The company exports 30% of its production.

This organization has a self-sufficiency in energy from the burning done on a boiler in one of its plants. In addition, the company has efficient techniques for reuse of water as it has no river in its vicinity. Although without any certification of an IMS, as shown in Table 8, the company conducts its activities incorporated into an IMS.

It is noted from Table 8, a concern with the environmental issue, since it is the branch of activity requires an adequate forest management because of the magnitude in terms of environmental impact it causes.

The company is relatively April, before the other, new, which may explain the absence of other certifications of their management systems. The environmental manager of the organization said in April that there is an interest in this practice, but the initial focus was to obtain the ISO 14001:2004 and FSC that are the most important branch of business activity.

It already has self-sufficiency, the company has a project in the short term to sell 7 MW of surplus production, which is enough to power 25,000 homes.

Like other enterprises, the four opted not to disclose quantitative indicators from the EMS based on ISO 14001:2004, but noted significant qualitative benefits accruing to the company, among the principal, can be cited:

- Improved control of water consumption
- Improved control of gas emissions to air
- Improved control of emissions of particulate matter into the atmosphere
- Control the generation of solid and liquid waste;
- Dissemination of a culture of environmental responsibility in the company;
- Improved image of the company in the presence of stakeholders, and
- Alignment of values and company policy to environmental principles.

Table 9. Main projects of CP in company

Local application	Project description
Reuse water	Exchange of rotating drums for special washers. Reuse of cooling water pumps for boiler feed water recovery, reuse of waste reverse osmosis as part of the water supply for industrial factory, water reuse and heat energy produced in the digester and the reduction in dryer consumption of drinking water
STS	Biological treatment by activated sludge waste from pulp production
Electrostatic precipitators	For collection, processing, burning and control of odorous gases and chimneys with online control gas output
Boiler	Burning of surplus biomass production of cellulose for energy production

According to environmental manager, the company is an offshoot of a corporate group that already owns in his profile, an environmentally responsible conduct of its business, thereby facilitating the process of ISO 14001:2004 certification, with emphasis as little difficulty, only adjustments to the company's records and documentation requirements of the standard.

Even without taking the nomenclature and methodology of the CP, the company earned 4 adopts procedures for environmental and economic impacts in its production system, and also be regarded as a successful case of CP by CETESB still follows 19 of the 20 steps proposed by UNIDO.

The commitment to the adequacy of the production system to obtain environmental improvements part of senior management, ie, the practice is inherent in the politics of IMS and also the company's EMS.

There is a team environment, which is the same as the EMS, which identifies the opportunities and objectives, evaluating the viability and moving to the area responsible for the sector that will be implemented.

Table 9 shows the major projects implemented CP Company 4.

For Table 9, we observe important projects of CP by the company, according to your industry, promoting alternatives to minimize waste and particularly the reuse of water is a serious problem for the pulp industry.

This company also failed to publish figures of their indicators, but acknowledges that some environmental benefits promoted by adopting procedures for CP, they are:

- Decrease in water use;
- Reduction of landfill waste gases, liquids and gases;
- Reduction of sulfur emissions to the atmosphere by the use of electrostatic precipitators, and
- Power generation through the burning of surplus biomass production of cellulose.

The Table 10 aims to illustrate important points of business to which it may be possible to structure a relationship between the organizations studied information presented throughout this chapter. This framework serves to expose a qualitative analysis and succinctly, the main information obtained from the companies mentioned in the chapter.

Table 10. Production of the companies grouped

Organizations	Possessed before the ISO EMS	Has management support for CP	Signatory's official CP	Number of steps followed by CP	Economics benefits accruing	Environmental benefits accruing
1	YES	YES	NO	13	YES	YES
2	YES	YES	YES	17	YES	YES
3	YES	YES	NO	11	YES	YES
4	YES	YES	NO	19	YES	YES

It became possible to observe that all companies had already implemented before the EMS ISO 14001:2004 certification, which greatly facilitated the certification process by the standard, only requiring adaptations to the requirements of ISO documentation and records. All have full management support for implementation of CP projects in their units, which greatly facilitates the practice.

Except for the second company, has officially declared no signatory of the CP, but note that all adopt more than 50% of the steps proposed by UNIDO for the implementation of CP. In addition, all companies have realized significant benefits from economic and environmental origin, which represent the main focus proposed by the application of CP.

Conclusion

The EMS based on ISO 14001:2004 enjoyed significant qualitative benefits to companies that adopt it, because it is not possible for quantitative analysis at the request of business secrecy, it became clear that there are important benefits for organizations. Some of these indicators are derived from the relation between the EMS based on ISO 14001:2004 and adopting procedures for CP. It was possible to see the profile of the CP in Brazil, since the survey sample was comprised of successful cases in this practice in the country. Thus it was noted that even companies that consider themselves officially signed the CP, not exactly follow the same methodology proposed by UNIDO, but achieve the objectives expected by the international body.

An important consideration to make is that CP actually receives important benefits to the business sector and also for the environment, however, note that the methodology proposed for this worldwide practice, could be revised, thereby aiming to facilitation of its applicability and thus increase the spread of this instrument on the market.

The results presented in this chapter can be used for future qualitative research on the subject, requiring more detailed quantitative analysis to demonstrate the correlation between EMS and CP in enterprises.

Acknowledgments

Thanks to CAPES (Coordination for the Improvement of Higher Education), to FAPESP (Foundation for Research Support of São Paulo) and CNPq (National Council for Scientific and Technological Development) for the important financial support for the research.

REFERENCES

Berland, N.; Loison, M. C. Fabricating management practices: "Responsible Care" and corporate social responsibility. *Society and Business Review*, v. 3, n. 1, p. 41-56, 2008.

IBGE – Brazilian Institute of Geography and Statistics. Available at http://www.ibge.gov.br. Accessed 01/10/2010.

ISO14001:2004 – Environmental management systems: requirements with guidance for use. *International Organization for Standardization* (2004). Geneva, Switzerland.

Martin, M. J.; Rigola, M. Incorporating cleaner production and environmental management systems in environmental science education ate University of Girona. *International Journal of Sustainability in Higher Education*, v. 2, n. 4, p. 329-338, 2001.

Matthews, D. H. Environmental management systems for internal corporate environmental benchmarking. *Benchmarking: An International Journal*, v. 10, n. 2, p. 95-106, 2003.

Melnyk, S. A., Sroufe, R. P., Calantone, R. L.; Montabon, F. L. Assessing the effectiveness of US voluntary environmental programmes: an empirical study. *International Journal of Production Research*, v. 40, n. 8, p. 1853-78, 2002.

Rosen, C. M. Environmental strategy and competitive advantage: an introduction. *Californian Management Review*, v. 43, n. 3, p. 9-16, 2001.

Seiffert, M. E. B. ISO 14001 - *Sistemas de gestão ambiental*: implantação objetiva e econômica. 2. ed. São Paulo: Atlas, 2006.

UNEP – United Nations Environment Programme. Current changes in approaches to environmental policy: cleaner and leaner production. Available at: http://www.unep.org. Accessed 14/11/2010.

In: Environmental Management
Editor: Henry C. Dupont

ISBN: 978-1-61324-733-4
© 2012 Nova Science Publishers, Inc.

Chapter 14

TOWARDS A WATERSHED APPROACH IN NON-POINT SOURCE POLLUTION CONTROL IN THE LAKE TAI BASIN, CHINA

Xiaoying Yang, Zheng Zheng and Xingzhang Luo
Department of Environmental Science and Engineering,
Fudan University, Shanghai, China

ABSTRACT

While accounting for 0.4% of its land area and 2.9% of its population, the Lake Tai basin generates more than 14% of China's Gross Domestic Production. Accompanied with its fast economic development is serious water environment deterioration in the Lake Tai basin. The lake is becoming increasingly eutrophied and has frequently suffered from cyanobacterial blooms in recent years. Although tremendous investment has been made to control pollutant discharge to improve its water quality, the Lake Tai's eutrophication trend has not been reversed due to the past emphasis on point pollution sources and the lack of effective measures for non-point source pollution control in the region.

A watershed approach is proposed to deal with the serious non-point source pollution issues in the region with four guiding principles: (1) Control from the source; (2) Reduction along transport; (3) Emphasis on waste reuse and nutrient recycling; and (4) Intensive treatment at key locations. Research as well as field applications that have been conducted to implement this watershed approach are introduced. Existing problems and their implications for future research needs as well as policy making are also discussed.

Keywords: Non-point source pollution, rural sewage treatment, rural waste recycling, watershed approach, Lake Tai.

1. INTRODUCTION

Lake Tai (119 Tai uction N, 30, Tai ucti E) is the third largest freshwater lake in China, and located in the highly developed and densely populated Yangtze River Delta. The lake is a shallow lake with a mean depth of 1.9 m and a maximum depth of 2.6 m. It extends a distance of 68.5 km from north to south and 56 km from east to west. The lake bottom features a flat terrain with an average topographic gradient of 0°0'19.66" and a mean elevation of 1.1 m above sea level (Qin et al. 2007).

The drainage basin of the Lake Tai spans over 36,500 km^2. Although it accounts for 0.4% of China's territory and 2.9% of its population, the Lake Tai basin generates more than 14% of China's Gross Domestic Production (GDP), and the basin's GDP per capita is 3.5 times as much as the state average (Zhang, Wang and Jin 2007). Lake Tai is vital to its surrounding region's socio-economic development, providing multiple services including water supply for residential, municipal, industrial, and agricultural needs, navigation, flood control, fishery, and tourism. Unfortunately, accompanied fast economic development and the resulting large amount of pollutant discharge, water environment of the Lake Tai basin has been deteriorating seriously since 1980's. The lake is becoming increasingly eutrophied and has frequently suffered from cyanobacterial blooms in recent decades.

Since the 1990's, China has taken a series of measures to control pollutant discharge to improve the Lake Tai's water quality. In 1996, the Chinese government set the following two targets on pollution control in the region: (1) Water quality of the Lake Tai's tributaries should meet the Category III National Surface Water Quality Standard (GB3838-2002) by 2000; and (2) The Lake Tai should be clean again by 2010. So far, point pollution sources such as industrial and municipal wastewater in the region have been largely put under control. Nevertheless, neither of the above two water pollution control targets has been met, and the Lake Tai's eutrophication trend has not been reversed either. In 2007, a severe algae bloom in the lake caused a drinking water contamination crisis, leaving 4.43 million people without safe drinking water in Wuxi City, Jiangsu Province. It was reported that the concentration of dimethyltrisulfide in a water sample collected on 4 June 2007 from the drinking-water intake was 11,399 mg/l, which is high enough to cause strong septic and marshy odors (Yang et al. 2008).

Current serious water pollution situation in the Lake Tai basin demonstrates the need to re-evaluate the effectiveness of previous water pollution control strategies, which put the emphasis solely on the treatment of pollution from industries and urban sewage. It is increasingly realized that the previously largely ignored nonpoint pollution sources are discharging significant amount of pollutants into the ambient environment (Chen et al. 2010). How to effectively control the non-point source pollution in the Lake Tai Basin is the key to improve the lake's water quality conditions.

2. TOWARD A WATERSHED APPROACH FOR NON-POINT SOURCE POLLUTION CONTROL

Compared with discrete point sources, non-point sources are diffusive and variable in the following aspects (U.S.EPA 2003):

1. Pollutants discharged from nonpoint sources enter surface and/or ground waters in a scattered manner at intermittent intervals.
2. Pollutant generation arises over an extensive land area and moves overland before it reaches surface waters or infiltrates into ground waters.
3. The extent of NPS pollution is related to uncontrollable climatic events, geographic and geologic conditions and it varies greatly from place to place and from year to year.
4. The extent of NPS pollution is often more difficult or expensive to monitor at the point(s) of origin, as compared to monitoring of point sources.

Many previous studies have identified agricultural production as the dominant non-point pollution source (Donner 2003, Turner and Rabalais 2003, U.S.EPA 2009). Nevertheless, the composition of non-point pollution sources is more complicated in the Lake Tai Basin. With a total population of 36 million, the Lake Tai Basin has a high population density of 978 persons per km^2, which is seven times of the national average. More than 50 percent of the population is scattered in rural villages, which are often located along the rivers as they have been in the long past. The sewage from the rural households is largely untreated, and much has been discharged directly into the nearby rivers. For example, residential wastewater, agricultural production, and animal feeding operations are identified as the major pollution sources to the Caoqiao River, a tributary to the Lake Tai. It is estimated that residential wastewater contributes 53% of COD, 35% of NH_3-N, 21% of TN, and 18% of TP load; agricultural production contributes 24% of COD, 37% of NH_3-N, 64% of TN, and 38% of TP load; and animal feed operations contribute 10% of the COD, 17% of NH_3-N, 12% of the TN, and 41% of TP load (Zhang et al. 2009).

To deal with the serious non-point source pollution situations in the Lake Tai Basin, a multidisciplinary watershed approach is proposed with four guiding principles: (1) control from the source; (2) reduction during transport; (3) intensive treatment at critical locations; and (4) emphasis on waste reuse and nutrient recycling. Technologies that have been developed to deal with various types of non-point pollution sources during implementation of this watershed approach are introduced in the rest of this section.

2.1. Rural Sewage Treatment

Although the technologies for centralized municipal sewage treatment have been well developed, they are not suitable for wide application in rural area because of the high costs needed for sewage collection, facility construction, and operation, as well as the requirement of rigorous watch and maintenance. Alternative technologies are needed for rural sewage treatment that ideally should be tailored to the characteristics of rural sewage on the one hand, and require low cost and little expertise for maintenance on the other hand. A lot of research has been conducted to develop alternative onsite wastewater treatment technologies (Oakley, Arthur and Autumn 2010). So far, most of them are limited to lab tests or pilot experiments in China. In the following, two promising technologies are introduced, which have been maturing over the past decade, and now are among the few that have seen wide applications in the Lake Tai region.

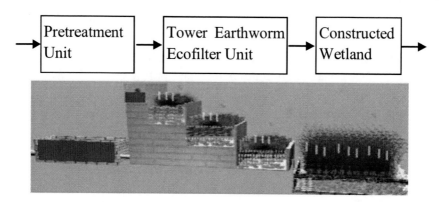

Figure 1. Configuration of the Tower Earthworm Ecofilter System.

2.1.1. Tower Earthworm Ecofilter System

The Tower Earthworm Ecofilter System consists of three major components: pretreatment unit, tower earthworm ecofilter, and constructed wetland (Figure 1). In the pretreatment unit, suspended solids and grease are removed so as to mitigate clogging in the ecofilter unit. The pretreatment unit may also facilitate the degradation of organic compounds. Depending on the influent characteristics as well as environment conditions, the unit could take many forms such as septic tanks, sedimentation basins, acidic hydrolysis tanks, and anaerobic ponds.

The ecofilter unit is where the majority of pollutant removal activities occur. Usually, it is composed of a sequence of linked sub-units, each containing a 30 cm layer of artificial soils and several supporting layers of substrates that are made of materials such as sands, gravels, cobbles, sawdust, and slag. In the ecofilter unit, pollutants are removed through various physical, chemical and biological processes. Earthworms in the soil layers not only increase void ratios to improve soil aeration and reduce substrate clogging, but also increase the specific surface area of the substrates for absorption and stimulate the conversion of nutrients and organic compounds (Sinha, Bharambe and Chaudhari 2008).

If needed, treated wastewater from the ecofilter unit is diverted to a constructed wetland for further removal of nitrogen and phosphorous. Normally, an underflow wetland system is used, which can be configured with 3 or 4 aquatic plant species.

Allowing a hydraulic load of 1-2 $m^3/(m^2 \cdot d)$, the Tower Earthworm Ecofilter System requires less space than many counterpart technologies. Both lab experiments and demonstration projects in the Lake Tai Basin have shown that the system is able to remove organic compounds and nutrients effectively, with more than 70% of removal rates for COD, NH_4^+-N, TN, and TP (Fang et al. 2010, Li et al. 2009). In addition, building and operating the earthworm ecofilter tower are fairly affordable in China's rural regions. It is estimated that on the average the construction cost is around 255 dollars for 1 t/d of treatment capacity (excluding the costs associated with building sewers and constructed wetland), while operation cost is around 5 cents per ton. Due to modest need for maintenance, the system could be operated independently by local residents. Finally, although built for wastewater treatment, the system could be beautifully designed and incorporate with the surrounding landscape. Figure 2 gives two field application examples of the Tower Earthworm Ecofilter System.

Figure 2. Field application examples of the Tower Earthworm Ecofilter System.

2.1.2. Soil Capillary Infiltration Ditch System

Tower Earthworm Ecofilter System is mainly used to treat wastewater for relatively concentrated settlement. For remotely located rural households, the Soil Capillary Infiltration Ditch system is more applicable. In the system, sewage from rural households is dispersed underground through perforated pipes or clay. During the process, nutrients are absorbed by soil organisms and plants. Above the system, the land surface could still be used for various purposes such as gardening (Figure 3).

The Soil Capillary Infiltration Ditch System is capable of treating sewage for several households. Pilot project results indicate that the system has a nitrogen removal rate of above 75% for nitrogen and above 85% for phosphorous. The system is quite affordable with construction cost around 165 dollars per house and operation cost less than 4.5 cents per ton of water treated. Figure 4 gives a field application example of the Soil Capillary Infiltration Ditch System.

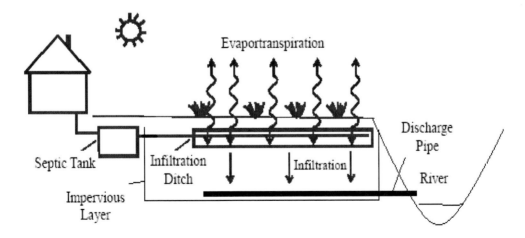

Figure 3. Configuration of the Soil Capillary Infiltration Ditch System.

Figure 4. A field application example of the Soil Capillary Infiltration Ditch System.

2.2. Rural Waste Reuse and Recycling

Agricultural production in China is heavily dependent on the application of energy-intensive inputs such as fertilizers and pesticides. According to National Bureau of Statistics of China, China's fertilizer consumption has increased rapidly from 8.8 million tons in 1978 to 51.1 million tons in 2007. Now China is the world's largest fertilizer consumer, accounting for one third of the world's total consumption. On the average, the rate of fertilizer application in China is around 450 kg/hm^2, which is twice the upper limit of 225 kg/hm^2 generally accepted by developed countries. The fertilizer application rate is even higher in the Lake Tai basin. In the Lake Tai basin, the majority of the cropland in the region belongs to the paddy soil under rice-wheat crop rotation (from mid-June to late October for rice, and from early November to next May for wheat). It is estimated that the farmers in the region are generally applying at a rate greater than 400 kg N ha^{-1} per crop, sometimes even greater than 600 kg N ha^{-1} per crop (Ju et al. 2004, Zhao et al. 2009).

The production of synthetic fertilizers, especially nitrogen fertilizer using the Haber–Bosch Process, consumes large amounts of fossil fuels. It is estimated that production of Nitrogen fertilizer alone accounts for around 5 percent of global natural gas consumption (Woods et al. 2010). While tremendous amount of energy is consumed in fertilizer production to supply crops with the needed N, P, K, and other nutrients, millions of tons of nutrients have been discarded as waste annually. It is estimated that around 754.7 million tons of crop straws and 3971.6 million tons of animal manure are generated annually in China's rural regions (Zhang, Tan and Gersberg 2010). More than 15% of crop straws are just discarded and burned in field, while another 40% is burned as residential fuels. Because of the lack of treatment capabilities, less than 5% of livestock manure is sufficiently treated and recycled in China, although it is high in nutrients required for crop growth. For example, on the average, chicken manure is composed of 25.5% of organic matter, 1.6% of N, 1.5% of P, and 0.9% of

K. There is an urgent need for the development of sustainable approaches to recover nutrients and energy from rural wastes as well as minimizing the associated secondary pollution.

Rural wastes such as crop residuals and aquatic plants are not degradable easily due to their high content of lignocellulosic. Through anaerobic co-digestion of the lignocellulosic and biodegradable materials such as animal manure, the biological conversion rate can be improved significantly. During anaerobic digestion, organic matters contained in the wastes are converted to biogas through the interactions of a very heterogeneous mixture of bacteria. Composition of the produced biogas is strongly dependent on the type of digestion process as well as the nature of the fermented organic matter. Nevertheless, its methane content generally ranges between 50 and 60%, and carbon dioxide ranges between 35 and 45% (Neves, Oliveira and Alves 2009). The low methane content of the raw biogas limits its possible uses, and it needs to be refined so as to reach an energy value comparable to that of the other energy sources. However, the high cost of biogas refinement has been a bottleneck to compromise its competitiveness with other energy sources.

Using algae with fast growth rates to absorb carbon dioxide from biogas so as to increase its methane content could prove to be a cost-effective way of improving the quality of the raw biogas. It is estimated that algae absorption can increase the methane content to over 85%, and the resulting high-quality biogas can act as the replacement for natural gas. Meanwhile, the produced algae biomass are not discarded as wastes but recycled back for another round of anaerobic digestion so as to achieve the maximization of waste recycling rates and the reduction of carbon dioxide emissions simultaneously (Figure 5).

In addition, the anaerobic digestion process results in a mineralization of organically bounded nutrients, in particular nitrogen, and converts them into inorganic forms that are more accessible to plants. For instance, it is reported that biogas residual contains 25% more ammonium (NH_4^+-N) than untreated liquid manure. Because essential nutrients (N, P, K) as well as trace elements required by plants are preserved in the biogas residue, the biogas residue has great potentials to act as the substitute and/or complement to commercial mineral fertilizers after proper pretreatment.

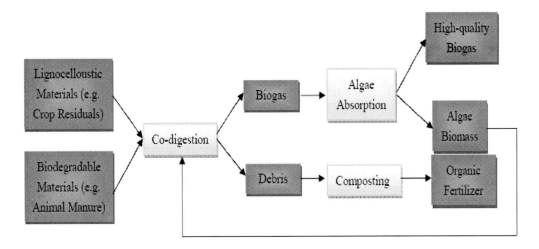

Figure 5. Flowchart of nutrient recovery and high-quality biogas production from rural wastes.

A number of studies have been conducted on how effectively biogas residue can substitute commercial mineral fertilizers in terms of crop yield, and the results support the significant potentials of biogas residue as a valuable substitute and/or supplement to commercial fertilizers to enhance yield for a broad range of plants including cereals and vegetables (Garg et al. 2005, Montemurro et al. 2008). Additionally, application of biogas residue can increase the soil organic matter content, which improves soil aggregation, facilitates soil drainage and aeration, suppresses plant disease, and stimulate soil microbial activities (Arthurson 2009).

2.3. "Fortified" Clean River Network

Compared with discrete pollutant sources, non-point source pollution is more difficult to control because of its diffusive nature. Without fixed discharge locations and transport path, it is hard to rely on the installation of conventional pollutant treatment devices to deal with non-point source pollution. Abatement of nonpoint sources should be more focused on land and runoff management practices, rather than on effluent treatment. Alternative approaches that make full use of the local landscape and waterways while fortified with biological, physical, or chemical structures could be more effective in this case. Some examples of the alternative approaches are described in the following.

2.3.1. Vegetated Filter Strips

Vegetated filter strips (VFS) can improve water quality by removing nutrients, sediment, suspended solids, and pesticides through various processes such as filtration, deposition, infiltration, absorption, adsorption, decomposition, and volatilization. Typical locations to build VFS include: below cropland or other fields; adjacent to wetlands, streams, ponds, or lakes; along roadways, parking lots, or other impervious areas; and above conservation practices such as terraces or diversions.

Vegetated filter strips are designed to be used under conditions in which runoff passes in a uniform sheet flow, and the key elements to ensure VFS to work are (U.S.EPA 1993):

(1) Slope. VFS function best on gentle slopes of less than 5 percent; slopes greater than 15 percent render VFS ineffective because surface runoff flow will not be sheet-like and uniform.
(2) Plant type. The best species for VFS are those which will produce dense vegetative cover resistant to overland flow. Native or at least noninvasive species should be used to avoid negatively impacting adjacent natural areas.
(3) Length. Contact time between runoff and vegetation in the VFS increases with increasing VFS length. Some sources recommend a minimum length of about 50 feet, while others recommend the VFS length to be at least as long as the runoff contributing area (Dillaha et al. 1989, Schueler 1987).
(4) Detention time. Consideration should be given to increasing the detention time of runoff as it passes through the VFS, such as the design of small rills that run parallel to the leading edge of the vegetated filter strip, which would serve to trap water as runoff passes through.

(5) Monitoring. VFS should be inspected periodically to determine whether concentrated flows are bypassing or overwhelming the system, and whether sediment is accumulating in quantities that would reduce its effectiveness.

Studies of VFS show that they can be an effective management practice for the control of nonpoint pollution from silvicultural, urban, construction, and agricultural sources. The effectiveness of VFS varies with topography, vegetative cover, climate, soil type, land use practices, and implementation. Study results indicate that VFS are most effective at sediment removal, with rates generally greater than 70 percent. The published results on the effectiveness of VFS in nutrient removal are more variable, but nitrogen and phosphorus removal rates are typically greater than 50 percent (U.S.EPA 1993).

2.3.2. Constructed Wetlands

Constructed wetlands are typically engineered complexes of saturated substrates, emergent and submerging vegetation, animal life, and water that simulate wetlands for human use and benefits. Constructed wetlands typically have four principal components that may assist in pollutant removal: (1) substrates with various rates of hydraulic conductivity; (2) plants adapted to water-saturated anaerobic substrates; (3) a water column (water flowing through or above the substrate); and (4) aerobic and anaerobic microbial populations (Hammer, Pullin and Watson. 1989).

Several factors must be considered in the design and construction of an artificial wetland to ensure the maximum performance of the facility for pollutant removal (U.S.EPA 1993):

(1) Hydrology. The most important variable in constructed wetland design is hydrology. If proper hydrologic conditions are developed, the chemical and biological conditions will, to a degree, respond accordingly.
(2) Soils. The underlying soils in a wetland vary in their ability to support vegetation, to prevent surface water percolation, and to provide active exchange sites for adsorption of constituents like phosphorus and metals.
(3) Vegetation. The types of vegetation used in constructed wetlands depend on the region and climate of the constructed wetland. When possible, use native plant species or noninvasive species to avoid negative impacts to nearby natural wetland areas.
(4) Geometry. The size and shape of the constructed wetland can influence the detention time of the wetland, the rate of water moving through the system, and subsequently the system's pollutant removal effectiveness.
(5) Pretreatment. In the constructed wetland system, forebays can be constructed to trap sediment before runoff enters the vegetated area. Baffles and diversions could be strategically placed to prevent trapped sediments from re-suspension prior to being cleaned out.
(6) Maintenance. Constructed wetlands need to be maintained for optimal performance as well as avoid any negative impacts to wildlife and surrounding areas. For example, non-native or undesirable plant species must be kept out of adjacent wetlands or riparian areas. Contamination of sediments due to toxics entering the constructed wetland must be controlled, and excess sediments should be removed from the system.

Many studies have been conducted to evaluate the effectiveness of artificial wetlands for removing pollutants from surface water runoff. In general, constructed wetland systems are effective at removing suspended solids and pollutants that attach to solids and soil particles. Typical removal rates for suspended solids were greater than 90 percent. Removal rates for total phosphorus ranged from 50 percent to 90 percent. Nitrogen removal was highly variable and ranged from 10 percent to 76 percent for total nitrogen (U.S.EPA 1993).

2.3.3. Restoration of Riparian Areas

Riparian areas consist of a complex organization of biotic and abiotic elements. They are effective in removing suspended solids, nutrients, and other contaminants from upland runoff, as well as maintaining stream channel temperature. Some studies also suggest that riparian vegetation acts as a nutrient sink, taking up and storing nutrients.

Restoration of riparian areas refers to the recovery of a range of functions that existed previously by reestablishing the area's hydrology, vegetation, and structure characteristics. The restoration can lead to multiple benefits such as the abatement of non-point source pollution and provide wildlife habitats.

Several practices could be adopted for riparian area restoration: (1) Restore the area's hydrologic regime. (2) Restore native plant species through either natural succession or selected planting. (3) Plan restoration as part of naturally occurring aquatic ecosystems. When selecting sites and designing restoration, priority should be put to restore the regions with high aquatic and riparian habitat diversity and high productivity; to maximize connectedness (between different aquatic and riparian habitat types); and to provide refuge or migration corridors along rivers between larger patches of uplands (U.S.EPA 1993).

2.3.4. Permeable Dam

Permeable dams are built with gravels and stones to block the flow of water in the rivers that are not used for the purposes of flood control and navigation. The water blockage allows the storage of surface runoff and leads to a heightened water level before the dam, which enables water to flow through subsequent water treatment facilities through gravity. Since it takes some time for water to flow through the permeable dam, the system has a detention period for water purification. In addition, vegetation can be planted on the dam, and the synergy between vegetation roots and soil microorganisms facilitate the degradation of organic compounds and nutrients similar to the "root zone" of constructed wetlands.

A permeable dam was constructed in the Yixing City of Jiangsu Province with 40% of gravels (with a size of 5-10 mm) and 60% of stones (with a size of 2-4 cm). Its length is 4.6 m at the top and 15.6 m at the bottom, with a width of 12.7 m and a height of 2.2 m. With an infiltration coefficient of 0.1 m/s, it collects and treats surface runoff from an upstream contribution area of 2.0 km^2. A four-month experiment indicates that its TN and TP removal rates are 15.6% and 23.4%, respectively (Tian and Zhang 2006).

2.3.5. Building of a "Fortified" Clean River Network

There is no single, ideal management practice system for controlling a particular pollutant in all situations. Rather, the system should be designed based on multiple factors such as the source and the type of pollutants; local agricultural, climatic, and environmental conditions; and the willingness and ability of the stakeholders to implement and maintain the practices. In addition, water quality problems can not usually be solved with one management practice since single practice is rarely able to provide the full range and extent of control needed at a site.

In view of the Lake Tai Basin's characteristic of dense river networks, multiple practices can be integrated to function together while incorporating local landscape and waterways, thereby building a "fortified" clean river network system that addresses treatment needs associated with pollutant generation from various sources, transport, and remediation.

Figure 6 gives an example of such system called "pre-dam." The pre-dam system is composed of four units: the collection and adjustment unit, the interception and deposition unit, the intensive purification unit, and the reuse unit. As seen from the figure, the design of the pre-dam system integrates a variety of physical and biological wastewater treatment practices while making full use of the local natural landscape and waterways. The whole system could be visualized as a "self-functioning" river network that is capable of removing pollutants along its path-way.

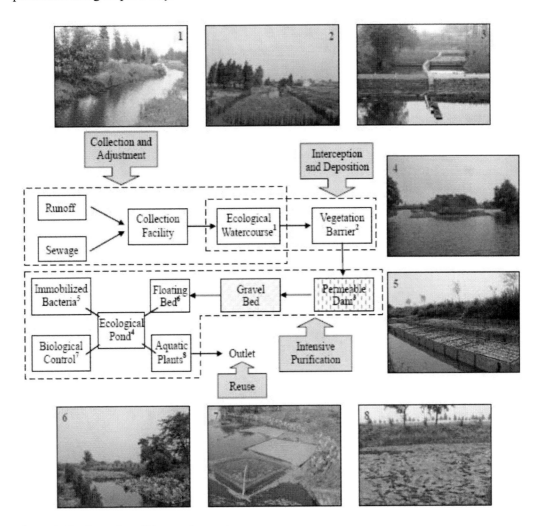

Figure 6. Configuration of the pre-dam system.

CONCLUSION

Like many other places in the world, non-point source pollution is the major cause of water environment deterioration in the Lake Tai basin of eastern China. The task of non-point source pollution control is more complicated in this region due to its large rural population base, improper application of fertilizers and pesticides, and mushrooming concentrated animal feeding operations. A watershed approach is proposed to target the various non-point pollution sources in the region under four principles: control from the source, reduction during transport, intensive treatment at key locations, and emphasis on waste reuse and nutrient recycling.

Technologies in rural sewage treatment, rural waste disposal, and surface runoff treatment have been developed in striving to implement the watershed approach in the Lake Tai region. Although these technologies are sound in principles, there is a still a long way before their full implementation in the Lake Tai region. Major obstacles to their implementation include the lack of participation from stakeholders and insufficient capacity in monitoring and maintenance. For example, a face-to-face interview in the Lake Tai basin showed that local rural villagers have low willingness to pay for the treatment of domestic sewage, with more than 80% of the interviewed villagers are only willing to pay 8 dollars or less annually for the treatment (Yang, Luo and Zheng 2011). Although facilities have been built to produce methane from rural wastes through anaerobic digestion, many of them have not operated normally because of the difficulty with collecting raw materials such as crop residues and animal manure. At the same time, although prohibited, many of the local farmers are still burning crop residues in field secretly. Furthermore, structures built for non-point pollution abatement often fall out of shape and stop functioning normally due to the lack of routine maintenance and management. Therefore, there is still much work ahead in improving non-point source pollution treatment technologies, as well as motivating the involvement of local stakeholders and building capacity to ensure effective and continuous implementation of the technologies.

REFERENCES

Arthurson, V. (2009) Closing the global energy and nutrient cycles through application of biogas residue to agricultural land - Potential benefits and drawbacks. *Energies,* 2, 226-242.

Chen, X., F. Wo, C. Chen and K. Fang (2010) Seasonal changes in the concentrations of nitrogen and phosphorus in farmland drainage and groundwater of the Taihu Lake region of China. *Environmental Monitoring and Assessment,* 169, 159-168.

Dillaha, T. A., R. B. Renear, S. Mostaghimi and D. Lee (1989) Vegetative filter strips for agricultural nonpoint source pollution control. *Transactions of the American Society of Agricultural Engineers,* 32, 513-519.

Donner, S. (2003) The impact of cropland cover on river nutrient levels in the Mississppi River Basin. *Global Ecology and Biogeography,* 12, 341-355.

Fang, C. X., Z. Zheng, X. Z. Luo and F. H. Guo (2010) Effect of hydraulic load on domestic wastewater treatment and removal mechanism of phosphorus in earthworm ecofilter. *Fresenius Environmental Bulletin,* 19, 1099-1108.

Garg, R. N., H. Pathak, D. K. Das and R. K. Tomar (2005) Use of flyash and biogas slurry for improving wheat yield and physical properties of soil. *Environmental Monitoring and Assessment,* 107, 1-9.

Hammer, D. A., B. P. Pullin and J. T. Watson. 1989. Constructed wetlands for livestock waste treatment. Knoxville, TN.

Ju, X., X. Liu, Z. F. and M. Roelcke (2004) Nitrogen fertilization, soil nitrate accumulation, and policy recommendations in several agricultural regions of China. *Ambio,* 33, 300-305.

Li, J. Z., X. Z. Luo, Z. Zheng, P. P. Li and Z. Meng (2009) Design of tower earthworm ecofilter system for treatment of centralized rural domestic sewage (translated). *China Water and Wastewater,* 25, 35-38.

Montemurro, F., S. Canali, G. Convertini, D. Ferri, F. Tittarelli and C. Vitti (2008) Anaerobic digestates application on fodder crops: effects on plant and soil. *Agrochimica,* 52, 297-312.

Neves, L., R. Oliveira and M. M. Alves (2009) Co-digestion of cow manure, food waste and intermittent input of fat. *Bioresource Technology,* 100, 1957-1962.

Oakley, S. M., J. G. Arthur and J. O. Autumn (2010) Nitrogen control through decentralized wastewater treatment: Process performance and alternative management strategies. *Ecological Engineering,* 36, 1520-1531.

Qin, B., P. Xu, Q. Wu, L. Luo and Y. Zhang (2007) Environmental issues of Lake Taihu, China. *Hydrobiologia,* 581, 3-14.

Schueler, T. 1987. Controlling urban runoff: A practical manual for planning and designing urban BMPs. Washington D.C.

Sinha, R. K., G. Bharambe and U. Chaudhari (2008) Sewage treatment by vermifiltration with synchronous treatment of sludge by earthworms: a low-cost sustainable technology over conventional systems with potential for decentralization. *Environmentalist,* 28, 409–420.

Tian, M. and Y. C. Zhang (2006) Experimental study on permeable dam technique to controlrural non-point pollution in Taihu basin (translated). *Acta Scientiae Circumstantiae,* 26, 1665-1670.

Turner, R. E. and N. N. Rabalais (2003) Linking landscape and water quality in the Mississippi river basin for 200 years. *Bioscience,* 53, 563–572.

U.S.EPA. 1993. Guidance specifying management measures for sources of nonpoint pollution in coastal waters. EPA 840-B-92-002. Washington D.C.

U.S.EPA. 2003. National management measures for the control of nonpoint pollution from agriculture. EPA-841-B-03-004. Washington D.C.

U.S.EPA. 2009. National water quality inventory: 2004 report to congress. EPA 841-R-08-001. Washington D.C.

Woods, J., A. Williams, J. K. Hughes, M. Black and R. Murphy (2010) Energy and the food system. *Philosophical Transactions of the Royal Society B,* 365, 2991-3006.

Yang, M., J. W. Yu, Z. L. Li, Z. H. Guo, M. Burch and T. F. Lin (2008) Taihu Lake not to blame for Wuxi's woes. *Science,* 319, 158.

Yang, X., X. Luo and Z. Zheng. 2011. Tower Earthworm Ecofilter System for rural sewage treatment in China. In *5th International Conference on Bioinformatics and Biomedical Engineering*. Wuhan, China.

Zhang, D. Q., S. K. Tan and R. M. Gersberg (2010) Municipal solid waste management in China: Status, problems and challenges. *Journal of Environmental Management*, 91, 1623-1633.

Zhang, L., W. Liu, B. You and B. Bian (2009) Characteristics of pollutant sources of Caoqiao River in Taihu Lake Basin (translated). *Research of Environmental Sciences*, 22, 1150–1155.

Zhang, T. X., X. R. Wang and X. Jin (2007) Variations of alkaline phosphatase activity and fractions in sediments of a shallow Chinese eutrophic lake (Lake Taihu). *Environmental Pollution*, 150, 288-294.

Zhao, X., Y. Xie, Z. Xiong, X. Yan, G. Xing and Z. Zhu (2009) Nitrogen fate and environmental consequence in paddy soil under rice-wheat rotation in the Taihu lake region, China. *Plant Soil*, 319, 225-234.

In: Environmental Management
Editor: Henry C. Dupont

ISBN: 978-1-61324-733-4
© 2012 Nova Science Publishers, Inc.

Chapter 15

SOCIO-ENVIRONMENTAL MARKETING AS AN ENVIRONMENTAL MANAGEMENT SYSTEM TO PROTECT ENDANGERED SPECIES

Juan José Mier-Terán[*,a]*, María José Montero-Simó*[‡,b]
and Rafael A. Araque-Padilla[‡,b]

[a]University of Cádiz, Spain
[b]University of Córdoba, Spain

ABSTRACT

The environmental management systems are normally used to obtain certain environmental objectives. That is why they are established as either business or institutional obligations aimed at identifying objectives that improve the environment and incorporate procedures in management to attain these objectives. Moreover, given that prevention is always better than cure, they ensure that the aforementioned is complied with.

A common objective in these systems is to change behaviour that is thought to be negative with regard to the environment, whether it is that of those working for the aforesaid organisations or that of the final users of the services. Social marketing has proven to be a very suitable technology to change behaviour, and with regard to environmental issues, the specialisation of Environmental Marketing is a good example of how it is applied to a specific subject matter. In this case its basic principles and the techniques that are applied are aimed at modifying negative behaviour towards environmental problems. In this context, the basis and foundations of the Socio-environmental Marketing are shown as a management tool used to change behaviour in the sustainable management of protected areas. Furthermore a specific case is presented whereby this management system is used to protect an endangered species, the Iberian Lynx.

---

[*] E-mail: juanjose.mier-teran@uca.es.
[‡] E-mail: jmontero@etea.com.
[‡] E-mail: raraque@etea.com.

1. INTRODUCTION

Since the beginning of the 20th century, marketing in business has continuously developed and for the majority of companies that compete in increasingly more complex and fragmented markets, it does, beyond all doubt, influence the consumer's purchasing behaviour. Nevertheless, it wasn't until the beginning of the seventies that this technology starts to be applied to different profit making business spheres of activity. Indeed, this is when Kotler and Zaltman (1971) first defined the social marketing concept. They stated that it consisted in "the design, implementation and control of programmes designed to affect how social ideas are accepted, which involve planned concepts about product, price, promotion, place and market research". This started a debate within the scientific community about whether marketing should be applied in the non-profit making area or not. After some years, this debate clearly evolved towards the wider concept of marketing that includes both the profit making and non-profit making spheres of activity. It can therefore be used by companies and other types of non-commercial organisations to obtain their own objectives. By doing so a product is not only considered as being a good or a service, but also as an idea, a place, a person, an institution and the "exchange values" (instead of trade) are thought of as being the fundamental nucleus of marketing. The last reason why it is so well accepted is linked to the fact that, in practice, many organisations already used marketing tools (in the majority of cases advertising) to attain their own objectives. Greenpeace, WWF, Friends of the Earth, Ecologists in Action or the Ministry of the Environment are just some examples of the non-profit making and private institutions and organisations from the environmental world that as previously stated, include marketing tools in their strategies. However, the majority of campaigns that are organised are not very effective, if at all, when they are only focused on promotion, and do not take into consideration other important aspects of marketing, such as market research, segmentation and positioning or product, price and place strategies.

Marketing strategies must bear in mind all the elements that affect the behaviour of individuals in the purchasing process. They should not only be channelled through advertising. These precepts are just as valid in the non-profit making sphere of activity; it's more effective to work with a marketing strategy designed to modify the selected behaviour by combining the aforesaid variables properly than when this is done exclusively through advertising campaigns that are not very effective when they are carried out in sporadically, if at all.

2. SOCIAL MARKETING AND SOCIO-ENVIRONMENTAL MARKETING CONCEPT

Social marketing is currently understood as being "the use of marketing principles and techniques to make the target public willingly accept, reject, modify or abandon a particular form of behaviour so that it benefits the individual, a group or society as a whole" (Kotler; Roberto and Lee, 2002). The sphere of activity covers very different sectors including health, safety, social involvement, nutrition or the environment. With regard to the latter, campaigns aimed at changing behaviour towards issues like energy and water saving, protecting wildlife, forest fires, recycling or using alternative forms of transport have been designed and implemented. In short, any issue in which it is possible to attain a certain objective by a voluntary change of behaviour or where there are value exchanges, can be dealt with from the social marketing point of view.

Given that social marketing is exclusively aimed at obtaining benefits for individuals (human centred point of view), where our outlook is focused on identifying the human race as just one more living creature (although obviously with different characteristics) among all the others that live on the earth. At the same time, considering the distinct character of using social marketing for environmental issues, a new concept that has been called "Socio-environmental Marketing" can be considered appropriate to address the problem of the endangered species. This is understood as "using social marketing principles and techniques that are intended to modify or promote human behaviour, on their own free will, that affects a natural resource, a species, an ecological community, an ecosystem, a protected area or the environment as a whole. By doing so it supports the biodiversity and the earth's sustainable development" (Mier-Terán, 2004).

It is however important to point out that social marketing is not valid for all environmental situations or problems. Andreasen (2002) explains that although social marketing social can be used for many types of situations, it isn't always the best option. There are two criteria that should preside over the use of social marketing, which, according to this writer, are effectiveness and adaptation. With regard to the first criterion, it is important to analyse the campaigns that have already been carried out successfully; all the same, as Hornik points out (2001) that the majority of social marketing campaigns involve a significant number of activities that are intended to influence the target public. This makes it very complicated to identify specific effects. The latter criterion refers to whether using social marketing to solve a problem is the best option or if it is ethically right to use it.

3. BASIS OF THE SOCIO-ENVIRONMENTAL MARKETING MODEL

Why are there people who behave in a certain way with regards to the environment that is considered to be appropriate and then others who don't? There are many models and theories to explain environmental behaviour (Rajecki, 1982; Ajzen and Fishbein, 1980); Castro, 2000; Fietkau and Kessel, 1981; Blake, 1999; Kollmus and Agyeman, 2002). However there are three factors in this model that explain more clearly why certain people don't commit themselves to adopting these forms of behaviour.

- There are people who are not sure how to be pro-environmental or know what benefits can be had from acting this way. For example, those who don't realise how important it is to drive slowly along certain roads to preserve a certain species such as the Iberian Lynx or who are unaware of the benefits to be had by doing so, to preserve the biodiversity in the ecosystems. Or people who hunt this species because they don't know how important it is to preserve it.
- There are people who, in spite of knowing how to behave properly, have considerable difficulty or put up barriers that prevent them from doing so. For example, those who know the outcome of driving fast along roads where there could

be a lynx, but who refuse to go at a moderate speed because they don't want to be late for work, or they just don't want to waste their time by driving moderately. There are also people who in spite of knowing the problems facing the species, set traps because they think the problem of other predators is more serious.

- The people who are aware of the outcome of the behaviour and who do not consider the barriers to be important, but nevertheless, they feel that they are better off behaving as they do at the moment, just because it is easier for them. For example, those who, in spite of knowing the dangers of driving so fast that are not in a hurry drive fast because they feel like it; they enjoy the speed, or those that set non-selective traps because they feel it is the most convenient way to control predators on their estate.

To actually manage to influence people's environmental behaviour, firstly the barriers perceived and the benefits from this behaviour must be understood. There are four key ideas related to the change in environmental behaviour:

- The majority of people tend to act naturally, behaving in a way that is the most beneficial with the minimum amount of barriers.
- The barriers and benefits perceived vary substantially among individuals. What is beneficial for someone might be a barrier for someone else.
- Behaviours compete with each other. People chose between different behaviours that are possible; behaving in a certain way normally means rejecting another form of behaviour.
- Behaviours are not always predictable and on occasions, they do not completely coincide with people's attitudes or predispositions.

We therefore need to know about the barriers, and benefits perceived for each of the behaviours analysed. Moreover, if the programmes are going to be effective, tools need to be used that enable us to increase the benefits perceived and reduce the barriers of the behaviours proposed. This occurs at the same time as the competitive behaviours barriers are increased and its benefits are reduced. Socio-environmental marketing is ideal to carry out this aforementioned process and it provides us with a group of tools created to modify human behaviour that negatively affects the environment. There are four basic points involved in developing the strategy that we propose:

- The desired behaviour to be promoted: For example, different activities can be carried out to preserve or improve the lynx population. Each one could involve a different behaviour (breeding in captivity, improvement of the populations of prey, improved habitats, action taken on roads where the lynx cross over, reduce the number of non-selective traps, etc.). Out of all the aforementioned, those that are the most suitable according to their characteristics have to be chosen. To determine this two questions have to be answered:

    a) What are the possibilities of the socio-environmental marketing activities actually changing this behaviour?

b) What benefits and barriers are associated to each of these?

- The group in which the behaviour is to be promoted. In principle, this group will consist of those who adopt a form of behaviour that is contrary to the one that is proposed. Within this group however, there will be certain individuals who are more prone to change than others.
- The conditions needed so that people change their behaviour. Matters that can help make the individual adopt the proposed behaviour or abandon their current form of behaviour that is considered to be negative should be identified.
- The pressure from the reference groups that guide the target public's behaviour. In the environmental field, it is common to find groups that lead opinions about how one should act in a specific situation.

There will be benefits and barriers for both the proposed and the competitive forms of behaviour. In principle, it is logical to think that the competitive behaviour will have more benefits and fewer barriers as this is the one that is currently being adopted by target individuals from the programme.

The objective of our programme is therefore based on modifying the ratio between benefits and barriers in favour of the proposed behaviour, in such a way that this appeals more to the target public. There are four ways to do this:

- Increase the benefits of the proposed behaviour.
- Reduce the barriers of the proposed behaviour.
- Reduce the benefits of the competitive forms of behaviour.
- Increase the barriers of the competitive forms of behaviours.

The possibilities are not mutually exclusive; they can all be developed at the same time. Now the question involves finding out how to design the strategies that will enable the four aforesaid points to be carried out. The institutions and organisations working in the environmental field use different strategies to tackle these matters. Some of them are legal; laws are used to govern the appropriate behaviour. In practice these measures are not always popular or efficient and normally the fact that there is a lack of control measures means that a large part of the population does not comply with these laws (have people stopped using poison because it has been legally prohibited?, do we drive at the permitted speed limit or less?, are only selective traps that are allowed set?, do people only hunt the right species during the legally permitted season?).

Another type of strategy is information. It is very common in the environmental field; it is what is traditionally known as "awareness campaigns". Information can be used efficiently to modify behaviour; for example, the campaigns to use condoms to prevent AIDS have been successful. However, information itself in the majority of cases is just not enough to change these forms of behaviour. It has to be accompanied by other activities that are not aimed at just increasing people's knowledge about a specific matter (are there less forest fires after the campaigns organised?). Despite the fact that on a long term this action is having an effect, in many cases, this type of strategy is only implemented because it is the easiest.

The educational strategy (the traditional environmental education) has also proved to be inefficient in terms of changing behaviour. Numerous examples show that this strategy can affect attitudes, although these are not direct determining factors of behaviour. In the environmental field, for example, there are surveys that make it clear that attitude concerning the environment is well above the behaviour. People adopt positive attitudes towards protecting our environment, although in practise they contribute very little to actually protect it.

The economic strategy has proved to be ineffective in the majority of situations concerning the environment too. This is because in the majority of cases the economic incentives are not valued positively by the target public because they feel as if their behaviour is being bought.

The socio-environmental marketing strategy incorporates all of the aforementioned and at the same time it could even complement them. It is not intended to replace the existing approaches, but rather enhance them with a marketing perspective. It aims to affect behaviour, instead of information or attitudes. It is based on focusing on the consumer as the starting point. It suggests that only by having extensive knowledge of the needs, wishes and attitudes of environmental behaviour and of the barriers and benefits associated to these, will it be possible to actually influence them. The scheme that we propose and the stages involved in developing a socio-environmental marketing strategy are as follows:

4. STAGES OF A SOCIO-ENVIRONMENTAL MARKETING PROGRAMME

To develop a socio-environmental marketing programme, four basic stages that provide knowledge about the forms of behaviour associated with environmental problems and the subsequent development of strategies to modify and/or eliminate this behaviour must be taken into consideration:

4.1. Behaviour that is Promoted

4.1.1. Identifying Behaviour that Should be Promoted and Competitive Forms of Behaviour

The following socio-environmental marketing tools need to be used to understand current behaviour towards the problem that is the object of the study and the definition of the proposed form of behaviour to be followed:

a) Research on socio-environmental marketing. This is essential in the initial stages (and throughout the whole process); it clarifies which forms of behaviour need to be worked on. In this way and despite the great variety of tools that are available at the moment, those that are the most important in socio-environmental marketing are the following:

- Review of the existing literature. It helps to find out the background of the problem and it enables the researcher to feed off the experience already gained and the results.
- In depth interviews. These enable us to find out about the different opinions and points of views raised on the same matter. They provide more knowledge about reasons that explain the current forms of behaviour.
- Field observation. This is very useful to directly see the problems. It enables us to detect and record people's behaviour in their own surroundings. The main advantage of this technique is that the interaction between the research and what is being researched is minimal, which eliminates the mistakes made in the interviews.
- Group dynamics. These enable the group's synergy to focus on identifying and dealing with the problems. It helps to find out about different forms of behaviour at the same time as discussing the arguments for and against these. Variants such as the so-called "Strategic Workshops" are useful.

b) Identifying behaviour. This is part of the first stage and basically it is the need to clearly identify the problem that is to be tackled and/or the behaviour that is to be changed or promoted, as well as the competitive forms of behaviour. The majority of the problems that we face in the environmental field are not associated to just one form of behaviour but to various; this means they need to be identified and specified so that the strategies that are suitable for each one of them can be designed. Therefore, it has to be established in scientific terms, that is to say, what forms of behaviour are causing the problem and which ones are being proposed as an alternative have to be clearly explained. The finishing touches must be made to the ideas that have led to establishing the need to carry out a socio-environmental marketing programme and they must be structured more methodically. This step is crucial, given that in non-profit making organisational attempts are often made to tackle problems without having defined them beforehand. The main advantage, as pointed out by Ackoff (1953), is that a problem that has been correctly established is partially resolved. One way of doing this is to show the problem as a relationship between variables, formulating it correctly, without any ambiguities and establishing the empirical way to solve it. There are different points that can be defined to clarify the object of the study which help us in this process and these are:

- The matter to be studied. This is the subject of the programme.
- The programme's possible approaches. These are the different ways or means of tackling the subject matter to be studied. The criteria to choose a certain approach depends on the following factors: potential of the behaviour change, social demand, completed campaigns, organisational capacity and the campaign's economic potential.
- Purpose of the campaign. This refers to the potential impact of the campaign.

Having defined these points, the fact that the socio-environmental marketing is enough to successfully tackle the points raised needs to be confirmed.

4.1.2. Knowledge about the Organisation that must Implement the Programme. Training/Awareness of the Human Resources in the Organisation

Business training for the majority of people working in the organisations that could potentially be in charge of socio-environmental marketing campaigns is not dominated by the marketing mentality. This applies especially in the context of the public authorities concerned with environmental issues where there is a strong tendency towards the natural sciences. In our criteria, a socio-environmental marketing programme cannot be implemented if those involved in the process have not assimilated what is normally known as "marketing philosophy" beforehand. Nowadays, with technological progress, this is not limited to the consumer approach, it goes far beyond this. Those working in these organisations that have inspirational marketing principles need to become involved so that, before developing the programme, all the components of the organisation are in line with these principles and the socio-environmental marketing techniques and methodology to deal with the problems are shared.

4.1.3. Identifying the Perceived Barriers and Benefits for the Forms of Behaviour Analysed

At this stage, the research idea still exists but now there is clear tendency towards the forms of behaviour that will be analysed. The barriers can be internal, belong to the individual or be external (socials, structural, etc.). In this way, the aforementioned research techniques will be of great use to identify and understand them, especially the quantitative ones. The following points have to be defined:

- Benefits and barriers of the existing behaviour.
- Benefits and barriers of the competitive forms of behaviour.
- Possible strategies.

4.2. The Group in Which the Behaviour is to be Promoted

The knowledge of the target public is a key factor in the development of the programme as it enables the resources that are used for the campaign to be optimised. The socio-environmental strategies will be different depending on the identified groups.

4.2.1. Market Segmentation

One of the key aspects in designing socio-environmental marketing strategies is the need to establish different groups, which different strategies are aimed at, among the population of change (Montero Simó, 2003). In this way the groups that must be, homogeneous within but heterogeneous among themselves *are identified*. This is what we normally define as segmentation. The behaviour of individuals can vary significantly from one to another and that is why strategies must be designed to adapt to the characteristics of each one. In this way the investment made is optimized. It is then that the socio-demographic variables (gender, age, income, sector, etc.) and psychographic variables (life styles, personality, etc.) become interesting because they enable us to identify significant differences from the point of view of the variables analysed and define the profile of each of the groups.

4.2.2. Segment Analysis and Study

Having defined and quantified the segments, their characteristics then have to be analysed in order to establish the essential elements in the work to be carried out for each one of them. Our proposal for the socio-environmental marketing campaigns is based on the studies from Prochaska, Norcross and DiClemente (1994) that describe a six stage model that is needed to change behaviour, these are the following:

- Precontemplation: In this state people do not intend to change their behaviour and usually they deny that there is a problem. e.g.: people who don't believe setting non-selective traps is a significant problem for the lynx.
- Contemplation: Individuals who admit that there is a problem and start to seriously think about the solution. E.g. Individuals that think that setting non-selective traps does pose a threat for the species, although they still haven't decided if they are going to stop setting traps.
- Preparation: At this stage the majority of people get ready to do something about it in the next few months, and they are making the finishing touches before starting to change their behaviour. e.g.: those who have decided that they are not going to set any more non-selective traps.
- Action: This is when the majority of people modify their behaviour in public and in their groups. E.g. individuals comment on the decision they have taken in their group.
- Maintenance: Work is done to consolidate the benefits obtained during the action and an effort is made not to go back to behaving as before. E.g.: people that have not set non-selective traps for quite a time and are proud of it.
- Termination: This is the last objective; the problem does not arise again. e.g.: individuals never set non-selective traps again.

Essentially the stage at which the target public is has to be identified in order to be able to move on to the higher stages, although it is difficult to go from one stage to another one much higher up without going through the intermediate stages.

4.3. Conditions that Favour the Change

To be able to change behaviour, not only do we need to have extensive knowledge of this behaviour and the reasons why it has to be adopted, we must also find out about the internal and external factors that can contribute to changing the behaviour. These can either favour the benefits it has or eliminate the barriers that arise. The analysis of these factors is crucial in building the subsequent strategies.

To modify behaviour, we need to know what surrounding factors can have an influence. Among the external factors, it is worth mentioning: the external public and the cultural, technological, demographic, natural, economic and legal political forces. The internal factors refer to the aspects concerning the capacities of the organisation. This is all used to build a

SWOT analysis (strengths / weaknesses – opportunities / threats). This enables us to identify the organisation's competitive situation in the context in which it is expected to develop. Therefore the following points need to be defined:

- Elements that can positively or negatively influence the campaign.
- Strengths and weaknesses of the organization / the surrounding threats and opportunities.

4.4. Developing the Programme

4.4.1. Defining the Objectives

On completing the analysis stage, we will have quite a clear idea of the current situation, which means we are ready to establish what the objectives of the campaign should be. These refer to what the organisation wants the target public to do, namely, the behaviour it wants them to adopt. In this way it is a good idea to establish three types of objectives:

- Behaviour objectives: They refer to what the organisation wants the target public to do.
- Knowledge objectives: This involves establishing what the organisation wants the target public to know about the campaign.
- Belief objectives: They define what the organisation wants the target public to believe.

4.4.2. Developing Strategies: Product, Price (Adoption Costs), Place (Supplying the Product) and Promotion

Having specified the behaviour on which action must be taken, the groups in which the action will be the most effective, their situation with regards to the proposed behaviour and the surrounding elements that could influence them, we are then ready to define the strategies that must be implemented to increase the benefits of the proposed behaviour and reduce the barriers. This is carried out at the same time as the benefits of the competitive forms of behaviour are decreased and their barriers are increased. This means that a strategy based on the classic marketing elements, namely: product, price, place and promotion is proposed. The meaning of each of these variables in socio-environmental marketing is as follows:

The product can be tangible or intangible (the latter is more common in socio-environmental marketing) and it involves the desired behaviour and the benefits associated with this behaviour (Kotler, Roberto and Lee, 2002). Different levels can be established for each product so that the benefits that are obtained are identified with the core product and the proposed behaviour with the actual product.

- Core product: Benefit that generates the product
- Actual product: It is the behaviour that it promoted
- Augmented product: They are the additional services and the elements that can make the product in question tangible.

The price is how much it costs the target public to adopt the new behaviour; this can be shown in monetary units or, what is more common, in effort, time and other intangible elements. Although on many occasions this is difficult to gauge, it has to be clearly defined, especially in the comparative sphere of activity, which means a balance has to be obtained between the benefits derived from the new behaviour and the costs entailed.

Place in socio-environmental marketing involves the way of making the adoption of the behaviour available to the consumer. This plan is extremely important given that many forms of behaviours proposed by social campaigns that do not adopt a socio-environmental marketing point of view are not carried out simply because the target public does not know (because nobody has told them) what they have to do, where and when they can do so. On many occasions this then creates a slight despondency in certain people who are willing to change their behaviour, but don't actually know how to do it in practise. Among other points the following has to be defined:

- Where will the behaviour be carried out?
- When and how can the behaviour be carried out?
- With regard to promotion, just like in commercial marketing, it is used to support the aforementioned variables. In socio-environmental marketing, advertising is usually more personalised, it uses personal direct media instead of mass media communication. Incentives are established by means of the promotion sales, whereas behaviour and public relations are thought of more in the sense of changing the external barriers. In the majority of cases the sales force is represented by environmental volunteers concerned with the campaign. The elements needed for this strategy are shown in the following points:
- Advertising: To influence behaviour firstly the campaign needs to get the attention of the target public that it is aimed at; having done this, the elements of persuasion are put into practice. Usually, the material used in the environmental campaigns is discrete and boring; the information that is put forward should be lively, specific and more personalised because it makes it easier to remember. That is why information about the audience is needed, as is a credible communication source (the more credible it is the more impact it has). The message has to be focused properly too. In this way, the messages that emphasise what can be lost if nothing is done is usually more persuasive. It is also important that the audience clearly understands the seriousness of the situation. Finally, personal contact acquires more importance in the communication activities that are carried out from this point of view.
- With regard to the promotions, they involve incentives in exchange for behaviour and they must work in the short term. The individuals that decide to change their behaviour should be rewarded for their efforts so that they feel as if they have done something positive.
- Public relations is included in the context of creating a favourable atmosphere for the socio-environmental marketing campaign and must involve looking for favourable behaviour with distinguished, well known people in society; they help to attain the objectives related to influencing the opinion leaders.
- Finally, the sales force will be in charge of directly passing on the socio-environmental marketing programme to the target public. It should be made up of

members of staff from the organisation and environmental volunteers that join the project. It is crucial that they know about the values that must be transmitted, and also the behaviour or the forms of behaviour that are promoted by the campaign.

Lastly and in order to make the campaign sustainable, the specific strategies on aspects that positively affect the aforesaid sustainability need to be developed and these are:

- Obligations: Obliging people is an element that helps the sustainable development strategy. It should start off with a small obligation: people respond better when the way they are asked to behave does not involve a substantial change with regard to their normal behaviour. Subsequently and having fulfilled the obligation, it is much easier to fulfil new ones. If they are fulfilled in writing they are even more effective just like if they are made public or communicated to groups. The person must be actively involved and it should not cost more that the benefits derived from the proposed behaviour. In short, it consists in "helping people to see their pro-environmental side" without having to resort to coercion.
- Reminders: The fulfilled obligations fade away and disappear over time that is why; people have to be reminded all the time. To do these elements can be used to help the target public remember the obligation they fulfilled. These elements must be self-explanatory; pictures and text are used to recall whatever they have been conceived for. They must be located near places where the proposed forms of behaviour are displayed and they are more effective when they made to be positive rather than when they are coercive or negative.
- Social norms: Finally, a certain type of "peer pressure" is needed; namely, a significant number of the people involved must behave as proposed to set an example for everyone else. (We can't expect others to do what we don't ourselves).

4.4.3. Strategy Pre-Test
One way to prevent a campaign from failing is to carry out a test before it is actually launched. This enables us to readjust those elements that are not suitable and correct the deviations/differences produced in the definitive campaign.

4.4.4. Implementation
Once the strategies have been designed and tested, it is now the moment to put them into practice with the selected target public. The organisation's ability to implement these is crucial at this stage. The results of the training stage have to show that the organisation/institution is prepared to attain the objectives defined for the campaign. If the organisation has been capable of understanding the inspirational principles of socio-environmental marketing, the results observed will be enhanced. The voluntary environmental work must play a crucial role in this stage. The questions that have to be answered at this stage are:

- What are the stages of the campaign?
- What happens at each stage?

4.4.5. Evaluation and Control

Finally, having developed the programme, whilst it is in progress control measures need to be established so that we can see whether the proposed objectives have been completely or only partially obtained. That is why, these have to be defined so that they are easy to gauge and quantify, and two questions have to be asked beforehand, which are as follows:

- How are the objectives going to be analysed?
- What measuring tools are going to be used?

5. CONSERVATION OF THE IBERIAN LYNX: A CASE WHERE THE SOCIO-ENVIRONMENTAL MARKETING MODEL IS APPLIED

Hereinafter a socio-environmental marketing proposal is made to develop a conservation campaign to protect the Iberian Lynx. This is an endangered species whose populations have been seriously threatened in the last thirty years although breeding in captivity programmes have been successful in the last few years.

5.1. IDENTIFYING FORMS OF BEHAVIOUR

Review of the bibliography concerning the studied species, especially concerning the work that have already been carried out and how effective it has been.

Matter to be studied	The decrease in the number of Iberian Lynx.
Possible approaches	Reduce the speed limit on roads that are frequented by the species. Decrease the amount of deaths caused by snares, traps and shooting. Support the breeding in captivity programmes. Improvement in species – prey populations. Improvement in the connections between towns Improve the habitats.
Goal	Reduce the rate of mortality of the species due to unnatural causes in 2011 by 50% with regard to 2010.

In-depth interviews with people involved with the subject matter that is being studied (experts on the species, competent authorities, estate owners, hunters, people who set traps, etc.)

Observation: Travel around the areas where the Lynx live to get a better understanding of the situation and the problems

Behaviour that is promoted: Set selective traps to capture predators (basically foxes)

Competitive forms of behaviour: Set traps

5.2. Knowledge about the Organisation that must Implement the Programme

Give a course on the social marketing principles and techniques
Organise a hands-on workshop to put the social marketing strategies into practice

5.3. Identifying the Benefits and the Barriers of the Behaviour

Personal interviews/surveys are held to get in-depth information about the benefits/barriers produced by using selective traps, snares, gins, etc.

	Proposed Forms of Behaviour	Competitive Forms of Behaviour	
	Set selective traps and release any Lynx that gets trapped in one	Set selective traps that kill Lynx if they get stuck in one	Set snares, gins or other non-selective methods
Benefits	We save an endangered species from extinction. We help the future generations survive and live	Stop the Lynx eating rabbits from the estate	Control predators (foxes) to stop them killing rabbits and partridges. It's cheaper
Barriers	It's more expensive and less effective in terms of controlling predators. It takes longer to set the traps	Sanctions, fines Peer pressure	

5.4. Market Segmentation

Study the different segments of the population that set non-selective traps, gins, snares, etc. Define their profile based on psychographic and socio-demographic criteria so as to design a campaign that is adapted to each of the identified profiles.

5.5. Analysis and Study of the Segments

Target public state transition matrix

	Precontemplation	Contemplation	Preparation	Action	Maintenance
Decided not to set any more non-selective traps six months ago	No	No	No	No	Yes
Have taken some type of decision on the matter in the last six months	No	No	No	Yes	Yes
Intend to do something about it next month	No	No	Yes	Yes	Yes
Will try to do something about it in the next six months	No	Yes	Yes	Yes	Yes

5.6. Analysis of the Surroundings

SWOT Matrix

Strengths	Weaknesses
LIFE project resources. Environmental volunteers. Protection strategies for the Imperial Eagle, Black Vulture, Rabbit, Partridge. More knowledge about the species.	Inability to stop snares and gins being set. Limited action in terms of private estates. Lack of environmental officials.

Threats	Opportunities
Population decline. Remote villages. Habitat fragmentation. Only two feasible populations. Snares and gins. Roads without crossings. Sick rabbits. Conflicts between administrations. Liberal policies in some Autonomous Communities concerning the use of snares and traps.	Positive peer pressure. More interest in the species. Unique species in the world. Breeding in captivity. Cultural change in new generations. More commitment from the sectors involved.

5.7. Defining Objectives

Behaviour objectives	Only set selective traps and release any lynx that gets stuck in one.
Knowledge objectives	The lynx is a species in danger of extinction. The lynx can get stuck in non-selective traps. These traps can kill or lame the lynx.
Belief objectives	Setting selective traps can help to preserve the lynx populations. The lynx is not dangerous for the rabbit populations on the estates.

5.8. Developing Strategies

Given the elements studied, the established strategy would be based on creating a selective trap, designed by the organisation that promotes the behaviour. This would be supplied to the owners of the estates where the lynx live for a lower price than the normal selective trap.

Core product	Save the Iberian lynx from extinction.
Actual product	Set selective traps instead of snares or gins.
Augmented product	Selective trap designed by the organisation promoting the campaign.
Monetary value	Part of the price of the traps that must be paid by the user.
Non-monetary price	Effort made to set selective traps.
Place	The hunting reserves where the selective traps are going to be used are determined. Where and how the traps must be set is decided upon. The organisation or the companies helping with the programme, distribute the selective traps out to those who request them.
Promtion	Objectives: Publicise the fact that selective traps can be bought cheaper. Benefits that guarantee: We save a species that is about to become extinct. The future generations will know this species alive.
Promotion sales	Buying selective traps is encouraged by making them cheaper. Volunteers help to set the traps if need be. There is some type of incentive available for those who inform the organisation whenever a lynx gets stuck in one of the traps.
Public relations	A prestigious person in the environmental field supports the campaign's idea of setting the selective traps that are being promoted.
Sales force	A team of volunteers is formed to help disseminate the campaign.

5.9. The Programme's Sustainability

Obligations:	An agreement is made with one of the estate owners where the lynx lives so that they use the selective traps.
Reminders:	Signs are put on the selective traps reminding users how important it is to save the species.
Social norms:	Hunting associations encourage the use of selective traps.

5.10. Strategy Pre-Test

Launch the campaign in specific geographical areas and analyse the results.

5.11. Implementation

First stage:	Design and build a prototype of a selective trap. Prepare and write the informative material that will be supplied with the trap. Define the areas where it is going to be put into practice. Information campaign for the hunting sector.
Second stage:	Tailor made personalised explanation about the campaign's objectives. Show photographs of the selective traps.
Third stage:	Setting the traps. Looking after them.

CONCLUSION

To preserve species that live on our planet, a multidisciplinary attempt at tackling the problems is needed and it has to have a much broader perspective than the existing one. More knowledge about the interrelations between the environment and other sectors is also essential. The majority of organisations that operate in the environmental field focus the decisions they take in this way. The socio-environmental marketing discipline is used as a base to propose methods aimed at changing the behaviour of humans on their own free will, as they know how ineffective coercive methods are on many occasions. The proposed model broadens the existing perspective used to tackle environmental problems, whereby the voluntary change in behaviour is established as the backbone of the action strategies aimed at obtaining the proposed objectives. The work methodology is based on real experience in social marketing, a science that has developed methods that are suitable to promote the necessary changes in the target populations under study so that the socially unacceptable behaviour can be changed.

REFERENCES

Ajzen, I. y Fishbein, M. (1980), *Understanding attitude and predicting social behaviour.* Englewood Cliff, New Jersey: Prentice-Hall.

Andreasen, A. R. (1994), "Social marketing: its definition and domain", *Journal of Public Policy & Marketing*, vol. 13, n° 1, (spring) 108–115.

Castro, R. de (2001), "Naturaleza y funciones de las actitudes ambientales", *Estudios de Psicología,* vol. 22, n° 1, 11–22.

Europark-España (2002), *Anuario EUROPARK-España del estado de los espacios naturales protegidos*, Madrid: Fundación BBVA.

McKenzie-Mohr, D. y Smith, W. (1999), *fostering sustainable Behaviour*. Gabriola Island, British Columbia, Canada: New Society.

McKenzie-Mohr, D. (2000): "Promoting Sustainable Behaviour: An introduction to Community-Based Social Marketing", *Journal of Social Issues*, vol. 56, n°. 3, 543–554.

Mier-Terán, J. J. (2004): "Marketing Socioambiental: Un nuevo paso en el desarrollo del Marketing Social", *Revista Internacional de Marketing Público y No Lucrativo*, vol. 1, n° 1, 139–153.

Montero Simó, M.J. (2003), El Marketing en las ONGD. *La gestión del cambio social, Bilbao*: Desclée de Brouwer.

Philips, A. (2002*), Management Guidelines for IUCN Category V Protected Areas: Protected Landscapes/Seascapes*. Gland, Switzerland and Cambridge, UK: IUCN xv + 122 pp.

Prochaska, J. O. y Diclemente, C.C. (1983), "Stages and Processes of Self-Change of Smoking: Toward an Integrative Model of Change," *Journal of Consulting and Clinical Psychology*, vol. 51, n° 3, 390–395.

INDEX

#

20th century, 231, 368

A

abatement, 361, 364
ABM framework, ix, 1, 2, 37
access, 52, 59, 61, 105, 215, 218, 282, 296
accessibility, 203, 230
accommodation, xiii, 229
accountability, 90
accounting, xv, 53, 105, 239, 351, 357
accreditation, 166
acetaldehyde, 139, 140, 146
acid, 118
acidic, 354
acquisition of knowledge, 187
activated carbon, 138, 142, 146, 147, 148, 149
activation energy, 132
activation parameters, 148
active centers, 118
active transport, 305
adaptation, 39, 210, 332, 335, 340, 369
additives, 58, 65
adhesives, x, 117
adjustment, 124, 127, 128, 132, 138, 362
adsorption, x, 117, 122, 126, 127, 129, 131, 132, 133, 134, 138, 139, 141, 143, 144, 145, 146, 147, 148, 359, 360
adsorption isotherms, 138, 143
adults, 244
advancement, 324
adverse effects, 208, 218
aesthetic, 14, 16, 18, 20, 25, 241, 255
aesthetics, 17, 35
Africa, 243
age, xiii, 229, 242, 245, 270, 327, 375

agencies, ix, 1, 2, 3, 4, 8, 9, 20, 26, 27, 28, 29, 35, 37, 38, 284
aggregation, 359
agricultural sector, 105
agriculture, 94, 98, 104, 109, 111, 112, 114, 272, 274, 366
AIDS, 372
air emissions, 67
air pollutants, 70, 100
air quality, 310
air temperature, 262
aldehydes, 119, 135
algae, 245, 352, 358
ammonium, 121, 358
amphibia, 280
anaerobic digestion, 357, 358, 364
annual rate, 29
ANOVA, 136
APA, 251
apex, 323
application techniques, 100
aptitude, 253
aquaculture, 121
aqueous matrix, x, 117
aqueous solutions, 124, 148, 149
aquifers, 105
architects, 191
Argentina, viii, xiii, 229, 231, 242, 243, 253, 257, 261, 271, 273, 275
aromatic rings, 118, 125
articulation, 68
asbestos, 77
Asian countries, 225
assessment, ix, 40, 46, 49, 62, 65, 66, 69, 73, 74, 78, 81, 82, 89, 94, 109, 164, 165, 172, 199, 200, 226, 234, 243, 253, 255, 272, 274, 292, 318, 319, 320, 324, 326, 327, 328
assessment models, 318
assessment tools, 94

assets, 194, 292
asymmetric information, 161
asymmetry, 166
atmosphere, 99, 100, 110, 185, 189, 241, 258, 259, 260, 270, 345, 347, 378
audit, xiii, 54, 57, 58, 59, 62, 75, 81, 82, 83, 84, 89, 92, 164, 168, 172, 175, 295, 296, 297, 300, 301, 305, 307, 308, 333
Austria, 48, 53
authority, 71, 77, 155, 157, 168, 170, 174, 175, 335
autonomy, 70
average revenue, 340
avoidance, 155, 211
awareness, xi, xii, xiii, 54, 70, 71, 72, 73, 83, 86, 89, 155, 168, 170, 175, 183, 187, 188, 189, 190, 192, 198, 199, 200, 201, 202, 207, 214, 229, 234, 296, 372

biotic, 361
birds, 280, 281, 286
blame, 366
boilers, 312
bonds, 125
bone, 268
bonuses, 290
brainstorming, 341
Brazil, viii, xiv, 122, 331, 332, 336, 337, 342, 343, 348
breathing, 103
breeding, 280, 371, 380
building blocks, 143
burn, 2, 12, 13, 21, 22, 25, 26, 27, 35
business function, 70
business management, 90, 159
businesses, 48, 58, 306

B

bacteria, 121, 357
ban, 234
banking, 290
Barents Sea, 328
barriers, 49, 52, 55, 56, 57, 91, 200, 204, 213, 296, 370, 371, 372, 375, 376, 377, 378, 381
base, 4, 84, 104, 163, 178, 239, 288, 363, 384
behavioral change, xi, 183
behaviors, x, 2, 6, 9
Beijing, 40
benchmarking, 155, 158, 348
benchmarks, 318
beneficiaries, 61
benefits, xii, xiv, xv, 5, 6, 7, 8, 9, 24, 35, 38, 50, 52, 54, 55, 57, 60, 61, 72, 97, 156, 157, 179, 208, 210, 212, 213, 214, 215, 216, 217, 218, 221, 222, 223, 224, 225, 228, 245, 265, 274, 283, 309, 315, 317, 319, 332, 335, 337, 339, 340, 341, 343, 344, 345, 347, 348, 360, 361, 364, 369, 370, 371, 372, 376, 377, 379, 381
bioaccumulation, 121
biodegradability, 264
biodegradable materials, 357
biodegradables, 307
biodiversity, 41, 65, 87, 98, 316, 323, 324, 369, 370
biogas, 357, 358, 364, 365
biological activity, 262
biological processes, 355
biomass, x, xiii, 117, 254, 257, 258, 259, 265, 266, 267, 268, 269, 270, 272, 273, 318, 319, 324, 341, 346, 347, 358
biomonitoring, 111
biosphere, 259, 329

C

cadmium, 147
calibration, 21, 77, 342
campaigns, 233, 368, 369, 372, 374, 375, 377, 378
CAP, 98
capillary, 113
capitalism, 311
carbon, xiii, 86, 95, 97, 114, 138, 148, 164, 257, 258, 259, 260, 261, 262, 263, 264, 265, 266, 268, 270, 271, 272, 273, 274, 275, 357, 358
carcinogen, 297
case studies, 6, 44, 212, 335
case study, xiv, 146, 214, 218, 227, 255, 295, 296, 345
cash crops, 105
catalyst, 291
categorization, 282, 292
cation, 258
cationic surfactants, 121, 122, 145
C-C, 125, 311
cell division, 106
cellulose, xiv, 331, 341, 346, 347
certificate, 212, 218, 221, 223, 224, 340
certification, x, xii, 47, 52, 58, 61, 80, 152, 163, 166, 167, 172, 173, 178, 179, 180, 207, 211, 212, 214, 215, 216, 218, 221, 223, 224, 225, 226, 228, 333, 336, 337, 338, 339, 340, 342, 343, 344, 345, 346, 347
challenges, 94, 284, 318, 366
changing environment, 83
chemical, x, xiv, 85, 97, 100, 107, 109, 117, 118, 119, 120, 121, 122, 127, 146, 148, 149, 212, 258, 261, 267, 275, 281, 296, 297, 307, 308, 331, 332, 335, 337, 338, 355, 359, 360

Index

chemical properties, 100
chemical structures, 359
chemicals, 212, 340, 341, 343
chemisorption, 134
Chicago, 148, 180
chicken, 357
children, 238, 242
Chile, 328
chimneys, 346
China, viii, xv, 152, 203, 208, 214, 215, 217, 226, 228, 311, 328, 351, 352, 354, 355, 357, 363, 364, 365, 366
Chinese government, 352
chromatography, 112, 113, 115
chromium, 146
circulation, 35, 57, 234
citizens, 12
City, 40, 260, 261, 279, 281, 284, 352, 362
clarity, 286
classes, 12, 13, 21, 118, 273
classification, 13, 161, 253
Cleaner Production, xiv, 91, 92, 178, 180, 203, 204, 228, 311, 331, 332, 334, 336
cleaning, 245, 309, 343
clients, 191, 211, 217, 221, 223, 224, 226
climate, ix, xiii, 1, 2, 3, 5, 8, 21, 26, 35, 36, 37, 38, 45, 104, 108, 231, 257, 258, 259, 260, 265, 270, 272, 273, 274, 305, 360, 361
cluster analysis, xiii, 229, 235, 238
clustering, 290
CO2, xiii, 88, 94, 95, 96, 99, 109, 257, 259, 266, 270
coastal management, 243, 255
coercion, 379
coffee, 232
cognition, 188
collaboration, 16, 20, 27, 164, 285
combined effect, 189
combustion, 86
commerce, 42, 227, 230
commercial, 2, 10, 22, 35, 37, 86, 123, 200, 225, 230, 268, 282, 308, 319, 320, 358, 368, 378
communication, 58, 59, 68, 70, 72, 75, 78, 79, 85, 90, 91, 155, 157, 158, 159, 164, 170, 172, 174, 176, 177, 221, 223, 334, 338, 378
community, ix, 1, 3, 8, 14, 16, 17, 22, 24, 26, 37, 38, 41, 44, 46, 48, 49, 118, 172, 202, 279, 289, 290, 311, 324, 328, 368, 369
compaction, 94, 96, 98, 111, 260
compatibility, 97, 163
competition, 152, 323, 324, 325
competitive advantage, 50, 55, 178, 215, 218, 296, 349
competitive behaviour, 371

competitive demand, 218
competitiveness, 49, 50, 312, 358
competitors, 51, 52, 162, 215, 343
complement, 358, 372
complex interactions, 2, 316
complex organizations, 68, 82
complexity, 6, 15, 39, 54, 76, 80, 277, 312, 318, 322, 327
compliance, ix, xi, 46, 51, 52, 58, 59, 62, 63, 64, 66, 72, 73, 75, 77, 81, 83, 152, 156, 157, 158, 163, 164, 168, 172, 173, 174, 176, 183, 199, 217, 225, 284, 296, 318, 337
composition, 2, 4, 65, 112, 120, 147, 254, 261, 353
compost, 272, 307
composting, 271, 273, 274, 300, 301, 307, 308, 310
compounds, 100, 118, 121, 122, 141, 143
computer, 3, 4, 6, 38, 103, 105, 327
computing, 324
conception, 230, 312
conceptual model, 5, 285
condensation, 119
conductance, 114
conductivity, 360
conference, 180, 204, 255, 327
configuration, 98
conflict, 108, 157, 224, 285
conformity, 61, 76, 77, 172, 174
Congo, 272
congress, 293, 366
consciousness, 184, 189, 210
consensus, xi, 98, 151, 154, 284, 318
consent, 339
conservation, ix, 17, 115, 231, 244, 260, 275, 278, 283, 284, 287, 291, 292, 293, 317, 318, 319, 321, 322, 323, 324, 325, 327, 329, 359, 380
constant rate, 264
constituents, 100, 360
constructed wetlands, 361, 362
construction, xi, xii, 13, 19, 21, 22, 28, 29, 31, 37, 183, 185, 186, 187, 188, 189, 190, 191, 192, 193, 194, 195, 196, 198, 199, 200, 201, 202, 203, 204, 205, 207, 208, 209, 210, 211, 212, 213, 214, 215, 216, 217, 218, 219, 220, 224, 225, 226, 227, 228, 337, 354, 355, 356, 360
consulting, 12, 53, 74, 192
consumers, xiv, 87, 161, 296, 331, 333
consumption, 2, 50, 87, 94, 95, 107, 108, 109, 113, 114, 162, 186, 187, 195, 304, 305, 306, 308, 341, 343, 344, 345, 346, 357
consumption rates, 2
contact time, 126
contaminant, 140
contaminated soil, 273

contamination, 65, 108, 173, 230, 267, 305, 352
control measures, 380
convergence, 92
conversion rate, 357
cooling, 304, 305, 346
cooperation, 61, 78, 89, 214, 217, 218
coordination, 71, 154, 159, 161, 167, 285, 286
coral reefs, 231, 255
correlation, 86, 87, 127, 132, 133, 137, 176, 275, 321, 348
correlation coefficient, 133, 137
correlations, 130
cost, xii, xv, 9, 14, 15, 18, 19, 20, 24, 25, 35, 42, 50, 51, 52, 53, 54, 57, 96, 98, 105, 109, 110, 118, 134, 146, 152, 156, 161, 163, 191, 200, 208, 214, 215, 217, 218, 223, 224, 225, 268, 305, 306, 309, 310, 332, 354, 355, 356, 358, 365, 379
cost saving, 57, 214, 217, 268
cotton, 274
counterbalance, 59
covering, 200, 337
CP, xiv, 331, 332, 335, 336, 338, 339, 340, 341, 342, 343, 344, 345, 346, 347, 348
CPC, 248
criticism, 167
crop, x, xiii, 93, 94, 96, 97, 99, 105, 106, 107, 108, 111, 113, 257, 258, 259, 260, 265, 267, 269, 271, 272, 274, 281, 357, 358, 364
crop production, 99, 106, 108, 260, 271
crop residue, xiii, 257, 258, 259, 364
crops, x, xiii, 93, 96, 100, 101, 105, 106, 107, 108, 111, 113, 257, 260, 268, 270, 274, 357, 365
cultivation, 96, 107, 113
cultural values, 195
culture, xi, 55, 85, 183, 184, 185, 186, 187, 188, 189, 190, 195, 196, 199, 200, 201, 202, 297, 309, 340, 343, 346
curriculum, 74
customers, 51, 88, 159, 162, 163, 167, 172, 337, 340
cyanosis, 121
cycles, 364
cycling, 260

D

damages, xi, 16, 22, 183
danger, 318, 382
data set, 12, 235
deaths, 380
decentralization, 365
decision makers, 3, 7
decision-making process, 11, 49, 65, 68, 71, 318
decomposition, 241, 259, 263, 264, 266, 359

deficiencies, 64
deficit, 106, 107
deforestation, 42, 307
deformation, 125
degradation, 97, 121, 148, 254, 258, 262, 266, 282, 306, 354, 362
degradation mechanism, 262
degradation process, 258
Delta, 317, 352
demographic characteristics, 4
Denmark, 53, 93, 113
Department of Agriculture, 40, 43, 45, 46, 93, 293
Department of the Interior, 292
deposition, 100, 344, 359, 362
deposits, 111
depth, 49, 61, 64, 67, 163, 177, 193, 261, 352, 373, 380, 381
desorption, 127, 132
detection, 101, 113, 188
detention, 360, 361, 362
developing countries, xiii, xiv, 104, 208, 210, 316, 331
deviation, 220, 221
diffusion, 56, 134, 178
digestion, 357, 365
direct foreign investment, 166
direct observation, 74
directives, 64
directors, 155
disaster, 40
discharges, 87
disclosure, 56, 336
discrimination, 161
discs, 98
diseases, 100
dispersion, 100, 101, 102, 109, 162
disposition, 242, 268
dissatisfaction, 212, 240
distilled water, 123
distribution, 60, 73, 101, 105, 243, 245, 258, 283, 322
diversification, 111
diversity, x, 2, 3, 6, 26, 39, 161, 245, 279, 286, 290, 321, 361
dogs, 234, 239, 240, 241, 250, 251
DOI, 292
dosage, 139
draft, 48
drainage, 105, 108, 245, 282, 352, 359, 364
draught, 96
drinking water, 346, 352
drought, 46
dry matter, 265, 266

drying, 106, 107, 112, 114
dumping, 118, 121, 210, 258
dyes, xi, 118, 121, 143, 144, 145, 148

E

earnings, 105
earthquakes, 86
earthworms, 365
Easter, 239, 240
EBFM, xiv, 315, 316, 317, 318, 319, 325, 326
ecological systems, ix, 1, 3, 278
ecology, 42, 227, 254, 326
economic activity, 153, 154, 209
economic development, xv, 104, 152, 284, 351, 352
economic efficiency, 153
economic growth, x, 2, 3, 5, 21, 26, 28, 29, 32, 36, 37, 44, 185
economic growth rate, 44
economic incentives, 55, 322, 372
economic performance, 333
economic values, 5, 339
economics, 40, 42, 94, 110, 113, 152, 304, 329
economies of scale, 162
ecosystem, ix, xiii, xiv, 4, 8, 40, 42, 44, 106, 277, 278, 280, 284, 285, 286, 287, 288, 290, 291, 292, 315, 316, 317, 318, 319, 320, 321, 322, 323, 324, 325, 326, 327, 328, 329, 369
Ecosystem-Based Fisheries Management, xiv, 315, 316
education, 74, 187, 203, 230, 232, 238, 290, 307, 309, 348, 372
educational opportunities, 281
educational research, 187
effluent, 339, 341, 359
effluents, 340
egg, 210
Egypt, 104, 115
elaboration, 328
electrical conductivity, 106
electricity, 86, 88, 304
electrophoresis, 113
elementary school, 238
elongation, 125
e-mail, 295
EMAS, vii, ix, x, 47, 48, 49, 50, 51, 52, 53, 54, 55, 56, 57, 58, 59, 60, 61, 62, 68, 70, 71, 72, 73, 74, 75, 76, 77, 78, 79, 80, 81, 82, 85, 90, 91, 92, 153, 163, 164, 179, 211, 312
emergency, 7, 16, 65, 66, 72, 73, 74, 75, 76, 84, 86, 156, 158, 174, 177, 338
emergency management, 73, 76, 86
emergency preparedness, 73, 338

emergency response, 7
emission, 48, 86, 95, 101, 108, 109, 113, 166, 185, 259, 311, 340, 343, 344
employees, 53, 59, 67, 68, 69, 70, 71, 72, 74, 75, 77, 78, 80, 85, 88, 155, 156, 157, 172, 174, 211, 238, 297, 298, 301, 304, 305, 306, 307, 308, 309, 310, 336, 338, 339, 340, 342, 345
employment, xiv, 18, 185, 209, 331
employment opportunities, 209
encouragement, 61
endangered species, xvi, 279, 281, 283, 284, 291, 293, 325, 368, 369, 380, 381
energy, 50, 63, 65, 87, 88, 94, 95, 96, 113, 132, 134, 185, 187, 200, 203, 204, 209, 259, 297, 301, 303, 304, 305, 306, 307, 308, 320, 327, 343, 345, 346, 357, 358, 364, 369
energy consumption, 88, 301, 305, 307, 343
energy efficiency, 87, 203, 204
energy input, 95, 96
enforcement, 5, 23, 214, 285, 317
engineering, 64, 180, 219, 226
England, 178, 254
enlargement, 57, 163
environmental aspects, xiii, 48, 55, 58, 60, 62, 63, 64, 65, 66, 67, 68, 69, 70, 73, 74, 76, 77, 80, 81, 83, 85, 86, 87, 89, 90, 91, 155, 167, 173, 184, 186, 295, 296, 298, 340
environmental audit, 81, 85, 152, 334
environmental awareness, 72, 210, 213, 214, 220, 222, 225, 231, 232, 338
environmental change, 39, 43
environmental conditions, 63, 100, 290, 362
environmental effects, 66, 153, 154, 158, 174, 325
environmental factors, 66, 156
environmental impact, xiii, 48, 59, 62, 63, 64, 66, 68, 72, 73, 74, 75, 82, 84, 86, 89, 93, 94, 97, 99, 109, 147, 158, 163, 166, 177, 200, 208, 215, 220, 222, 292, 296, 297, 304, 306, 307, 308, 309, 332, 337, 339, 345
environmental issues, xiv, xv, 62, 63, 67, 70, 71, 72, 78, 86, 87, 152, 157, 158, 164, 172, 174, 208, 217, 226, 244, 245, 295, 296, 304, 309, 333, 337, 340, 367, 369, 374
environmental management, ix, xi, xii, xiii, xiv, xv, 48, 52, 58, 59, 60, 61, 66, 67, 68, 69, 70, 71, 72, 73, 74, 76, 78, 81, 84, 89, 90, 91, 92, 151, 152, 153, 156, 165, 167, 173, 174, 176, 177, 178, 179, 180, 181, 207, 208, 209, 210, 211, 214, 217, 218, 220, 222, 224, 225, 226, 227, 245, 295, 296, 313, 331, 332, 333, 335, 348, 367
environmental policy, x, xiii, 47, 48, 62, 67, 68, 69, 70, 72, 73, 74, 75, 89, 91, 92, 154, 155, 156, 157,

159, 173, 174, 175, 177, 295, 296, 333, 334, 337, 349
environmental protection, x, xiv, 67, 81, 93, 200, 212, 214, 217, 225, 331, 338
Environmental Protection Agency, 292
environmental quality, 235, 245, 253
environmental regulations, 172, 211
environmental resources, 291
environmental sustainability, 211
environmental variables, 318, 319
EPA, 353, 359, 360, 361, 362, 365, 366
equilibrium, 121, 124, 126, 127, 129, 131, 133, 134, 142, 145, 147, 262
equipment, 72, 75, 77, 96, 97, 155, 186, 210, 212, 302, 304
equity, 191
erosion, 94, 230, 243, 244, 258, 260, 281
EU, 48, 49, 54, 55, 59, 64, 91, 98, 99, 110, 162
eucalyptus, xiii, 257, 268, 269, 270
Europe, 48, 95, 104, 152, 163, 166, 178, 231, 274
European Commission, x, 47, 49, 53, 59, 60, 61, 90, 91, 92, 99
European Community, 91
European Eco Management and Audit Scheme, ix
European Parliament, 56, 154
European Union, 91, 153, 209, 210, 211
Europeanisation, 92
evacuation, 7, 8, 14, 17, 38, 43, 44, 45
evaporation, 105, 123, 316
evapotranspiration, 106, 107
everyday life, 86
evidence, xiv, 44, 51, 53, 54, 60, 81, 87, 174, 311, 315, 321
evolution, 46, 49, 63, 67, 86, 186, 203
exclusion, 2, 40, 75
execution, 196, 199, 202
experimental condition, 118
experimental design, 136, 266
expertise, 54, 82, 354
exploitation, 255, 320, 323
exposure, 3, 9, 99, 100, 101, 103, 121
external environment, 77, 107
externalities, 160, 161, 162
extinction, 283, 325, 381, 382
extraction, 120, 148, 185, 341
extracts, x, 114, 117, 119, 120, 122, 124, 139, 142, 148

F

fabrication, 149
factor analysis, xii, 207, 221, 224
families, 231, 238, 240, 242, 305

family characteristics, 242
farmers, 106, 114, 357, 364
farmland, 364
fauna, 230, 232, 243, 244, 251
Federal Register, 10, 46
feedstock, 121
feelings, 230
fertility, 260, 270, 271, 274
fertilization, 365
fertilizers, 109, 307, 357, 358, 363
fiber, 212, 273, 342
field crops, 106
field trials, 97
filtration, 122, 359
financial, x, xv, 47, 51, 52, 53, 60, 68, 69, 71, 153, 174, 195, 199, 200, 202, 214, 298, 312, 332, 333, 339, 344, 345, 348
financial institutions, 60
financial planning, 72
financial resources, 52, 53, 69, 339
financial support, 195, 202, 348
fire resistant materials, 16
fire suppression, ix, 1, 2, 11, 38, 46
fires, 7, 8, 15, 42, 43, 280, 369
fish, 67, 121, 244, 245, 280, 281, 317, 321, 328
Fish and Wildlife Service, 278, 279, 283, 285, 287, 288, 289, 293
fisheries, xiv, 315, 316, 317, 319, 320, 322, 324, 325, 326, 327, 328, 330
fishing, 316, 317, 319, 320, 321, 322, 323, 324, 325, 326, 329
fitness, 305, 333
flammability, 11, 13, 118
flexibility, 4, 80, 89, 119, 179, 284, 296
flocculation, 122
flooding, 280, 281, 305
floods, 86
flora, 244, 251, 282
fluctuations, 114
foams, 123, 142, 146, 147
focus groups, 16, 17, 18, 28, 297, 308
food, 104, 115, 122, 327, 365, 366
force, 17, 57, 66, 96, 165, 185, 210, 378, 383
forecasting, 324
foreign companies, 210
foreign direct investment, 210
foreign exchange, 105
foreign firms, 218
foreign investment, 210
forest fire, 43, 369, 372
forest management, ix, 1, 2, 5, 7, 29, 35, 36, 37, 38, 40, 340, 345

formaldehyde, 118, 119, 120, 123, 124, 125, 135, 136, 137, 139, 140, 146
formation, 119, 273
formula, 122
foundations, xv, 164, 367
France, vii, 47, 92, 110, 113, 115
free will, 369, 384
freedom, 22
freshwater, 282, 329, 352
fruits, 105
FTIR, 120, 123, 124, 125
fuel consumption, 94, 95, 99, 341
fuel loads, 21, 27
fuel management, 42
full capacity, 48
funding, 26, 52, 55, 61, 307, 310
funds, 26, 312
fungi, 121
fuzzy sets, 19

G

garbage, 234, 239, 250
gasification, 142
GDP, 332, 352
gel, 123, 124, 134, 143, 144, 146, 147
gelation, xi, 117, 118, 119, 120, 121, 123, 124, 136, 138, 139, 140, 143, 146
geography, 92
geology, 63
Germany, 48, 54, 110, 111, 113, 162, 315
global warming, 86, 258, 344
globalization, 227
GNP, 26, 28, 30, 32, 33
goods and services, 65, 74, 161, 162
governance, 68
government policy, 201
governments, 22, 153, 164, 284
grades, 324
graph, 98
grass, xiii, 99, 257, 280
grasses, 292
gravity, 86, 362
grazing, 280, 282
Great Britain, 110
Greece, 93, 114
green buildings, 199
greenhouse, xi, 88, 94, 100, 101, 102, 103, 107, 108, 109, 111, 112, 113, 114, 115, 164, 166, 183, 203, 258, 262, 265, 266, 307, 308, 344
greenhouse gases, 88, 164, 258, 344
Gross Domestic Product, xv, 332, 351, 352
gross national product, 209

groundwater, 104, 108, 121, 267, 279, 281, 364
grouping, xiii, 201, 229
growth, xiv, 3, 4, 5, 8, 13, 18, 28, 29, 32, 36, 37, 40, 41, 90, 108, 111, 114, 166, 184, 187, 209, 265, 266, 268, 270, 271, 331, 333, 357, 358
guidance, 60, 98, 110, 165, 186, 287, 291, 292, 348
guidelines, 48, 49, 57, 59, 80, 155, 162, 165, 217
guiding principles, xv, 199, 279, 286, 351, 353

H

habitat, ix, 7, 278, 279, 280, 281, 283, 284, 286, 288, 290, 291, 292, 293, 329, 361
harmful effects, 103, 121
harmony, 56
harvesting, 37, 97, 98, 111, 320
hazardous substances, 75
hazardous waste, 65, 88
hazards, ix, 1, 7, 43, 118
headache, 103
health, xiii, 8, 103, 109, 191, 270, 277, 297, 298, 304, 305, 306, 309, 310, 338, 343, 369
health ratings, 297
health risks, 270
heart rate, 121
heavy metals, xi, 118, 121, 122, 145, 148, 274, 309
height, 234, 362
heterogeneity, 5, 42, 49, 85, 107
highlands, 271
history, xi, 7, 85, 151, 152, 254, 279, 292, 293, 312, 329, 344
homeowners, ix, 1, 3, 7, 8, 11, 12, 13, 14, 16, 17, 19, 23, 37, 38, 39
homes, 2, 7, 12, 16, 22, 345
homogeneity, 50, 161, 167
Hong Kong, vii, 180, 183, 185, 186, 191, 192, 193, 194, 198, 199, 201, 202, 203, 204, 205, 209, 213, 214, 217, 227
hotels, 230
hourly wage, 25
housing, 10, 23, 204
human, ix, 1, 2, 3, 5, 7, 8, 9, 37, 38, 39, 46, 55, 56, 72, 98, 100, 101, 109, 111, 121, 153, 157, 174, 184, 185, 190, 200, 208, 228, 230, 254, 278, 288, 316, 333, 360, 369, 371
human actions, 2, 3, 7, 8
human behavior, x, 2, 38
human capital, 55
human development, 7, 289
human dimensions, 3
human exposure, ix, 1, 101
human health, 100, 109
human nature, 157

human resources, 56, 72, 153
humidity, 100
Hungary, 54
hunting, 281, 305, 307, 383
husbandry, 259
hybrid, 271
hydrogen, 125
hydrolysis, 354
hypothesis, 127, 132, 138, 139, 143, 144

I

Iberian Lynx, xvi, 368, 370, 380
ideal, xii, 18, 19, 36, 208, 230, 242, 362, 371
identification, 64, 65, 66, 77, 80, 140, 156, 158, 160, 175, 176, 291, 336, 344
identity, 67, 242, 244
ignitability, 40
image, 50, 51, 98, 214, 217, 220, 222, 333, 337, 340, 343, 346
imagery, 106, 113
imbalances, 108, 258
immigration, 231
immobilization, 140
impact assessment, 165
improvements, 71, 77, 157, 158, 177, 201, 296, 333, 338, 343, 344, 346
incidence, 332, 333
income, 14, 98, 345, 375
independence, 18, 82
independent variable, 136
indirect effect, 167, 318
individuals, 4, 17, 43, 195, 368, 369, 370, 371, 375, 376, 378
induction, 155
industrial revolution, 178, 311
industrial sectors, 86, 208, 211
industrialized countries, 258
industry, xi, xii, xiv, 48, 79, 81, 88, 97, 114, 121, 155, 162, 183, 185, 186, 188, 189, 193, 199, 200, 201, 202, 204, 205, 207, 208, 213, 215, 217, 227, 228, 254, 307, 331, 333, 345, 346
inflation, 35
information technology, 94
infrastructure, xiii, 63, 208, 229, 230, 231, 234, 235, 239, 240, 241, 242, 244, 245, 248
initiation, 66, 199
institutions, 13, 48, 55, 56, 79, 310, 368, 372
integration, xiv, 39, 59, 60, 90, 154, 156, 159, 164, 279, 297, 315, 326
integrity, xiii, 185, 277, 278, 290, 297
interaction effect, 136
interest groups, 85, 177

interface, 10, 40, 45, 46, 161
internal change, 83
internal environment, 70
international standards, 90
interrelations, 384
intervention, 69
invertebrates, 286
investment, xv, 53, 161, 200, 209, 210, 345, 351, 375
investors, 194, 296
ions, 147, 148
Iran, 272
irrigation, 105, 106, 107, 108, 109, 112, 113, 114, 260, 316
ISC, 129
Islam, 147
islands, 254
ISO 14001, vii, ix, xi, xii, xiv, 49, 51, 58, 59, 70, 71, 72, 75, 76, 77, 78, 79, 80, 81, 82, 91, 92, 151, 152, 153, 162, 163, 164, 165, 166, 167, 168, 169, 170, 171, 172, 173, 174, 175, 176, 177, 178, 179, 180, 207, 211, 212, 213, 214, 215, 216, 217, 218, 219, 220, 221, 222, 223, 224, 225, 226, 227, 228, 310, 312, 331, 332, 333, 334, 336, 337, 338, 339, 340, 342, 343, 344, 345, 346, 347, 349
isomers, 100
isotherms, 138, 147, 149
Israel, 104, 114
issues, ix, xv, 16, 40, 45, 48, 52, 61, 65, 67, 69, 70, 71, 73, 87, 156, 158, 186, 189, 200, 235, 244, 279, 284, 288, 290, 304, 309, 310, 333, 351, 365, 369
Italy, 47, 53, 61, 112, 242, 254

J

Japan, 152, 166, 272
jaundice, 121
justification, 51, 290

K

KBr, 123
kill, 381, 382
kinetic model, 126, 263
kinetic parameters, 264
kinetics, 124, 129, 142, 147, 148, 149, 263, 274

L

labeling, 164, 165
lack of control, 372
lakes, 121, 359

landfills, 259, 268, 307
landings, 329
landscape, 3, 18, 42, 46, 63, 98, 230, 235, 243, 253, 278, 279, 291, 355, 359, 362, 365
language barrier, 81
Latin America, 253
Law of the Sea Convention, 316
laws, xiv, 58, 62, 63, 64, 67, 86, 152, 210, 211, 296, 311, 331, 372
laws and regulations, 62, 63, 296
leaching, 267, 271, 274
lead, 8, 17, 70, 74, 98, 123, 136, 140, 148, 156, 162, 211, 265, 282, 296, 325, 361, 371
leadership, 155, 156, 157, 168, 175, 177
leaks, 308
learning, 4, 73
legal protection, 284
legislation, 58, 61, 64, 66, 100, 101, 152, 155, 157, 158, 159, 163, 177, 201
legislative authority, 87
leisure, 230
LIFE, 382
life cycle, 60, 91, 158, 164, 187, 200, 272
light, 12, 18, 28, 33, 62, 300, 301, 305
liquid chromatography, 113
liquids, 148, 347
livestock, 281, 282, 357, 365
loans, 65
local authorities, 232
local community, 51, 89, 289
local government, 22, 60
localization, 289, 291
logging, 26
logistics, 110, 338, 343
Luo, viii, 351, 364, 365, 366
Luxemburg, 110, 113

M

machinery, 77, 94, 95, 96, 97, 99, 109, 110, 212, 307
magnesium, 267
magnitude, 66, 132, 134, 186, 268, 270, 345
majority, 3, 8, 16, 57, 94, 193, 241, 304, 321, 354, 357, 368, 369, 370, 372, 374, 376, 378, 384
Malaysia, 180
mammal, 281, 317
mammals, 327
man, 230
manpower, 342
manufactured goods, 65, 211
manufacturing, 28, 29, 37, 47, 60, 68, 83, 86, 179, 180, 306, 335, 338, 343
manure, 258, 259, 265, 272, 357, 358, 364, 365

mapping, 74
marine fish, 329
marital status, xiii, 229
marketing, x, xv, 47, 49, 79, 214, 216, 217, 218, 312, 367, 368, 369, 371, 372, 373, 374, 375, 377, 378, 379, 380, 381, 384
marketing strategy, 368, 372
mass, 88, 143, 160, 163, 261, 270, 321, 378
mass media, 378
materials, x, 2, 11, 12, 13, 15, 19, 20, 21, 22, 23, 24, 26, 63, 68, 88, 117, 120, 134, 138, 142, 146, 185, 186, 187, 191, 194, 204, 209, 211, 212, 259, 260, 263, 270, 293, 297, 306, 307, 343, 354
matrix, x, 117, 118, 235, 381
matter, xv, 56, 138, 142, 152, 186, 264, 265, 266, 341, 343, 345, 357, 367, 372, 373, 374, 380, 381
measurement, xii, 76, 77, 87, 113, 155, 158, 168, 171, 172, 174, 175, 184, 186
media, 69, 378
median, 177
Mediterranean, 43, 104, 105, 106, 107, 108, 109, 110, 255, 271, 273
Mediterranean countries, 104, 105, 108
membership, 61
membranes, 148
memorizing, 81
memory, 4, 103
mercury, 147, 305
messages, 79, 378
metal ion, 147
metal ions, 147
metals, 148, 360
methodology, 76, 77, 84, 135, 168, 172, 187, 189, 226, 336, 338, 341, 346, 348, 374, 384
methylation, 120
methylene blue, 124, 148, 149
Mexico, viii, xiv, 7, 229, 253, 271, 273, 315, 316, 321, 326, 327, 329, 330
Miami, 146
microbial communities, 261
microorganisms, 264, 362
microscopy, 123
migration, 280, 319, 362
mineralization, 263, 264, 266, 268, 271, 272, 274, 358
mission, 99, 110, 164, 199, 204
Missouri, 1, 44
mixing, 123
modelling, 40, 41, 44, 101, 113
models, ix, xiv, 1, 4, 5, 6, 7, 8, 18, 35, 37, 38, 39, 41, 42, 43, 44, 45, 97, 106, 127, 129, 131, 152, 263, 308, 315, 316, 317, 318, 319, 320, 321, 322, 323, 324, 325, 326, 327, 328, 370

modifications, 24, 53, 67, 97
moisture, 107, 112, 264
molecular weight, 144
molecules, 119, 125, 131, 273
mollusks, 322
Montana, ix, 1, 3, 8, 13, 26, 35, 37, 43, 44
morale, xiii, 295
Morocco, 104
mortality, 318, 321, 325, 380
motif, 98
motivation, xi, 51, 94, 153, 157, 177, 183, 187, 188, 189, 190, 192, 195, 199, 202, 214, 239, 240
multidimensional, 244
multinational companies, 85
multiple factors, 362
municipal solid waste, 272
music, 239, 240, 241, 250

operations, 26, 66, 67, 68, 71, 72, 76, 80, 94, 95, 96, 97, 98, 99, 100, 110, 154, 158, 166, 173, 174, 176, 179, 186, 217, 353, 363
opportunities, xiv, 14, 48, 49, 78, 83, 90, 91, 152, 153, 173, 175, 214, 279, 295, 296, 338, 341, 344, 346, 376
optimal performance, 361
optimization, 42, 50, 124, 135, 136, 138
organic compounds, 209, 354, 355, 362
organic matter, 94, 258, 260, 264, 266, 267, 270, 271, 274, 275, 357, 359
organizational culture, 72, 339
organize, 63, 306
osmosis, 148, 346
overharvesting, 42
overlap, 221, 223
oxidation, 94, 262

N

national parks, 23, 86
natural gas, 303, 304, 357, 358
natural resources, xi, 65, 117, 118, 152, 185, 191, 208, 230, 290
natural science, 374
negative consequences, 305
negative effects, 94
neglect, 86, 88
Netherlands, 114, 204
networking, 61
neutral, 18, 192, 196
New York Stock Exchange, 337, 340
nitrates, 108
nitrogen, x, 93, 94, 108, 113, 272, 273, 355, 356, 357, 358, 360, 361, 364
nitrogen quotas, x, 93
North Africa, 104
North America, 311
null, 192, 197, 198, 221
nutrient, xv, 108, 113, 258, 260, 265, 351, 353, 358, 360, 361, 364
nutrition, 369

O

objectivity, 175
obstacles, 49, 55, 213, 214, 217, 364
oceans, 278, 327
officials, 382
OH, 125
oil, 121, 296
one sample t-test, 219
operating costs, 96, 221, 223

P

Pacific, 40, 46, 321
paints, 337
palladium, 147
parallel, 72, 95, 96, 101, 360
parents, 242, 244
participants, xi, 16, 56, 84, 183, 185, 186, 187, 188, 189, 190, 191, 192, 199, 200, 201, 305
patents, 118
pathways, 101
payback period, 298, 304, 305, 306, 307
peat, 148
penalties, 76, 296
penetrability, 258, 267
perchlorate, 148
percolation, 360
performance indicator, 77
permit, 58, 65, 89, 109, 279, 284
personal contact, 192, 378
personality, 155, 375
persuasion, 87, 378
pesticide, 98, 99, 100, 101, 102, 103
pests, 95, 100
petroleum, 121
pH, 129, 261
pharmaceutical, 121
phenol, 149
Philippines, 39
phosphates, 108
phosphorous, 355, 356
phosphorus, 113, 271, 360, 361, 364, 365
photographs, 141, 383
photosynthesis, 259
physical characteristics, 316

physical features, 230, 235
physical properties, 146, 258, 267, 365
planning decisions, 65
plant growth, 105, 270
plants, 40, 63, 85, 107, 108, 112, 114, 259, 265, 269, 280, 286, 345, 355, 357, 358, 359, 360
plastics, 337
platform, 153
playing, 195, 240, 242, 243
PLS, 239, 240, 249
PM, 88
poison, 372
policy, x, xv, 5, 8, 10, 12, 21, 22, 24, 25, 26, 29, 32, 36, 40, 41, 42, 45, 46, 47, 58, 68, 69, 83, 84, 105, 155, 156, 158, 160, 168, 169, 175, 176, 194, 195, 200, 201, 291, 310, 333, 341, 343, 346, 351, 365
policy making, xv, 194, 195, 201, 351
political force, 376
politics, 346
pollutants, x, 99, 101, 117, 121, 122, 145, 281, 353, 355, 361, 362, 363
pollution, xv, 100, 108, 118, 121, 122, 153, 157, 208, 211, 214, 240, 245, 255, 281, 282, 296, 301, 305, 306, 307, 308, 312, 332, 335, 340, 342, 351, 352, 353, 357, 359, 360, 361, 363, 364, 365, 366
polymer, 343
polymerization, 118, 119, 123, 124, 140, 142
polymers, 343
polyphenols, 143
ponds, 354, 359
pools, 261, 264
population, xv, 16, 79, 104, 166, 209, 316, 317, 320, 325, 329, 351, 352, 353, 371, 372, 375, 381
population density, 353
population growth, 104, 209
porosity, 134
portfolio, 63, 337
positive attitudes, 372
potential benefits, 52
poverty, 110
precedent, 235
precipitation, 122, 148, 258
predation, 323, 324, 325, 328
predators, 325, 370, 381
preparation, 8, 48, 95, 107, 123, 147, 219, 341
preparedness, 155, 156, 168, 171, 176
prescribed burning, 10, 28, 33
present value, 35
preservation, 80, 290
pressure groups, 153
prestige, 210
prevention, xv, 51, 98, 157, 212, 312, 332, 335, 367
primary function, 281

priming, 260
principal component analysis, 235
Principal Components Analysis, xiii, 229, 239, 240
principles, xv, 67, 71, 72, 73, 74, 79, 81, 83, 89, 90, 109, 153, 157, 161, 164, 165, 177, 191, 203, 208, 210, 279, 284, 285, 289, 291, 317, 336, 341, 343, 346, 363, 364, 367, 369, 374, 379, 381
private ownership, 279
probability, 2, 11, 12, 13, 15, 16, 17, 18, 19, 20, 21, 22, 28, 35, 66, 131, 269, 322
probability distribution, 28
problem solving, 327
process control, 321
producers, 161, 162
product life cycle, 65
production technology, 84
productive capacity, 342
professionals, 10, 16, 32, 38, 39, 82, 202
profit, 44, 230, 368, 374
profitability, 50, 93, 199, 296, 312
project, 3, 6, 39, 50, 54, 99, 110, 146, 156, 186, 191, 192, 199, 204, 278, 284, 285, 292, 310, 328, 339, 344, 345, 356, 378, 382
proposition, 166
protected areas, xv, 231, 322, 326, 367
protection, xii, 23, 33, 41, 152, 207, 208, 210, 212, 213, 217, 226, 244, 245, 251, 278, 338
proteins, 273
prototype, 41, 383
public administration, 67, 79
public awareness, 79, 290
public education, 286
public policy, 43
public sector, 48, 52
public service, 239, 240, 241, 250
public-private partnerships, 291
publishing, 56
pulp, 332, 335, 340, 341, 345, 346
purification, 68, 69, 362
purity, 123
PVS, 246

Q

qualitative research, 307, 348
quality assurance, 291
quality of life, 298, 342
quality standards, 161
quantification, 68
Queensland, 275
questionnaire, xii, 189, 190, 192, 195, 201, 202, 207, 211, 212, 218, 219, 220, 234
quotas, x, 93, 317, 319, 328

R

race, 369
radius, 98
rainfall, 105, 260
rangeland, 46
rate of return, 219
raw materials, 65, 118, 142, 153, 209, 306, 335, 339, 344, 364
reaction mechanism, 119
reaction temperature, 135
reactive sites, 119
reactivity, 119
reading, 61
reality, 152, 167, 179
reasoning, 6, 235
recall, 379
recognition, 48, 54, 55, 61, 155, 163, 166, 187, 191, 195, 201, 214, 221, 222, 337, 340
recommendations, 12, 13, 44, 58, 90, 187, 290, 365
reconstruction, 80
recovery, 8, 65, 293, 317, 322, 346, 358, 361
recovery plan, 293
recreation, 230, 231, 238, 240, 242, 254, 286
recreational, 14, 86, 230, 243, 253, 254, 279
rectification, 202
recurrence, 78, 89
recycling, xv, 60, 65, 108, 114, 204, 306, 307, 332, 335, 343, 345, 351, 353, 358, 364, 369
redundancy, 235
reference frame, 63
regionalization, 323
regions of the world, 260
regression, 127, 129, 130, 131, 132, 135, 136, 138, 143, 144, 149
regression analysis, 149
regression equation, 135
regression method, 149
regulations, xiv, 5, 15, 16, 20, 23, 49, 64, 95, 201, 210, 215, 217, 225, 289, 295, 296, 316, 317
regulatory agencies, 172
regulatory controls, 201
regulatory framework, 153
regulatory requirements, 48
reinforcement learning, 40
relaxation, 230, 238, 242
relevance, 54, 56, 74, 82, 83, 87, 227, 230, 345
reliability, 82, 318
relief, 52, 55, 61
remedial actions, 187
remediation, 118, 362
remote sensing, 106
renewable energy, 297, 304, 306, 307

repair, 317
repetitions, 98
reptile species, 281
reputation, 167, 296
requirements, 22, 45, 51, 55, 56, 57, 58, 59, 61, 64, 65, 68, 69, 70, 71, 72, 74, 75, 77, 78, 80, 84, 86, 89, 94, 95, 98, 104, 106, 113, 157, 158, 163, 164, 165, 166, 168, 169, 170, 171, 173, 174, 176, 184, 194, 200, 285, 286, 289, 291, 296, 325, 333, 334, 338, 340, 342, 346, 347, 348
research institutions, 49
researchers, 3, 5, 7, 38, 118, 120, 138, 267, 325
reserves, 260, 383
residuals, 357
residues, x, 100, 102, 109, 113, 115, 117, 259, 364
resilience, 43
resins, 142
resistance, 97, 118, 157, 159
resolution, 41, 106
resorcinol, 119
resource management, 316, 318
resources, ix, x, 3, 16, 41, 47, 50, 52, 54, 55, 71, 83, 87, 89, 90, 104, 110, 153, 154, 155, 156, 157, 158, 159, 170, 174, 178, 185, 186, 195, 217, 232, 243, 288, 290, 291, 292, 306, 308, 316, 317, 319, 320, 322, 375, 382
response, 4, 6, 8, 38, 54, 70, 73, 135, 136, 137, 156, 168, 171, 176, 227, 274, 298, 304, 323, 338
responsiveness, 70, 74, 178, 310
restaurants, 232, 245
restoration, 42, 282, 361
restrictions, 12, 67, 82, 97, 322, 324
retail, 63
revenue, 53, 305
rewards, 55
rights, iv
rings, 119, 125
risk, ix, 1, 2, 3, 5, 7, 8, 9, 10, 11, 14, 15, 16, 17, 18, 33, 39, 40, 41, 43, 44, 46, 51, 74, 100, 121, 152, 214, 217, 254, 267, 278, 280, 282, 283, 296, 310, 325
risk assessment, 8, 46
risk management, 51
robotics, 110
root, 105, 106, 107, 108, 112, 258, 267, 273, 362
root system, 107, 112
rotations, 260
routes, 98, 153
Royal Society, 366
rules, ix, 1, 2, 3, 4, 5, 6, 9, 15, 19, 20, 21, 27, 38, 39, 48, 57, 61, 75, 81, 83, 85, 344
runoff, 281, 359, 360, 361, 362, 364, 365
rural development, 2

S

safety, 8, 10, 12, 89, 161, 162, 191, 200, 208, 305, 343, 369
Safety Management System, 339, 342
salinity, 108, 280, 316
saliva, 103
salt concentration, 108
sanctions, 210
saturation, 166
savings, 50, 95, 96, 99, 109, 110, 113, 214, 260, 305
sawdust, 354
scanning electron microscopy, 123
scarcity, 104, 110
scholarship, 326
science, 8, 39, 41, 43, 285, 317, 348, 384
scope, x, 82, 117, 118, 139, 161, 162, 163, 164, 168, 173, 174, 175, 333
sea level, 231, 352
security, 59, 104, 230, 240, 245, 250
sediment, 244, 359, 360, 361
sedimentation, 354
sediments, 259, 361, 366
seed, 96, 149
seeding, 99
selectivity, 320, 324, 328
self-assessment, 62
self-confidence, 220, 223
self-sufficiency, 345
seminars, 74
semi-structured interviews, 189, 193, 195
sensitivity, 106, 152, 189, 240
sensors, 105, 106, 108, 111, 113
septic tank, 354
service firms, 312
services, xiii, xv, 48, 59, 60, 62, 63, 64, 65, 66, 75, 86, 88, 163, 173, 194, 200, 220, 229, 231, 234, 235, 239, 240, 241, 242, 243, 245, 250, 278, 290, 335, 352, 367, 377
settlements, 9, 86, 228
sewage, 86, 147, 258, 260, 261, 263, 264, 265, 266, 267, 268, 269, 270, 271, 272, 273, 274, 341, 351, 352, 353, 354, 355, 356, 364, 365, 366
shade, 230
shape, 99, 119, 361, 364
shock, 121
shoot, 107, 108
shoreline, 255
shortage, 105, 108
showing, 97, 153, 263, 265, 270, 340
shrimp, 317, 319, 321, 322, 325, 328, 329
shrubland, 12
side effects, 79
SIGMA, 122, 123
signalling, 107, 112
signals, 107, 114, 167
significance level, 198
silica, 134
simulation, 4, 5, 8, 12, 14, 18, 22, 27, 31, 38, 39, 40, 41, 42, 43, 44, 45, 112
simulations, ix, 1, 4, 5, 8, 17, 33, 35, 40, 45, 46, 322, 324
Singapore, 212, 213, 227
skin, 101
slag, 355
sludge, xiii, 69, 147, 257, 258, 259, 260, 261, 262, 263, 264, 265, 266, 267, 268, 269, 270, 271, 272, 273, 274, 275, 341, 346, 365
small businesses, 178
small firms, 53
SOC, xiii, 257, 258, 259, 260, 265, 267, 270
social behaviour, 187, 384
social consensus, 79
social control, 85
social environment, 232
social group, 185
social life, 43, 240, 241
social phenomena, 41
social relations, 48
social relationships, 48
social responsibility, 166, 348
social sciences, 41
social status, 242
social structure, 231
society, x, 3, 93, 164, 184, 185, 186, 202, 230, 282, 308, 317, 325, 337, 338, 369, 378
sodium hydroxide, 123
software, 6, 135, 161, 219, 235
soil erosion, 281
soil organic carbon, xiii, 257, 258, 271, 272, 273
soil particles, 361
soil type, 95, 281, 360
solid waste, xi, 183, 185, 208, 282, 338, 341, 366
solution, 18, 19, 36, 48, 81, 99, 108, 114, 118, 123, 126, 147, 305, 308, 375
sorption, 143, 147, 148, 149, 264
South Africa, 242, 253
South China Sea, 328
sowing, 95, 96, 265
Spain, 101, 117, 151, 152, 162, 163, 242, 255, 367
specialisation, xv, 367
specialists, 82
speciation, 275
species, xiv, 2, 121, 146, 210, 260, 265, 268, 270, 272, 278, 279, 280, 282, 283, 284, 286, 287, 289, 290, 293, 315, 316, 317, 318, 319, 320, 321, 322,

324, 325, 326, 327, 330, 355, 359, 361, 369, 370, 372, 376, 380, 382, 383, 384
specific knowledge, 6
specific surface, 355
specifications, 75, 195
spending, xv, 231, 332
sponge, 343
Spring, 40, 43, 91
SPSS software, 221
stability, 185, 211, 273
stabilization, 265, 274, 275
staff development, 73
stakeholder groups, xi, 183, 190, 192, 193, 194, 195, 196, 198, 200, 201, 202
stakeholders, xi, 5, 17, 31, 51, 55, 67, 68, 75, 78, 79, 84, 86, 88, 91, 152, 153, 158, 159, 172, 179, 183, 185, 186, 190, 194, 195, 198, 199, 200, 201, 202, 277, 284, 312, 318, 346, 362, 364
standard deviation, 269
standardization, 210, 214, 217, 220, 222, 224, 225, 340
state, 8, 23, 26, 64, 85, 94, 109, 120, 166, 188, 217, 241, 253, 278, 281, 282, 283, 291, 317, 318, 326, 327, 332, 335, 336, 345, 352, 375, 381
states, 58, 60, 64, 210
Statistical Package for the Social Sciences, 227
statistics, 197, 199, 211
stimulus, 71
storage, xiii, 77, 257, 258, 259, 261, 268, 270, 271, 300, 301, 306, 362
storms, 239, 240, 241, 244
strategic planning, 83, 156, 177
strategic position, 343
stress, 104, 105, 106, 114, 305
stretching, 125
structure, xi, 4, 5, 6, 7, 8, 11, 12, 13, 18, 19, 20, 21, 23, 24, 38, 43, 58, 63, 69, 70, 83, 84, 85, 95, 118, 119, 120, 121, 122, 151, 152, 154, 156, 169, 172, 253, 319, 320, 323, 324, 327, 333, 347, 361
structuring, 83
style, xiii, 277
subgroups, 142
subsidy, 98
substitutes, 306
substrate, 108, 264, 355, 360
succession, 34, 361
sugar industry, 97
sugarcane, 97, 111
sulfate, 115
sulfur, 347
sulphur, 209, 267
Sun, 231
supervision, x, 74, 75, 83, 93, 157

supervisors, xi, 183, 191, 201
supplier, 65, 162
suppliers, 74, 75, 76, 79, 88, 155, 159, 172, 191, 200, 337, 340
supply chain, 90, 160, 199, 200, 202
suppression, xi, 7, 183
surface area, 123
surfactant, 121, 122, 123, 129, 130, 135, 136, 138, 139, 141, 143, 147, 148
surfactants, xi, 118, 121, 147
surplus, 267, 345, 346, 347
surveillance, 57, 76, 77
sustainability, ix, xi, xiv, 79, 115, 117, 118, 183, 184, 185, 186, 187, 188, 189, 190, 192, 194, 195, 196, 199, 200, 201, 205, 272, 278, 295, 296, 297, 298, 304, 305, 306, 307, 308, 309, 310, 322, 379
sustainable development, xii, 48, 152, 154, 184, 186, 189, 190, 191, 194, 199, 201, 202, 204, 207, 208, 210, 220, 222, 228, 297, 369, 379
Sweden, 92
Switzerland, 54, 326, 348, 384
symptoms, 265
synthesis, 41, 111, 118, 135, 138, 189

T

tactics, 7, 11
Taiwan, 311
tanks, 354
tannins, x, 117, 118, 119, 120, 122, 140, 143
target, 35, 57, 79, 87, 118, 129, 139, 140, 192, 304, 305, 306, 308, 309, 316, 317, 320, 325, 343, 363, 369, 371, 372, 375, 376, 377, 378, 379, 384
target population, 317, 384
target populations, 317, 384
TCA, xii, 207, 219, 227
teams, 199, 282
technical efficiency, 323
techniques, x, xv, 39, 41, 94, 95, 100, 101, 106, 107, 108, 109, 117, 148, 165, 345, 367, 369, 374, 375, 381
technological change, 307
technological progress, 374
technologies, 94, 105, 109, 186, 204, 342, 354, 364
technology, xi, xv, 48, 72, 83, 95, 96, 99, 117, 153, 215, 217, 290, 296, 365, 367, 368
telecommunications, 105
telephone, 161, 219
temperature, 100, 123, 132, 133, 135, 136, 137, 146, 230, 249, 258, 260, 305, 316, 361
terraces, 359
terrestrial ecosystems, 272
territorial, 63

territory, 86, 352
testing, 8, 74, 167, 342
texture, 106, 264, 270, 273
thermal activation, 134
thermodynamics, 149
thinning, 10, 28, 33
third dimension, xii, 208, 225
threats, ix, xiv, 1, 12, 230, 282, 286, 295, 296, 376
time frame, 186
time periods, 319
time series, 321, 322, 324
tissue, 274
total energy, 304, 305
tourism, 48, 210, 230, 231, 232, 240, 253, 254, 255, 352
toxic waste, 282
toxicity, 121, 265
toxin, 297
trace elements, 261, 275, 358
tracks, 96, 98
trade, 43, 51, 54, 65, 160, 162, 226, 309, 341, 368
traditions, 185
training, xiii, 58, 59, 69, 70, 72, 73, 74, 76, 77, 80, 82, 83, 89, 155, 157, 158, 159, 168, 170, 174, 175, 177, 295, 296, 300, 301, 307, 309, 334, 339, 374, 379
transaction costs, 160, 161, 163, 167
transformation, 153, 166, 272
transparency, 59, 84, 86, 90, 166
transport, xv, 63, 64, 65, 338, 343, 351, 353, 359, 362, 364, 369
transportation, 65, 282, 286, 297, 300, 301, 304, 305, 306, 307
treatment, ix, x, xv, 13, 14, 15, 18, 19, 27, 28, 29, 30, 31, 32, 33, 34, 45, 78, 86, 117, 121, 122, 141, 146, 148, 149, 261, 265, 266, 269, 272, 281, 332, 341, 346, 351, 352, 353, 354, 355, 357, 359, 362, 364, 365, 366
treatment agents, 121
trial, 96, 262
trustworthiness, 82
Turkey, xii, 112, 207, 208, 209, 210, 211, 215, 216, 218, 219, 220, 224, 225, 226, 227, 255
Turkish Contractors Association, xii, 207, 219

U

U.S. Department of the Interior, 45, 293
UGC, xiv, 315, 316, 317, 318, 319, 320, 321, 322, 323, 324, 325, 326
UK, 272, 384
UN, 258, 335
UNESCO, 184, 204

uniform, 4, 8, 265, 359
United Nations, xv, 152, 184, 204, 228, 313, 328, 332, 335, 349
United Nations Industrial Development Organization, xv, 332, 335
United States, 11, 12, 41, 44, 45, 152, 178
updating, 64, 77, 84, 316
Upper Gulf of California, xiv, 315, 316, 320, 323, 328, 329
urban, xiii, 2, 10, 40, 42, 63, 199, 203, 226, 229, 230, 232, 244, 245, 253, 279, 286, 352, 360, 365
urban settlement, 63
urbanisation, 231, 244
USA, ix, 1, 12, 41, 42, 46, 111, 113, 203, 214, 215
USDA, 2, 9, 12, 13, 45, 46, 261

V

vacuum, 123
validation, 40, 43, 84, 166, 171, 180
valuation, 172
variables, xii, xiii, 5, 9, 19, 21, 25, 26, 36, 50, 52, 54, 124, 135, 139, 140, 146, 207, 222, 229, 235, 239, 241, 320, 368, 374, 375, 377, 378
variations, 35, 105, 112, 162, 321, 324
vegetables, 105, 112, 359
vegetation, 4, 7, 9, 10, 12, 13, 14, 32, 35, 37, 230, 234, 258, 259, 273, 279, 280, 359, 360, 361, 362
vegetative cover, 359, 360
vehicles, 233, 250, 251, 306
velocity, 101
ventilation, 100, 101, 305
vibration, 65, 125
vision, 58, 98, 155
vocabulary, 164
volatile organic compounds, 343
volatilization, 359
voluntary commitment, x, 47
Volunteers, 383
vomiting, 104, 121
vulnerability, 8, 39, 282

W

Wales, 254
Washington, 39, 42, 43, 45, 46, 111, 115, 293, 327, 328, 329, 365, 366
waste, ix, x, xv, 50, 64, 65, 68, 70, 77, 87, 88, 117, 134, 146, 147, 186, 187, 195, 204, 205, 208, 209, 211, 212, 213, 217, 226, 245, 258, 259, 267, 296, 297, 300, 301, 307, 308, 312, 332, 335, 337, 338, 339, 341, 345, 346, 347, 351, 353, 357, 358, 364, 365, 370

waste disposal, 335, 364
waste management, 205, 211, 259
waste treatment, 147, 365
wastewater, 148, 281, 338, 352, 353, 354, 355, 362, 364, 365
water quality, xv, 245, 281, 351, 352, 353, 359, 362, 365, 366
water resources, 94, 104
watershed, xv, 351, 353, 363, 364
waterways, 359, 362, 363
WD, 165, 166
weakness, 8, 66
web, 109, 318
welfare, 104
wetlands, ix, 278, 279, 280, 281, 282, 285, 286, 289, 290, 291, 292, 359, 360, 361, 365
wetting, 105, 106, 114
whales, 328
wildfire, ix, 1, 2, 3, 5, 7, 8, 9, 10, 11, 12, 13, 14, 15, 16, 17, 18, 20, 22, 23, 24, 25, 26, 27, 28, 29, 33, 35, 37, 38, 39, 41, 42, 43, 44, 45, 46
wildland-urban interface, 39, 40, 41, 44, 45
wildlife, 24, 230, 271, 279, 280, 292, 361, 369
wildlife conservation, 271, 292
windows, 304, 305, 310
Wisconsin, 39

withdrawal, 104
wood, 28, 29, 31, 36, 120, 268, 311, 341, 342, 344
wood products, 28, 29, 31, 36
woodland, 260
workers, 72, 73, 74, 81, 100, 101, 103, 109, 112, 118, 191
workforce, 209
workplace, 58, 313
worldwide, xii, 207, 230, 258, 337, 342, 348

X

xylem, 112

Y

yield, 95, 96, 105, 106, 107, 112, 260, 265, 266, 267, 269, 271, 274, 275, 318, 319, 358, 365
young adults, 238
young people, xiii, 229, 243

Z

zinc, 271